An Overview of General Relativity and Space-Time

This textbook equips Masters' students studying Physics and Astronomy with the necessary mathematical tools to understand the basics of General Relativity and its applications. It begins by reviewing classical mechanics with a more geometrically oriented language, continues with Special Relativity and, then onto a discussion on the pseudo-Riemannian space-times. Applications span from the inner and outer Schwarzschild solutions to gravitational wave, black holes, spherical relativistic hydrodynamics, and Cosmology. The goal is to limit the abstract formalization of the problems, to favor a hands-on approach with a number of exercises, without renouncing to a pedagogical derivation of the main mathematical tools and findings.

Series in Astronomy and Astrophysics

The *Series in Astronomy and Astrophysics* includes books on all aspects of theoretical and experimental astronomy and astrophysics. Books in the series range in level from textbooks and handbooks to more advanced expositions of current research.

Series Editors:
M Birkinshaw, University of Bristol, UK
J Silk, University of Oxford, UK
G Fuller, University of Manchester, UK

Recent books in the series

Dark Sky, Dark Matter
J M Overduin and P S Wesson

Dust in the Galactic Environment, 2nd Edition
D C B Whittet

The Physics of Interstellar Dust
E Krügel

Very High Energy Gamma-Ray Astronomy
T C Weekes

Numerical Methods in Astrophysics: An Introduction
P Bodenheimer, G P Laughlin, M Rózyczka, H W Yorke

An Introduction to the Physics of Interstellar Dust
Endrik Krugel

Astrobiology: An Introduction
Alan Longstaff

Fundamentals of Radio Astronomy: Observational Methods
Jonathan M Marr, Ronald L Snell, and Stanley E Kurtz

Stellar Explosions: Hydrodynamics and Nucleosynthesis
Jordi José

Cosmology for Physicists
David Lyth

Cosmology
Nicola Vittorio

Cosmology and the Early Universe
Pasquale Di Bari

Fundamentals of Radio Astronomy: Astrophysics
Ronald L. Snell, Stanley E. Kurtz, and Jonathan M. Marr

Introduction to Cosmic Inflation and Dark Energy
Konstantinos Dimopoulos

Physical Principles of Astronomical Instrumentation
Matthew Griffin, Peter A. R. Ade, Carole Tucker

A Guide to Close Binary Systems
Edwin Budding and Osman Demircan

An Overview of General Relativity and Space-Time
Nicola Vittorio

An Overview of General Relativity and Space-Time

Nicola Vittorio

CRC Press
Taylor & Francis Group
Boca Raton London New York

CRC Press is an imprint of the
Taylor & Francis Group, an **informa** business

First edition published 2022
by CRC Press
4 Park Square, Milton Park, Abingdon, Oxon, OX14 4RN

and by CRC Press
6000 Broken Sound Parkway NW, Suite 300, Boca Raton, FL 33487-2742

CRC Press is an imprint of Informa UK Limited

ISBN: 978-0-367-69288-9 (hbk)
ISBN: 978-0-367-68304-7 (pbk)
ISBN: 978-1-003-14125-9 (ebk)

DOI: 10.1201/9781003141259

Typeset in Nimbus
by KnowledgeWorks Global Ltd.

Dedication

To Liù and Ludovico

Contents

Part II From Curvature to Observations

List of Figures

List of Boxes

Part I

From Forces to Curvature

1 Space and Time: The Classical View

1.1 INTRODUCTION

As it is well known, in Classical Mechanics *space* and *time* are two distinct categories. However, they share the same properties of *homogeneity* and *isotropy*. A priori, there are no special places where to be or special directions along which to move. Likewise, there is no preferred time for an event to occur; if a system evolves through a number of intermediate equilibrium states, it is always possible to go through the same intermediate states, but in a reversed order. The goal of this chapter is to start introducing some of the geometrical formalism used in the rest of the book, remaining for the moment within the framework of Classical Mechanics.

1.2 METRIC SPACE

The position in space of a point P is completely defined by its position vector, $\vec{\ell}$, bound to the origin O of the chosen coordinate system.

$$\vec{\ell} = \xi^1\hat{e}_1 + \xi^2\hat{e}_2 + \xi^3\hat{e}_3 = \sum_{k=1}^{3} \xi^k\hat{e}_k = \xi^k\hat{e}_k \tag{1.1}$$

Here ξ^k are the *contravariant* components of $\vec{\ell}$, whereas $\hat{e}_k$ are the corresponding *covariant* components of the basis vectors. By convention, the indexes of contravariant and covariant components are shown as superscripts and subscripts, respectively. The Einstein *summation convention* implies summing over any couple of repeated (or *dummy*) indexes, provided that one of the two is contravariant and the other covariant [see Exercise A.1]. The last equality in Eq.(1.1) exploits this convention: we omit the summation symbol and will assume hereafter that Latin indexes always run from 1 to 3.

Now, consider two points, P and Q, of coordinates ξ^k and $\xi^k + d\xi^k$, respectively. The infinitesimal displacement vector $d\vec{\ell}$, connecting Q to P, can be written in terms of its contravariant components $d\xi^k$:

$$d\vec{\ell} = d\xi^k\hat{e}_k \tag{1.2}$$

The distance between these two points is given by the magnitude of the displacement vector. The corresponding squared quantity writes

$$d\ell^2 = d\vec{\ell}\cdot d\vec{\ell} = (d\xi^i\hat{e}_i)\cdot(d\xi^j\hat{e}_j) = (\hat{e}_i\cdot\hat{e}_j)d\xi^i d\xi^j \tag{1.3}$$

where the symbol "·" indicates the *dot* or *inner* product of vectors. Eq.(1.3) can be written in a more compact and elegant way:

$$d\ell^2 = \gamma_{ij}d\xi^i d\xi^j \tag{1.4}$$

DOI: 10.1201/9781003141259-1

after defining the matrix

$$\gamma_{ij} \equiv \hat{e}_i \cdot \hat{e}_j \tag{1.5}$$

Eq.(1.4) is a very important one. It defines the *metric* of the space. Once we know the *metric coefficients* [*c.f.* Eq.(1.5)], the line element $d\ell$ yields the relative distance of two given points as a non-negative, real number. Remember that the vectors have an absolute meaning, their magnitude and direction being independent from the chosen coordinate system. On the contrary, the vector components *do* depend on the specific coordinate system we are using. So, the *lhs* of Eq.(1.4) is an invariant, while the explicit form of the *rhs* depends on the chosen coordinate system. Using orthonormal (Cartesian) coordinates [$\xi^i \equiv \{x,y,z\}$] simplifies things. In fact, in this case $\gamma_{ij} = \delta_{ij}$, and the line element can be written in its standard, well-known Euclidean form.

$$d\ell^2 = dx^2 + dy^2 + dz^2 \tag{1.6}$$

If we had chosen cylindrical coordinates [$\xi^i \equiv \{r,\theta,z\}$], we would have obtained a different expression for the *rhs* of Eq.(1.6) [see Exercise A.2]. Hereafter, unless otherwise specified, we will use a Cartesian coordinate system.

1.3 HOMOGENEITY AND ISOTROPY OF SPACE

Homogeneity

The space is *homogeneous*: according to the Copernican Principle, all positions must be equivalent. This means that the choice of the origin of the coordinate system shouldn't matter. Consider two coordinate systems, $\mathscr{K}$ and $\mathscr{K}'$. The latter is obtained by translating $\mathscr{K}$ while keeping the orthonormal basis vectors parallel to themselves. Given a point in space, its position vectors in $\mathscr{K}$ and $\mathscr{K}'$ are then related: $\vec{\ell} = \vec{\ell}' + \vec{d}$, where $\vec{d}$ is the (constant) position vector of $\mathscr{O}'$ *w.r.t.* $\mathscr{K}$ [see Figure 1.1a]. Clearly, the metric remains unchanged by such a translation. In fact, the basis vectors in $\mathscr{K}$ and $\mathscr{K}'$ have the same orientation. It follows that $d\xi^i = d\xi'^i$ [*c.f.* Eq.(1.2)] and $\gamma_{ij} = \gamma'_{ij}$ [*c.f.* Eq.(1.5)]. Thus,

$$d\ell^2 = \gamma_{ij} d\xi^i d\xi^j = \gamma'_{mn} d\xi'^m d\xi'^n \tag{1.7}$$

As expected, the metric is invariant under a *translation* of the reference frame.

Isotropy

The space is also *isotropic*: there is no preferred direction in which to move, and all the directions must be equivalent. This implies that the orientation in space of the basis vectors shouldn't matter. To see this point, consider again two different coordinate systems, $\mathscr{K}$ and $\mathscr{K}'$, the latter being obtained by rotating $\mathscr{K}$ by an angle ϕ around one of its basis vector. In this way $\mathscr{K}$ and $\mathscr{K}'$ have the same origin, but different orientation of the coordinate axes [see Figure 1.1b]. In general terms, we can write the orthogonal transformation that relates the new and old coordinates as follows:

$$\xi^k = L^k{}_m \xi'^m; \qquad \xi'^m = M^m{}_n \xi^n; \tag{1.8}$$

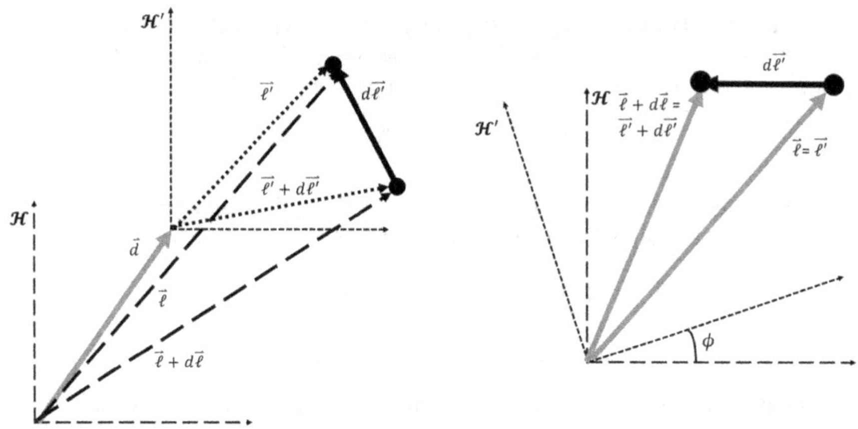

Figure 1.1 Space homogeneity: the line element $d\ell$ is unaffected by a rigid translation of the reference frame (left panel). Space isotropy: the line element $d\ell$ is unaffected by the orientation of the coordinate axes (right panel).

Here ξ^k and ξ'^m are the contravariant components of the position vector in $\mathscr{K}$ and $\mathscr{K}'$, while

$$L^k{}_m = \frac{\partial \xi^k}{\partial \xi'^m}; \qquad M^m{}_n = \frac{\partial \xi'^m}{\partial \xi^n} \tag{1.9}$$

are the orthogonal rotation matrixes. It is important to realize that the transformations of Eq.(1.8) are *linear* transformations, as both the rotation matrixes, $\mathbf{M}$ and $\mathbf{L} = \mathbf{M}^{-1}$, can be expressed in terms of a (constant) rotation angle, ϕ. If $\mathscr{K}'$ is obtained by a rotation of $\mathscr{K}$ around its ξ^3-axis, then we can write

$$\mathbf{L} \equiv \begin{pmatrix} \cos\phi & \sin\phi & 0 \\ -\sin\phi & \cos\phi & 0 \\ 0 & 0 & 1 \end{pmatrix}; \qquad \mathbf{M} \equiv \begin{pmatrix} \cos\phi & -\sin\phi & 0 \\ \sin\phi & \cos\phi & 0 \\ 0 & 0 & 1 \end{pmatrix}; \tag{1.10}$$

describing clock-($\mathbf{L}$) and counterclock-($\mathbf{M}$) rotations, respectively [see Figure 1.1b]. Using Eq.(1.8), we can easily find how the metric coefficients transform under such a rotation

$$d\ell^2 = \gamma_{ij} d\xi^i d\xi^j = \gamma_{ij} L^i{}_m d\xi'^m L^j{}_n d\xi'^n = \gamma'_{mn} d\xi'^m d\xi'^n \tag{1.11}$$

where

$$\boxed{\gamma'_{mn} = \gamma_{ij} L^i{}_m L^j{}_n} \tag{1.12}$$

This equation can be written in a compact way: $\gamma' = \mathbf{L}^T \boldsymbol{\gamma} \mathbf{L}$. In the case of an orthonormal vector basis, $\gamma_{ij} = \delta_{ij} = \gamma'_{ij}$ as $\mathbf{L}$ is an orthogonal matrix. Eq.(1.11) shows that, as expected, the metric is invariant under a rotation of the reference frame.

CONTRAVARIANT COMPONENTS: A FORMAL DEFINITION

A set of three quantities, V^k $(k = 1,3)$, are the contravariant components of a 3D vector if they transform on a change of coordinates as in Eq.(1.8):

$$V^k = L^k{}_m V'^m \qquad V'^k = M^k{}_m V^m \tag{1.13}$$

Let's make a step further along the same line, by writing the displacement vector in $\mathscr{K}$ and $\mathscr{K}'$, and by using the same notation of Eq.(1.8)

$$d\vec{\ell} = \begin{cases} d\xi'^j \hat{e}'_j = M^j{}_i d\xi^i \hat{e}'_j \ = d\xi^i \hat{e}_i \\ \\ d\xi^i \hat{e}_i \ = L^i{}_m d\xi'^m \hat{e}_i = d\xi'^m \hat{e}'_m \end{cases} \tag{1.14}$$

where, in the last equalities, we have found the transformation law of the basis vectors.

$$\hat{e}_i = M^j{}_i \hat{e}'_j; \qquad \hat{e}'_m = L^i{}_m \hat{e}_i; \tag{1.15}$$

We can use Eq.(1.15) to go beyond the simple case of the basis vectors and state the following:

COVARIANT COMPONENTS: A FORMAL DEFINITION

A set of three quantities, V_k $(k = 1,3)$, are the covariant components of a 3D vector if they transform on a change of coordinates as in Eq.(1.15):

$$V_m = M^k{}_m V'_k \qquad V'_m = L^k{}_m V_k \tag{1.16}$$

In conclusion, we have to use the rotation matrix **M** (**L**) or its inverse **L** (**M**) for finding the new (old) contravariant and covariant vector components, respectively. Note that the sum of two vectors is still a vector [see Exercise A.3].

1.4 COVARIANT OR CONTRAVARIANT VECTOR COMPONENTS

To clarify the difference between *covariant* and *contravariant* components of a vector, let's first define the procedure of lowering a contravariant index by using the metric coefficients:

$$d\xi_i \equiv \gamma_{ij} d\xi^j \tag{1.17}$$

As usual, we are summing over the repeated (Latin) *dummy* index j that goes from 1 to 3. The quantities $d\xi_i$ $(i = 1,3)$ define the *covariant* components of the displacement vector $d\vec{\ell}$ [see Exercise A.4].

Covariant and contravariant components of a vector can be different even if they refer to the same geometrical object. Consider the simple case of a 2D Euclidean space. If we choose an orthonormal basis, the metric coefficients are $\gamma_{ij} = \delta_{ij}$ [*c.f.* Eq.(1.5)]. It follows from Eq.(1.17) that the *covariant* and *contravariant* components of the displacement vector are the same and that

$$d\ell^2 = (d\xi^1)^2 + (d\xi^2)^2 = (d\xi_1)^2 + (d\xi_2)^2 \tag{1.18}$$

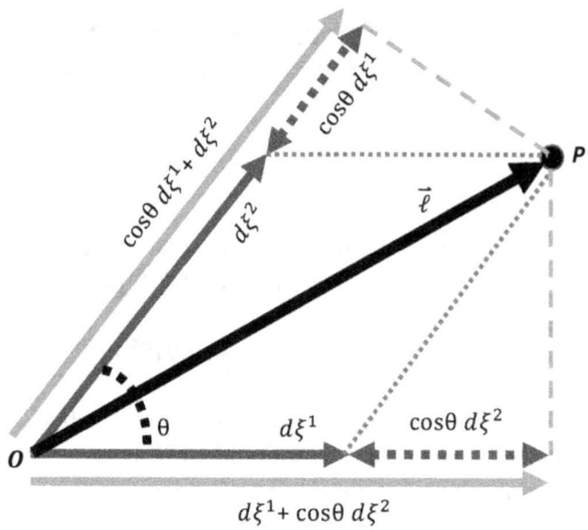

Figure 1.2 Contravariant (dark gray) and covariant (light gray) components of the position vector $\vec{\ell}$ (black) of the point P in a skew coordinate system

However, for the skew coordinate system of Figure 1.2, γ_{ij} is not diagonal anymore [*c.f.* Eq.(1.5)]:

$$\gamma_{ij} = \begin{pmatrix} 1 & \cos\theta \\ \cos\theta & 1 \end{pmatrix} \tag{1.19}$$

Then, according to Eq.(1.17)

$$\begin{aligned} d\xi_1 &= \gamma_{1j} d\xi^j = d\xi^1 + \cos\theta d\xi^2 \\ d\xi_2 &= \gamma_{2j} d\xi^j = \cos\theta d\xi^1 + d\xi^2 \end{aligned} \tag{1.20}$$

Figure 1.2 shows that the *contravariant* components of $d\vec{\ell}$ are those obtained by projecting the tip of the position vector, $\vec{\ell}$, using parallels to the coordinate axes (parallelogram rule). On the contrary, the *covariant* components of the same vector are those obtained by the intercepts of the normals to the coordinate axes. So, the same vectors can be described equally well in terms of its *covariant* or *contravariant* components, which differ in the general case of a non-orthonormal coordinate system. Note that Eq.(1.17) allows to write Eq.(1.4) in an even more compact form.

$$d\ell^2 = d\xi_i d\xi^i \tag{1.21}$$

Remember that the line element $d\ell$ has been defined as the magnitude of the displacement vector [*c.f.* Eq.(1.3)]. This allows us to formalize the definition of the *dot* or *inner* product of two vectors.

INNER PRODUCT OF VECTORS: DEFINITION

The inner product of two vectors, **A** and **B**, is given by the sum of the product of their covariant and contravariant components:

$$\mathbf{A}\cdot\mathbf{B} = A_i B^i = A^j B_j \tag{1.22}$$

Note that the inner product of two vectors is an invariant [see Exercise A.5]. On the basis of this definition, Eq.(1.21) together with Eq.(1.20) provides

$$d\ell^2 = (d\xi^1)^2 + (d\xi^2)^2 + 2\cos\theta\, d\xi^1 d\xi^2 \tag{1.23}$$

This is different from the familiar expression of Eq.(1.18) if $\theta \neq \pi/2$, but it is perfectly consistent with the well-known trigonometric Law of Cosines. Again, with reference to Eq.(1.4), remember that the *lhs* is an invariant, whereas the explicit form of the *rhs* depends on the chosen coordinate system.

1.5 MOTION OF A FREE TEST-PARTICLE

In Classical Mechanics, the action of a test-particle is defined as the integral of a Lagrangian, L, between two times, t_1 and t_2, say

$$S = \int_{t_1}^{t_2} L(\vec{\ell}, \vec{v}, t)\,dt \tag{1.24}$$

The Lagrangian depends in general on the position and the velocity of the test particle, and on the time as well. The trajectory followed by the particle to go from position A (at time t_1) to position B (at time t_2) renders the action stationary: $\delta S = 0$. This leads to the Euler-Lagrangian equations.

$$\frac{d}{dt}\frac{\partial L}{\partial \dot{\xi}^i} = \frac{\partial L}{\partial \xi^i} \tag{1.25}$$

where ξ^i and $\dot{\xi}^i$ are the contravariant components of the test-particle position ($\vec{\ell}$) and velocity ($\vec{v} = d\vec{\ell}/dt$) vectors, respectively. Let's assume here a Cartesian coordinate system.

The homogeneity and isotropy of both space and time put constraints on the motion of a free-test particle. In fact, the Lagrangian cannot explicitly depend on time (there is no specific time for a free particle to be observed in a particular status of motion) nor on the position (there is no special place for a free-particle to be at a given time). Then, the Lagrangian of a free-particle can depend *only* on its velocity. But the velocity is a vector, and for the isotropy of space, the Lagrangian cannot depend on the direction of $\vec{v}$, but only on its magnitude:

$$L = \frac{1}{2}m|\vec{v}|^2 \equiv T \tag{1.26}$$

BOX 1.1 GEODESICS

From a geometrical point of view, we can state that a free particle moves from point A to point B following the portion of a geodesic that connects the two points. A geodesic is a curve characterized by a length, which has a stationary value *w.r.t.* arbitrary small variation of the curve parameters. This concept can be expressed in terms of a variational principle by requiring $\delta \int_A^B d\ell = 0$, or equivalently

$$\delta \int_A^B \frac{d\ell}{d\lambda} d\lambda = 0 \tag{B1.1.a}$$

Here λ is a parameter defined along the geodesic, and, as usual, we assume that $\delta\xi^k(A) = \delta\xi^k(B) = 0$. Given Eq.(1.4), we have

$$\frac{d\ell}{d\lambda} = \sqrt{\gamma_{ij}\frac{d\xi^i}{d\lambda}\frac{d\xi^j}{d\lambda}} \tag{B1.1.b}$$

Thus, Eq.(B1.1.a) can be written as

$$\int_A^B \frac{d\lambda}{d\ell}\left(\gamma_{ij}\frac{d\xi^i}{d\lambda}\frac{d\delta\xi^j}{d\lambda}\right) d\lambda = \int_A^B \gamma_{ij}\frac{d\xi^i}{d\ell}\frac{d\delta\xi^j}{d\ell} d\ell = -\int_A^B \gamma_{ij}\frac{d^2\xi^i}{d\ell^2}\delta\xi^j d\ell = 0 \tag{B1.1.c}$$

Because of the arbitrariness of $\delta\xi^j$, Eq.(B1.1.c) yields

$$\frac{d^2\xi^i}{d\ell^2} = 0 \quad \Rightarrow \quad \frac{d\xi^i}{d\ell} = c^i \tag{B1.1.d}$$

where we have chosen a Cartesian coordinate system (*i.e.* $\gamma_{ij} = \delta_{ij}$) and the constants of integration, c^i, are given by the direction cosines of the unit vector tangent to the geodesic. Since $d\ell = |\vec{v}|dt$, Eq.(B1.1.d) is perfectly consistent with Eq.(1.27). Thus, a free particle moves along a portion of a geodesic, which in our case is a straight line. This illustrates another property of a geodesic: it is the straightest line that one can draw in a given space to join two arbitrary points.

where m is the mass of the test-particle, $|\vec{v}| = \sqrt{\dot{\xi}_k\dot{\xi}^k}$ the magnitude of its velocity and T its kinetic energy. Thus, Eq.(1.25) yields

$$\frac{d^2\xi^i}{dt^2} = 0 \quad \Rightarrow \quad \frac{d\xi^i}{dt} = c^i|\vec{v}| \tag{1.27}$$

Here the integration constants have been written in terms of the direction cosines, c^i, and of the constant magnitude of the velocity vector, $|\vec{v}|$. All this brings to the well-known First Newton's Law: *a free-particle maintains constant its velocity in magnitude and direction. If a free-particle is at rest, it will remain at rest.*

Let's conclude this section by noting that Eq.(1.27) has a very simple and important, geometrical interpretation: *a test-particle not subjected to external forces moves along geodesics of the space [see Box 1.1 and Eq.(B1.1.d)].*

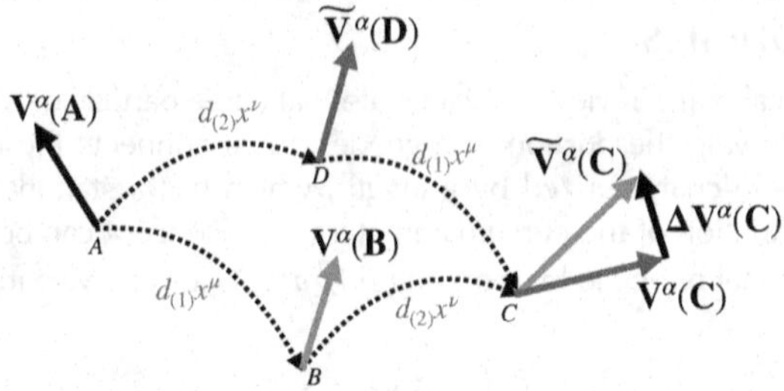

Figure 1.3 The position vectors $\vec{\xi}$ and $\vec{\xi}'$ of the particle P *w.r.t.* two inertial frames, $\mathscr{K}$ and $\mathscr{K}'$, in uniform relative motion

1.6 INERTIAL FRAMES AND GALILEO'S RELATIVITY PRINCIPLE

We can clearly revert the argument given in the last section and say that the First Newton's Law is valid only in those reference frames where the space appears to be homogeneous and isotropic. These reference frames are called *inertial* reference frames and slide one *w.r.t.* the others with constant velocities. Let's consider two of these inertial reference frames, $\mathscr{K}$ and $\mathscr{K}'$. Both of them have a Cartesian coordinate system, with each of the basis vectors of $\mathscr{K}$ parallel to the corresponding ones of $\mathscr{K}'$. So, we can write $\xi^i = \xi'^i + \xi^i_{\mathscr{O}'}(t)$ (see Figure 1.3), where ξ^i and $\xi^i_{\mathscr{O}'}$ are the contravariant components of two position vectors, those of the particle and of the origin $\mathscr{O}'$ in $\mathscr{K}$. Likewise, ξ'^i are the contravariant components of the particle position in $\mathscr{K}'$. Note that unlike the case of Eq.(1.3), here $\xi^i_{\mathscr{O}'}$ changes with time. This brings us to the *velocity composition law* by Galileo: *the velocity of a test-particle in $\mathscr{K}$ is given by its velocity in $\mathscr{K}'$ plus the dragging speed of $\mathscr{K}'$* w.r.t. *$\mathscr{K}$*. In formulae,

$$\frac{d\xi^i}{dt} = \frac{d\xi'^i}{dt} + \frac{d\xi^i_{\mathscr{O}'}}{dt} \tag{1.28}$$

There are two assumptions behind Eq.(1.28). First, most important, the time flows *at the same rate* both in $\mathscr{K}$ and in $\mathscr{K}'$. Secondly, $\mathscr{K}$ and $\mathscr{K}'$ are uniformly sliding one *w.r.t.* to the other: $d^2\xi^i_{\mathscr{O}'}/dt^2 = 0$. Then,

$$\frac{d^2\xi^i}{dt^2} = \frac{d^2\xi'^i}{dt^2} = 0 \tag{1.29}$$

As expected, the accelerations of a free test-particle vanish both in $\mathscr{K}$ and in $\mathscr{K}'$. This leads to the *Galileo*'s Relativity Principle: *the classical laws of motion are valid in all the inertial frames, as accelerations are unaffected by the uniform, relative motion of these frames.*

1.7 THE DERIVATIVE OF A VECTOR

In the previous section, we have evaluated the velocity of a test-particle by taking the derivative of its position vector. As we will see later in the book, the derivative

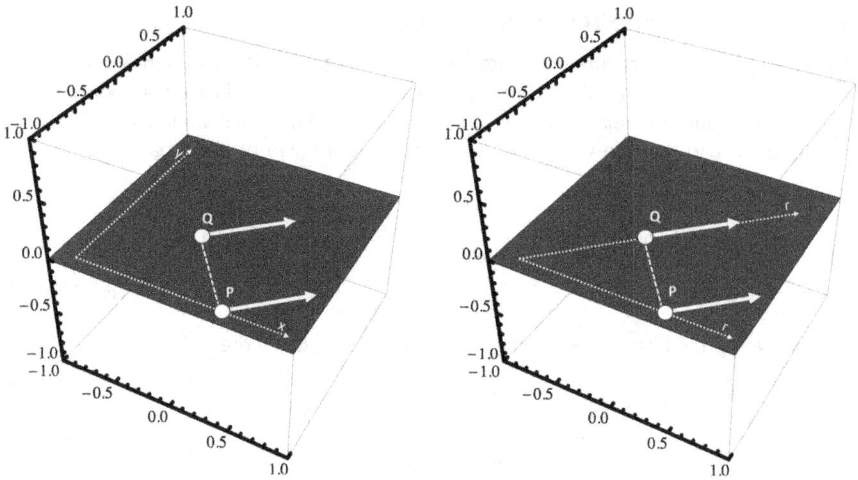

Figure 1.4 Left panel (a): parallel transport and Cartesian coordinates; Left panel (b): parallel transport and polar coordinates. See text.

of a vector is a very delicate operation that must be performed in a very careful way. For the sake of simplicity, let's restrict ourselves to a 2D vector field in a plane: $\vec{V} = V^i(\rho^k)\hat{e}_i$, ρ^k being the components of the position vector. Now, let's ask how this vector changes from $P(\rho^k)$ to $Q(\rho^k + d\rho^k)$. To assess the vector variation, we have first to *parallel transport* the vector $\vec{V}(\rho^k)$ from P to Q to obtain $\vec{V}'(\rho^k + d\rho^k)$. This is done by keeping constant the angle between the vector and the tangent to the geodesic connecting P to Q. In our case, the geodesic is a straight line and its unit tangent vector, $d\xi^k/d\ell$, points always in the same direction [see Box 1.1 and Eq.(B1.1.d); see also Figure 1.4a]. The infinitesimal variation of the vector can then be defined as $D\vec{V} = \vec{V}(\rho^k + d\rho^k) - \vec{V}'(\rho^k + d\rho^k)$, the two vectors being now both bound to Q. From a more formal point of view, we can write

$$D\vec{V} = \frac{\partial}{\partial \rho^k}\left(V^i\hat{e}_i\right) d\rho^k = \frac{\partial V^i}{\partial \rho^k} d\rho^k \hat{e}_i + V^i \frac{\partial \hat{e}_i}{\partial \rho^k} d\rho^k \tag{1.30}$$

that is,

$$D\vec{V} = \left(\frac{\partial V^i}{\partial \rho^k} d\rho^k + V^j \Gamma^i_{jk} d\rho^k\right)\hat{e}_i \tag{1.31}$$

where in the last equality we used the Christoffel symbols defined in Box 1.2 [*c.f.* Eq.(B1.2.d)]. This equation has a simple interpretation if we rewrite it as follows:

$$DV^i = dV^i - \delta V^i \tag{1.32}$$

BOX 1.2 THE CHRISTOFFEL SYMBOLS

Depending on the particular problem we are dealing with, it could sometimes be better to use polar rather than Cartesian coordinates, the two being connected by well-known, standard relations: $x = r\cos\phi$; $y = r\sin\phi$. An infinitesimal 2D displacement vector can then be written by using either Cartesian or polar basis vectors.

$$d\vec{\ell} = dx\ \hat{e}_x + dy\ \hat{e}_y = \left(\frac{\partial x}{\partial r}dr + \frac{\partial x}{\partial \phi}d\phi\right)\hat{x} + \left(\frac{\partial y}{\partial r}dr + \frac{\partial y}{\partial \phi}d\phi\right)\hat{y} \tag{B1.2.aa}$$

$$= dr\ \hat{e}_r + d\phi\ \hat{e}_\phi \tag{B1.2.ab}$$

By comparing the terms proportional to dr and $d\phi$, we can write

$$\begin{aligned} \hat{e}_r &= \left(\frac{\partial x}{\partial r}\hat{x} + \frac{\partial y}{\partial r}\hat{y}\right) = \cos\phi\hat{x} + \sin\phi\hat{y} \\ \hat{e}_\phi &= \left(\frac{\partial x}{\partial \phi}\hat{x} + \frac{\partial y}{\partial \phi}\hat{y}\right) = -r\sin\phi\hat{x} + r\cos\phi\hat{y} \end{aligned} \tag{B1.2.b}$$

Note that $\hat{e}_r$ is a unit vector, whereas $\hat{e}_\phi$ is not. When we move from a point $P(r,\phi)$ to a point $Q(r+dr, \phi+d\phi)$, the polar basis vectors are expected to change in directions.

$$\frac{\partial \hat{e}_r}{\partial r} = 0; \qquad \frac{\partial \hat{e}_r}{\partial \phi} = \frac{\partial \hat{e}_\phi}{\partial r} = \frac{\hat{e}_\phi}{r}; \qquad \frac{\partial \hat{e}_\phi}{\partial \phi} = -r\hat{e}_r \tag{B1.2.c}$$

The derivatives of the basis vectors can then be expressed in terms of the basis vectors themselves. In general, we can write

$$\frac{\partial \hat{e}_j}{\partial \rho^k} = \Gamma^i_{\ jk}\hat{e}_i \tag{B1.2.d}$$

where $d\rho^k \equiv \{dr, d\phi\}$ and the coefficients $\Gamma^i_{\ jk}$ are the so-called *affine connections* or *Christoffel symbols*. We will come back to them later in the book. By comparing Eq.(B1.2.c) and Eq.(B1.2.d) it is straightforward to show that there are only two non-vanishing Christoffel symbols.

$$\Gamma^1_{22} = -r; \qquad \Gamma^2_{12} = \frac{1}{r} \tag{B1.2.e}$$

A more formal definition of the Christoffel symbols can be given in terms of the coordinate transformation:

$$\Gamma^i_{\ jk} = \frac{\partial x^i}{\partial \xi^l}\frac{\partial \xi^l}{\partial x^j \partial x^k} \tag{B1.2.f}$$

where $\xi^l \equiv \{x, y\}$ and $x^i \equiv \{r, \phi\}$ are in this case Cartesian and polar coordinates, respectively. It is easy to show that the definition given in Eq.(B1.2.f) provides the results given in Eq.(B1.2.e) [see Exercise A.6]]

We can then state that the *intrinsic variation* of the i-th component of a vector, DV^i, is obtained by subtracting to its *total variation*, $dV^i = (\partial V^i/\partial \rho^k)d\rho^k$, the *spurious variation*, $\delta V^i = -V^j \Gamma^i_{jk} d\rho^k$, introduced by the parallel transport. The *spurious variation* clearly depends on the chosen coordinate system. For Cartesian coordinates, the Christoffel symbols vanish, as the basis vectors remain unchanged when moving from P to Q [*c.f.* Eq.(B1.2.d)]. This is not the case for polar coordinates [see Figure 1.4b]. In this case, $\rho^k \equiv \{r, \phi\}$ and we can use Eq.(B1.2.e) to write

$$DV^{(1)} = \frac{\partial V^{(1)}}{\partial \rho^k} d\rho^k - rV^{(2)} d\phi \qquad DV^{(2)} = \frac{\partial V^{(2)}}{\partial \rho^k} d\rho^k + \frac{V^{(1)}}{r} d\phi + \frac{V^{(2)}}{r} dr \quad (1.33)$$

where the superscripts (1) and (2) identify the contravariant components of $\vec{V}$ and $D\vec{V}$ along $\hat{e}_r$ and $\hat{e}_\phi$, respectively.

1.8 VELOCITY OF INTERACTIONS AND SECOND NEWTON'S LAW

Let's consider an isolated system composed by N interacting particles. In this case, the Lagrangian writes

$$L = \sum_{n=1}^{N} \frac{1}{2} m_{(n)} \dot{\xi}_{(n)i} \dot{\xi}^i_{(n)} - U\left[\xi^i_{(1)}, \ldots, \xi^j_{(N)}\right] \quad (1.34)$$

where the first term gives the kinetic energy of the system, while U is its potential energy. As it is well known, in Classical Mechanics U depends only on the particle positions. This implies that their interactions are *instantaneous* and, then, that they propagate at an infinity velocity. This assumption is intimately related to the existence of an *absolute time* (time flows at the same rate in all reference frames, and all the watches can always be synchronized) and to the *Galilean Principle of Relativity*. In fact, if the velocity of interaction was not infinite, then it should vary from one inertial reference frame to another one, implying that the physical evolution of the system would depend on the chosen reference frame. This would be in clear contradiction with the Galilean Principle of Relativity. The Euler-Lagrange equations for the n-th particle

$$\frac{d}{dt} \frac{\partial L}{\partial \dot{\xi}_{(n)i}} = \frac{\partial L}{\partial \xi_{(n)i}} \quad (1.35)$$

provide

$$\frac{dp_{(n)i}}{dt} = F_{(n)i} \quad (1.36)$$

Here $p_{(n)i} = m_{(n)} \dot{\xi}_{(n)i}$ and $F_{(n)i} = -\partial U / \partial \xi_{(n)_i}$ are the covariant components of the n-th particle momentum and of the force exerted on that particle by all the other ones. Clearly, changing the potential energy by a constant does not affect the dynamics of the system. Eq.(1.36) gives the Second Newton's Law: *the rate of change of momentum of a test-particle is directly proportional to the force applied on it, and it takes place in the direction of the applied force.*

1.9 HOMOGENEITY OF TIME: ENERGY CONSERVATION

As discussed in Section 1.5, the homogeneity of time implies that the Lagrangian cannot depend on time. Then, for a system of particles, we have

$$\frac{dL}{dt} = \sum_{n=1}^{N_p} \frac{\partial L}{\partial \xi^i_{(n)}} \dot{\xi}^i_{(n)} + \sum_{n=1}^{N_p} \frac{\partial L}{\partial \dot{\xi}^i_{(n)}} \ddot{\xi}^i_{(n)} = \sum_{n=1}^{N_p} \left(\frac{d}{dt} \frac{\partial L}{\partial \dot{\xi}^i_{(n)}} \dot{\xi}^i_{(n)} + \frac{\partial L}{\partial \dot{\xi}^i_{(n)}} \ddot{\xi}^i_{(n)} \right) \tag{1.37}$$

that is,

$$\frac{dL}{dt} = \frac{d}{dt} \sum_{n=1}^{N_p} \left(\frac{\partial L}{\partial \dot{\xi}^i_{(n)}} \right) \dot{\xi}^i_{(n)} \tag{1.38}$$

This implies that there is a conserved quantity, the energy of the particle system:

$$E \equiv \sum_{n=1}^{N_p} \left(\frac{\partial L}{\partial \dot{\xi}^i_{(n)}} \right) \dot{\xi}^i_{(n)} - L \tag{1.39}$$

In fact, for an isolated system, $L = T - U$ and we recover the well-known expression for the energy of the system: $E = T + U$. Thus, the conservation of energy for an isolated system directly derives from the homogeneity of time.

1.10 HOMOGENEITY OF SPACE: THE THIRD NEWTON'S LAW

As a consequence of the assumed homogeneity of space, the properties of an isolated system cannot change for a translation of the reference frame [*c.f.* Section 1.3]. Thus, if $\xi^i_{(n)} \to \xi^i_{(n)} + \delta\xi$, then $L \to L + \delta L$. To first order,

$$\delta L = \sum_{n=1}^{N_p} \left(\frac{\partial L}{\partial \xi^i_{(n)}} \right) \delta\xi \tag{1.40}$$

Because of the arbitrariness of $\delta\xi$ the condition $\delta L = 0$ implies that for an isolated system [*c.f.* Eq.(1.35) and Eq.(1.36)]

$$\sum_{n=1}^{N_p} \frac{\partial L}{\partial \xi^i_{(n)}} = -\sum_{i=1}^{N_p} \frac{\partial U}{\partial \xi^i_{(n)}} = \sum_{n=1}^{N_p} F_{(n)i} = 0 \tag{1.41}$$

where, again, $F_{(n)i}$ are the covariant components of the force exerted by the system on the n-th particle. This yields the Third *Newton*'s Law: *to every action, there is always opposed an equal reaction. The mutual actions of two bodies upon each other are always equal in magnitude, but oriented in opposite directions:* $F_{(1)i} = -F_{(2)i}$. Thus, the Third *Newton*'s Law directly derives from the homogeneity of space.

1.11 PLANETARY MOTIONS

One of the great achievement of Classical Mechanics is the successful description of the planetary motions. As we will see in Chapter 10, General Relativity challenges this result. Because of this, we want here to briefly remind the reader about the basic results of Classical Mechanics, as they will be later used as a benchmark in comparing the General Relativity predictions with the astronomical observations.

Let's write the Lagrangian of a test-particle subjected to a central gravitational field, in polar coordinates ($\xi^1 = r$, $\xi^2 = \theta$ and $\xi^3 = \phi$):

$$L = \frac{1}{2}\left(\dot{r}^2 + r^2\dot{\theta}^2 + r^2\sin^2\theta\dot{\phi}^2\right) + \frac{GM}{r} \tag{1.42}$$

The equations of motion are derived by the Euler-Lagrangian equations [*c.f.* Eq.(1.25)]. For $k = 2$, we get

$$\frac{d}{ds}\left(r^2\dot{\theta}\right) - r^2\sin\theta\cos\theta\dot{\phi}^2 = 0 \tag{1.43}$$

that admits the planar solution, $\theta = \pi/2$: the particle moves in the equatorial plane of our chosen polar reference frame. The equation for $k = 3$ (and $\theta = \pi/2$)

$$\frac{d}{ds}\left(r^2\sin^2\theta\dot{\phi}\right) = 0 \tag{1.44}$$

provides the conservation of the angular momentum per unit mass:

$$r^2\dot{\phi} = H \tag{1.45}$$

consistent with Second *Kepler*'s Law: *a line segment joining a planet and the Sun sweeps out equal areas during equal intervals of time.*

The energy per unit mass of the test-particle can be written as follows:

$$\mathscr{E} = \frac{1}{2}\left(\mathrm{v}_{\parallel}^2 + \mathrm{v}_{\perp}^2\right) - \frac{GM}{r} \tag{1.46}$$

Here $\mathrm{v}_{\parallel}^2 = \dot{r}^2$ and $\mathrm{v}_{\perp}^2 = H^2/r^2$ are the radial and transverse velocity components, whereas M is the mass of the central body, the Sun in our case. Since $\mathrm{v}_{\parallel}^2 \geq 0$, Eq.(1.46) yields

$$\mathscr{E} \geq \frac{1}{2}\frac{H^2}{r^2} - \frac{GM}{r} \equiv V_{eff}(r) \tag{1.47}$$

where $V_{eff}(r)$ is the effective potential shown in Figure 1.5a. The condition given in Eq.(1.47) is not fulfilled in the shaded area of Figure 1.5a, the forbidden region of our parameter space. The motion of a test-particle occurs at constant $\mathscr{E}$. Note that if the angular momentum is different from zero, V_{eff} has one minimum at $r_{min} = H^2/GM$. So, we might have circular orbits (for $r = r_{min}$); closed, elliptical orbits for $E < 0$; and open, hyperbolic orbits for $E > 0$. As long as $H \neq 0$, the test-particle cannot be gravitationally captured by the central body because of the potential barrier at small

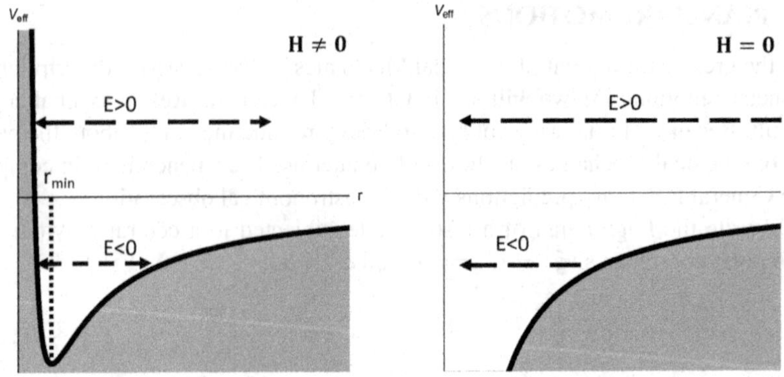

Figure 1.5 Left panel (a): the effective potential as a function of the radial distance for a non-vanishing angular momentum. We can have a circular orbit with radius $r_{min} = H^2/GM$, elliptical orbits with $E < 0$ and hyperbolic trajectories for $E > 0$. Right panel (b): the effective potential as a function of the radial distance for a vanishing angular momentum.

values of r. On the contrary, particles in radial motion, *i.e.*, with $H = 0$, toward the central body will be always captured (see Figure 1.5b).

To find an explicit solution for the radial motion, we could use Eq.(1.42) with $k = 1$. However, it is more convenient to use the energy conservation given in Eq.(1.46). Let's then define a new variable, $u[\phi(t)] = r^{-1}(\phi, t)$, and its derivative, $u' \equiv du/d\phi$. Then, $\dot{r} = -u'\dot{\phi}/u^2 = -u'H$. With these definitions, Eq.(1.46) becomes

$$\frac{1}{2}\left[(u')^2H^2 + u^2H^2\right] - GMu = \mathscr{E} \tag{1.48}$$

This is a non-linear differential equation for u. Instead of directly solving it, let's derive it *w.r.t.* ϕ to get

$$u'u''H^2 + uu'H^2 = GMu' \tag{1.49}$$

This equation clearly admits as a solution $u' = 0$, that is, a circular orbit with $r = r_{min}$ [*c.f.* Figure 1.5a]. Since we want to deal with planetary motion, we are more interested to elliptical rather than circular orbits. So, let's discard the solution $u' = 0$. Eq.(1.49) then provides

$$u'' + u = \frac{GM}{H^2} \tag{1.50}$$

This equation admits the following solution

$$u = \frac{GM}{H^2} + B\cos\phi \tag{1.51}$$

where B is an integration constant to be determined in terms of the initial conditions. Remembering that $u = r^{-1}$, we can rewrite Eq.(1.51) in a more familiar form

$$r = \frac{A}{1 + e\cos\phi} \tag{1.52}$$

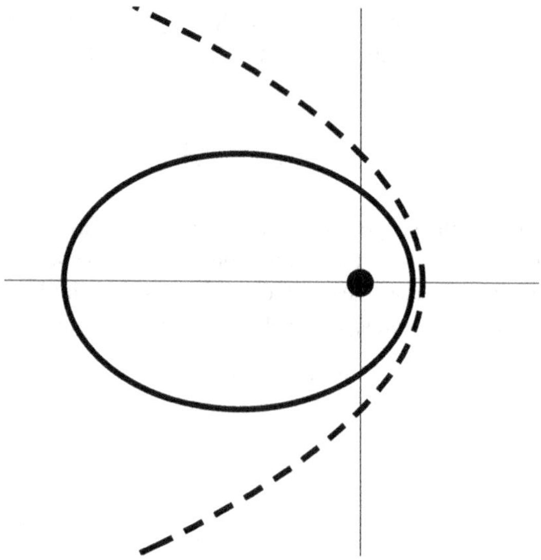

Figure 1.6 Elliptic ($e < 1$) and hyperbolic ($e > 1$) trajectories of a test-particle in a central gravitational field

where $A = H^2/GM$ is an amplitude and $e = H^2B/GM$ is the orbit eccentricity [see Figure 1.6]. If $e < 1$, Eq.(1.52) describes an ellipse, and this brings us to the First *Kepler*'s Law: *the orbit of a planet is an ellipse with the Sun at one of the two foci.*

1.12 NON-INERTIAL REFERENCE FRAMES

Let's conclude this chapter by discussing non-inertial reference frames. These frames play an important conceptual role in passing from Special to General Relativity. In view of this important step, let's briefly review how inertial forces are described in Classical Mechanics. Consider first the lab. reference frame, $\mathscr{K}$, with Cartesian coordinates ξ^k. In this frame, the Lagrangian of a free test-particle is given by Eq.(1.26) and the Euler-Lagrangian equations provide $d^2\xi^k/dt^2 = 0$, as it should be for a free particle [*c.f.* Eq.(1.27)]. Consider now a new reference frame, $\mathscr{X}$, and choose again an orthonormal coordinate system. The $\mathscr{X}$ frame rotates counterclockwise with a constant angular velocity ω around the ξ^3-axis in the lab frame $\mathscr{K}$. In this case, is as follows:

$$\begin{aligned} \xi^1 &= x\cos(\omega t) - y\sin(\omega t) \\ \xi^2 &= x\sin(\omega t) + y\cos(\omega t) \\ \xi^3 &= z \end{aligned} \tag{1.53}$$

where x, y and z are the coordinates of the test-particle in $\mathscr{X}$. Note that Eq.(1.53) describes a *non-linear* coordinate transformation, unlike that described by Eq.(1.8). In fact, the Jacobian matrix, $\partial\xi^i/\partial x^k$, is not a constant anymore, and it is now a function of the coordinates. Thus, we can conclude that non-linear coordinate transformations allow to pass from an inertial to a non-inertial frame. Let's verify this statement by substituting Eq.(1.53) in Eq.(1.26) to get the Lagrangian of the test-particle in the rotating frame $\mathscr{X}$.

$$L = \frac{1}{2}m\left(\dot{x}^2 + \dot{y}^2 + \dot{z}^2 + \omega^2 x^2 + \omega^2 y^2 - 2\omega y\dot{x} + 2\omega x\dot{y}\right) \tag{1.54}$$

Note that in the $\mathscr{X}$ frame the space is neither homogeneous (the Lagrangian depends on the position) nor isotropic (the Lagrangian depends on the single components of the velocity vector). This is not unexpected, as an observer in $\mathscr{X}$ sees a preferred direction, around which $\mathscr{X}$ rotates. The Euler-Lagrangian equations

$$\frac{d}{dt}\frac{\partial L}{\partial \dot{x}^k} = \frac{\partial L}{\partial x^k} \tag{1.55}$$

now provide

$$\begin{aligned} \frac{d^2x}{dt^2} &= \omega^2 x + 2\omega\frac{dy}{dt} \\ \frac{d^2y}{dt^2} &= \omega^2 y - 2\omega\frac{dx}{dt} \end{aligned} \tag{1.56}$$

These are the well-known, classical equations of motion of a free-particle subjected only to the inertial forces in $\mathscr{X}$: the centrifugal, $\omega^2\vec{r}$, and the Coriolis, $-2\vec{\omega}\times\vec{v}$, inertial forces. We will discuss again non-inertial reference frames in Chapter 3, in the framework of four-dimensional space-times.

2 From Space *and* Time to Space-Time

2.1 INTRODUCTION

In Classical Mechanics, the interactions among particles propagate at an infinity velocity. As discussed in Chapter 1, this conclusion rests on two assumptions: the absoluteness of time and the validity of the Galileo's Relativity Principle. The theory of Special Relativity, proposed in 1905 by Einstein [18], combines the Relativity Principle with the existence of a limiting speed for the propagation of the interactions. To obtain this goal, it is necessary to reconsider in depth the concept of time, abandoning the classical idea of its "absoluteness." This is the only way to have the same limiting speed in every inertial frame. The goal of this chapter is to review these concepts, together with the basic notions and formalism of Special Relativity.

2.2 A METRIC SPACE-TIME

To see why the idea of an absolute time has to be abandoned, consider two inertial frames, $\mathscr{K}$ and $\mathscr{K}'$, uniformly sliding one *w.r.t.* to the other. If the speed of light in the vacuum c has to be an invariant, then

$$c = \frac{dl}{dt} = \frac{dl'}{dt'} \tag{2.1}$$

where dl (dl') is the (infinitesimal) coordinate spatial interval travelled by a light signal in a time interval dt (dt') in $\mathscr{K}$ ($\mathscr{K}'$). There are no reasons to assume *a priori* that dl is equal to dl'. Thus, there can't be any reason to infer that dt is equal to dt'. In other words, we have now to take into account a constraint on the ratio between space and time, rather than on space and time separately. It follows that time has to be downgraded from its absolute role in Classical Mechanics to the level of all the other coordinates. Thus, the *time coordinate*, as the spatial ones, *must* depend on the chosen reference frame.

All this can be formalized by using the idea of *event*, but in a deeply different way *w.r.t.* Classical Mechanics. In fact, now an event is defined not in space *and* time, but rather in *space-time*. It follows that the event $\mathscr{E}$ is still characterized by a *when* (one time coordinate, $\xi^o = ct$) and by a *where* (three spatial coordinates, ξ^k with $k = 1, 2, \text{and} 3$). Then, an *event* can now be conveniently represented *geometrically* as a point of a *fictitious* four-dimensional space, the *space-time*. Then, in analogy with Eq.(1.1), we can indicate with ξ^α the *contravariant* components of the position vector of the event $\mathscr{E}(\xi^\alpha)$. So far, we have done nothing but replace the Latin indexes (running from 1 to 3) with Greek indexes (running from 0 to 3). This is the

DOI: 10.1201/9781003141259-2

convention we will adopt hereafter. Note that this extension, by itself, is necessary, but not sufficient for our task. We clearly need a particular space-time where the invariance of the speed of light can be easily enforced for all the inertial observers. This is done by using a hyperbolic geometry that combines the familiar 3D Euclidean space with 1D time dimension. As a result, we have the four-dimensional *Minkowski* space-time, fully described by the so-called *Minkowski* metric:

$$ds^2 = d\xi^{0^2} - d\xi_k d\xi^k = \eta_{\alpha\beta} d\xi^\alpha d\xi^\beta \tag{2.2}$$

where the matrix

$$\eta_{\alpha\beta} = \text{diag}(+1, -1, -1, -1) \tag{2.3}$$

identifies the Minkowski metric coefficients. Let's make two comments. First, the writing of Eq.(2.3) implies choosing an orthogonal coordinate system[1]. Secondly, unlike what was done in Eq.(1.6), we now *algebraically* sum in quadrature the differential of the coordinates. For this reason, we refer to the Minkowski space-time as to a pseudo-Euclidean, four-dimensional flat space. Let's anticipate here what we mean by flat space.

FLAT SPACE

Definition 2.1. *An N-dimensional space is said to be flat if it is possible to find a coordinate system, covering all the space, where the metric can be written in a diagonal form.*

$$\boxed{ds^2 = \varepsilon_\alpha dx^{\alpha 2}} \tag{2.4}$$

The coefficients ε_α can be either positive or negative, although of norm one. According to the Einstein convention, in Eq.(2.4), α runs from 1 (or 0) to N (or $N-1$).

Thus, because of its signature[2] [*c.f.* Eq.(2.3)], the line element of a Minkowski space-time is not necessarily associated to a *positive defined* "distance" between two nearby events, $\mathscr{E}_1(\xi^\alpha)$ and $\mathscr{E}_2(\xi^\alpha + d\xi^\alpha)$. In fact, the Minkowski metric can return positive ($ds^2 > 0$), as well as null ($ds^2 = 0$) or negative ($ds^2 < 0$) values, depending on $d\xi^{0^2}$ being larger, equal or smaller than $d\xi_k d\xi^k$. In these cases, we talk about time-, light- or space-like intervals, respectively. For light-like intervals, we can write

$$(d\xi^0)^2 - d\xi_k d\xi^k = 0 \tag{2.5}$$

in the $\mathscr{K}$ and

$$(d\xi'^0)^2 - d\xi'_k d\xi'^k = 0 \tag{2.6}$$

---

[1] The standard basis for the Minkowski space-time is provided by a set of four mutually orthogonal vectors $\{\hat{e}_0, \hat{e}_1, \hat{e}_2, \hat{e}_3\}$ such that $\hat{e}_0 \cdot \hat{e}_0 = 1 = -\hat{e}_1 \cdot \hat{e}^1 = -\hat{e}_2 \cdot \hat{e}_2 = -\hat{e}_3 \cdot \hat{e}_3$ and $\hat{e}_\alpha \cdot \hat{e}_\beta = 0$ for $\alpha \neq /\beta$. This is the convention that we will be using hereafter in this and the next Chapters.

[2] The signature of a quadratic form is defined in terms of the numbers of vanishing, positive and negative eigenvalues of the quadratic form. Thus, $\text{sign}(\eta_{\mu\nu}) = \{0, 1, 3\}$.

in the $\mathscr{K}'$ inertial frames. This is fully consistent with the requirement of Eq.(2.1). Note that this is still an ansatz, as we have not yet discussed which coordinate transformations links the writing of Eq.(2.5) in $\mathscr{K}$ with the writing of Eq.(2.6) in $\mathscr{K}'$. For the moment, let's stress here that the geometrical formulation of the invariance of the speed of light in the vacuum is structurally embedded in the functional form of the Minkowski metric of Eq.(2.2) and stands on the invariance of line element, ds. It follows that the classification of time-, light- and space-like intervals is an absolute one, regardless of the specific choice of the reference frame we want to use.

2.3 HOMOGENEITY AND ISOTROPY OF THE MINKOWSKI SPACE-TIME

In Classical Mechanics, there is a particular class of reference frames, the *inertial* ones, from where space *and* time share the same properties of being homogeneous and isotropic [*c.f.* Section 1.3]. Here we want to extend these considerations to the Minkowski space-time, showing that there exists a class of reference frames, the *inertial* ones, from where the Minkowski space-time appears to be both homogeneous and isotropic.

Homogeneity

For the Minkowski space-time to be *homogeneous*, all positions in space-time must be equivalent. This means that the choice of the origin of the coordinate system shouldn't really matter. Then, imagine to displace, without rotating, an inertial reference frame where the metric form is given by Eq.(2.2). As a result, the coordinates of all the points of the Minkowski space-time will change accordingly: $\xi^\alpha = \xi'^\alpha + d^\alpha$, where d^α are the (constant) contravariant components of the displacement vector. It is immediate to verify that the metric of Eq.(2.2) remains unchanged under such a rigid displacement as $\eta_{\alpha\beta} = \eta'_{\alpha\beta}$ and $d\xi^\alpha = d\xi'^\alpha$.

Isotropy

For the Minkowski space-time to be *isotropic*, all directions must be equivalent and the orientation of the coordinate axes shouldn't matter. Thus, the metric given in Eq.(2.2) should remain unchanged under a rotation of the coordinate system. In analogy with Eq.(1.8) and Eq.(1.9), let's write this coordinate transformation as follows:

$$\xi^\alpha = L^\alpha{}_\mu\, \xi'^\mu; \qquad \xi'^\mu = M^\mu{}_\alpha\, \xi^\alpha \tag{2.7}$$

where

$$L^\alpha{}_\mu \equiv \frac{\partial \xi^\alpha}{\partial \xi'^\mu}; \qquad M^\mu{}_\alpha \equiv \frac{\partial \xi'^\mu}{\partial \xi^\alpha} \tag{2.8}$$

can be interpreted as "rotation" matrixes in the four-dimensional Minkowski space-time. As such, they must depend only on a single constant parameter, the "rotation" angle. If so, Eq.(2.7) still describes *linear* transformations, where neither $L^\alpha{}_\mu$ nor $M^\mu{}_\alpha$ can depend on the space-time coordinates. The four-by-four matrixes of Eq.(2.8) are one the inverse of the other ($\mathbf{M} = \mathbf{L}^{-1}$), like the are three-by-three matrixes of Eq.(1.9). Then, on the basis of Eq.(2.7) and Eq.(2.8), we can write

$$ds^2 = \eta_{\alpha\beta} d\xi^\alpha d\xi^\beta = \eta_{\alpha\beta} L^\alpha{}_\mu d\xi'^\mu L^\beta{}_\nu d\xi'^\nu = \eta'_{\mu\nu} d\xi'^\mu d\xi'^\nu \tag{2.9}$$

BOX 2.1 VECTORS IN THE MINKOWSKI SPACE-TIME

As done in Section 1.3, we can use Eq.(2.7) to extend its use beyond the class of position vectors. We can then state the following:

• *Contravariant components of a four-vector*
A set of four quantities V^α $(\alpha = 0,3)$ are the contravariant components of a four-vector if they transform, on a change of coordinates, as in Eq.(2.7).

$$V^\alpha = L^\alpha{}_\mu V'^\mu; \qquad V'^\mu = M^\mu{}_\nu V'^\nu \tag{B2.1.a}$$

The lowering of contravariant indexes writes as follows [*c.f.* Eq.(1.17)]

$$d\xi_\alpha = \eta_{\alpha\beta} d\xi^\beta \tag{B2.1.b}$$

With this definition, Eq.(2.2) becomes [*c.f.* Eq.(1.21)].

$$ds^2 = \begin{cases} d\xi_\alpha d\xi^\alpha = d\xi_\alpha L^\alpha{}_\mu d\xi'^\mu \\ d\xi'_\mu d\xi'^\mu = d\xi'_\mu M^\mu{}_\alpha d\xi^\alpha \end{cases} \tag{B2.1.c}$$

We must then conclude that the covariant components of the infinitesimal displacement vector transform as follows:

$$d\xi_\alpha = M^\mu{}_\alpha d\xi'_\mu; \qquad d\xi'_\mu = L^\alpha{}_\mu d\xi_\alpha \tag{B2.1.d}$$

Going beyond the class of position vectors, we can state the following:

• *Covariant components of a four-vector*
A set of four quantities V_α $(\alpha = 0,3)$ are the *covariant components* of a four-vector if they transform, on a change of coordinates, as in Eq.(B2.1.d).

$$V_\alpha = M^\mu{}_\alpha V'_\mu; \qquad V'_\mu = L^\alpha{}_\mu V_\alpha \tag{B2.1.e}$$

Finally, we can extend what was done in Section 1.4, by defining the dot (or inner) product of four-vectors and by stating the following:

• *The inner product of the two four-vectors*
Consistently with Eq.(B2.1.c), the inner product of two four-vectors is given by the sum of the product of their covariant and contravariant components:

$$\mathbf{A} \cdot \mathbf{B} = A_\alpha B^\alpha = A^\beta B_\beta \tag{B2.1.f}$$

The dot product is of course an invariant [see Exercise A.7].

where

$$\eta'_{\mu\nu} = \eta_{\alpha\beta} L^{\alpha}{}_{\mu} L^{\beta}{}_{\nu} \tag{2.10}$$

The space-time is isotropic *if and only if*

$$\eta'_{\mu\nu} = \text{diag}\{+1, -1 - 1 - 1\} \tag{2.11}$$

This is the condition that allows us to specify the form of the "rotation" matrixes given in Eq.(2.8). This is what we want to discuss in the next section.

2.4 ORDINARY VS. HYPERBOLIC ROTATIONS

When we talk about "rotations" in space-time, we have to be more specific. Let's underline two important differences *w.r.t.* to the familiar rotations in a 3D Euclidean space. First, since we are dealing with a four-dimensional space-time, there are now six different possible "rotations," in six different coordinate planes: the (three) space planes, $\xi^i - \xi^j$ (with $i > j$ and $i \neq j$), and the (three) space-time planes, $\xi^i - \xi^0$. The first three are standard rotations around each of the three ξ^i-axes. Thus, a rotation around the ξ^3-axis is described by the following matrixes:

$$L^{\alpha}{}_{\mu} \equiv \begin{pmatrix} 1 & 0 & 0 & 0 \\ 0 & \cos\phi & \sin\phi & 0 \\ 0 & -\sin\phi & \cos\phi & 0 \\ 0 & 0 & 0 & 1 \end{pmatrix}; \quad M^{\mu}{}_{\alpha} \equiv \begin{pmatrix} 1 & 0 & 0 & 0 \\ 0 & \cos\phi & -\sin\phi & 0 \\ 0 & \sin\phi & \cos\phi & 0 \\ 0 & 0 & 0 & 1 \end{pmatrix} \tag{2.12}$$

with the rotation angle bound to be in the interval $0 \leq \phi \leq 2\pi$, as in Eq.(1.10) [3].

The second, and more important, difference has to do with the signature of the Minkowski metric [*c.f.* Eq.(2.3)], proper of a *hyperbolic* geometry. It follows that in order to fulfill the condition given in Eq.(2.11), we have to consider *hyperbolic* rotations in the $\xi^1 - \xi^0$, $\xi^2 - \xi^0$ and $\xi^3 - \xi^0$ planes. For example, a *hyperbolic* rotation in the $\xi^1 - \xi^0$ plane is described by the following rotation matrixes:

$$L^{\alpha}{}_{\mu} \equiv \begin{pmatrix} \cosh\psi & \sinh\psi & 0 & 0 \\ \sinh\psi & \cosh\psi & 0 & 0 \\ 0 & 0 & 1 & 0 \\ 0 & 0 & 0 & 1 \end{pmatrix}; \quad M^{\mu}{}_{\alpha} \equiv \begin{pmatrix} \cosh\psi & -\sinh\psi & 0 & 0 \\ -\sinh\psi & \cosh\psi & 0 & 0 \\ 0 & 0 & 1 & 0 \\ 0 & 0 & 0 & 1 \end{pmatrix} \tag{2.13}$$

where now $-\infty \leq \psi \leq \infty$[4].

Let's conclude this section with a final comment to visualize the difference between standard and hyperbolic rotations. In the standard case, a rotation by an angle

---

[3] We will leave as an exercise to find the equivalent of Eq.(2.12) for rotations around the ξ^2- and ξ^3-axes and to verify that these rotations leave unaltered the metric form of Eq.(2.2) [see Exercises A.8].

[4] We will leave as an exercise to find the equivalent of Eq.(2.13) for hyperbolic rotations in the $\xi^2 - \xi^0$ and $\xi^3 - \xi^0$ planes and to show that these rotations leave unaltered the metric given in Eq.(2.2) [see Exercise (A.9)].

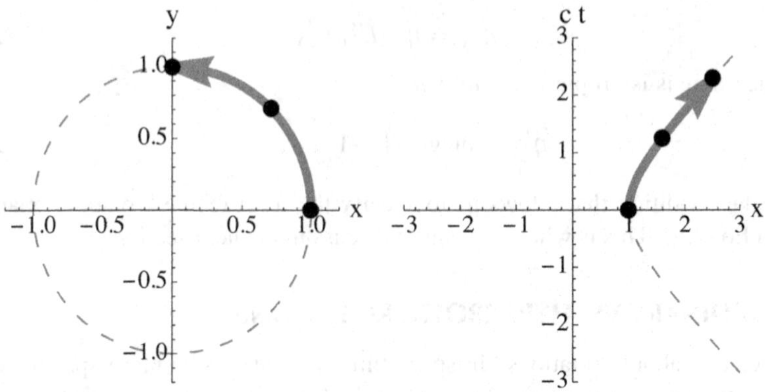

Figure 2.1 Ordinary vs. hyperbolic rotations. Left panel a): an ordinary rotation by an angle $\phi = \pi/2$ moves a point along a circumference. Right panel b): a hyperbolic rotation by an angle $\psi = \pi/2$ moves a point along the branch of a hyperbola.

ϕ moves a point along a circumference (see Figure 2.1a). On the contrary, a hyperbolic rotation by an angle ψ moves a point along the branches of a hyperbola (see Figure 2.1b). Note that the result of the two consecutive hyperbolic rotations generally depends upon the order with which they are performed [see Exercise (A.10)].

2.5 HYPERBOLIC ROTATIONS VS. DRAG VELOCITIES

There is another important difference between ordinary and hyperbolic rotations. Let's discuss this point by considering, as an example, the case of a hyperbolic rotation in the $x-ct$ plane. Then, by using Eq.(2.7) and Eq.(2.13), we can write the coordinate transformation in a more explicit form

$$\begin{cases} ct = \cosh\psi ct' + \sinh\psi x'; \\ x = \sinh\psi ct' + \cosh\psi x'; \\ y = y'; \\ z = z' \end{cases} \qquad \begin{cases} ct' = \cosh\psi ct - \sinh\psi x; \\ x' = -\sinh\psi ct + \cosh\psi x; \\ y' = y; \\ z' = z \end{cases} \tag{2.14}$$

We want to stress again that Eq.(2.14) provides *linear* coordinate transformations that preserve the metric form of a homogeneous and isotropic space-time [*c.f.* Eq.(2.2)]. But the space-time is observed to be homogeneous and isotropic only from inertial frames. We can then conclude that Eq.(2.14) allows us to move from one *inertial* reference frame to another (still *inertial*) one. Then, consider two inertial reference frames, $\mathscr{K}$ and $\mathscr{K}'$, sliding uniformly one *w.r.t.* the other along their common x-axes. The spatial origin $\mathscr{O}'$ of $\mathscr{K}'$ moves *w.r.t.* to $\mathscr{K}$ with a velocity of constant magnitude, $V = x/t$, oriented as the x-axis of $\mathscr{K}$. We can synchronize the watches of $\mathscr{K}$ and $\mathscr{K}'$ by imposing $t = t' = 0$ when the spatial origins $\mathscr{O}$ and $\mathscr{O}'$ coincide. A given event $\mathscr{E}$ has coordinates $\xi^\alpha \equiv \{ct, x, y, z\}$ in $\mathscr{K}$ and $\xi'^\alpha \equiv \{ct', x', y', z'\}$ in

$\mathcal{K}'$. For the spatial origin $\mathcal{O}'$ of $\mathcal{K}'$, we obviously have $x' = y' = z' = 0$. Thus, Eq.(2.14) yields

$$\tanh\psi = \frac{x}{ct} \equiv \frac{V}{c} \equiv \beta \tag{2.15}$$

This shows how a geometrical quantity, the hyperbolic rotation angle ψ, is related to a physical quantity, the uniform drag velocity V of $\mathcal{K}'$ *w.r.t.* $\mathcal{K}$ [see Exercise A.11]. Let's stress the important physical difference between ordinary and hyperbolic rotations. In the former the spatial coordinate axes of $\mathcal{K}'$ are rotated *w.r.t.* $\mathcal{K}$, but without any relative motion of the two frames[5]. Conversely, in the latter, there is not a rotation of the spatial coordinated axes, while there is a uniform, relative motion of the two frames[6]. From Eq.(2.15), it follows that $\cosh\psi = \gamma$ and $\sinh\psi = \beta\gamma$, where $\gamma \equiv (1-\beta^2)^{-1/2}$ is the *Lorentz factor* and $\beta = V/c$. Thus,

$$L^{\alpha}{}_{\mu} \equiv \begin{pmatrix} \gamma & \beta\gamma & 0 & 0 \\ \beta\gamma & \gamma & 0 & 0 \\ 0 & 0 & 1 & 0 \\ 0 & 0 & 0 & 1 \end{pmatrix}; \quad M^{\mu}{}_{\alpha} \equiv \begin{pmatrix} \gamma & -\beta\gamma & 0 & 0 \\ -\beta\gamma & \gamma & 0 & 0 \\ 0 & 0 & 1 & 0 \\ 0 & 0 & 0 & 1 \end{pmatrix} \tag{2.16}$$

and the transformation of Eq.(2.14) can be directly written in terms of the drag velocity.

$$\begin{cases} ct = \gamma(ct' + \beta x'); \\ x = \gamma(x' + \beta ct'); \\ y = y'; \\ z = z' \end{cases} \qquad \begin{cases} ct' = \gamma(ct - \beta x); \\ x' = \gamma(x - \beta ct); \\ y' = y; \\ z' = z \end{cases} \tag{2.17}$$

These are the so-called *Lorentz*'s transformations, connecting—we stress it again—two *inertial* reference frames. Coordinate transformations describing the relative uniform motion of the frames $\mathcal{K}$ and $\mathcal{K}'$ without any rotation of the spatial coordinate axes are called *boosts*. The matrixes in Eq.(2.16) are then called *boosts* (or *Lorentz*) matrixes. As expected, these transformations mix space and time coordinates. As already discussed (see Exercise A.10), Lorentz transformations commute *if and only if* they describe hyperbolic rotations around the same axis. Equivalently, two consecutive boosts commute *if and only if* they occur along the same direction. This is not the case for the two successive Galileo transformations, which always commute independently in the direction of the drag velocity $\vec{V}$ [see Exercise A.12]. Note also that the Lorentz transformations reduce to the classical Galileo transformations for $\beta \to 0$ (or $c \to \infty$) : $t = t'$; $x = x' + Vt'$ (or $x' = x - Vt$); $y = y'$; $z = z'$.

2.6 LORENTZ TRANSFORMATIONS: A GRAPHICAL APPROACH

In order to visualize hyperbolic rotations in a Minkowski space-time, consider the $x-ct$ plane of the lab. reference frame, $\mathcal{K}$. A *world line* is a series of events experienced

[5] If the first frame is inertial, it will be so also the second one, as no relative motion is present.

[6] Again, if the first frame is inertial, it will be so also the second one, as their relative motion is uniform.

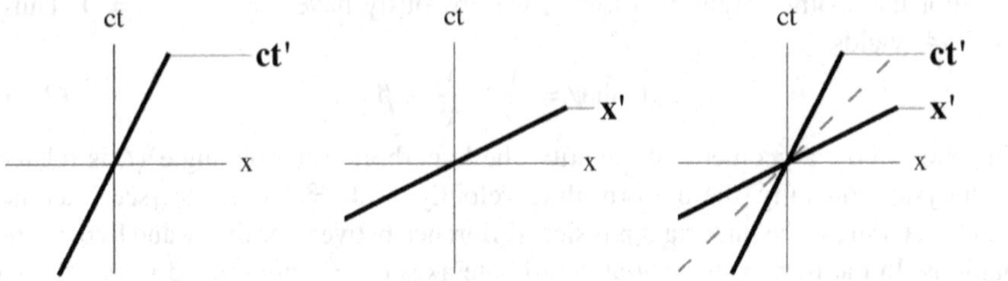

Figure 2.2 The time and space axes of the $\mathscr{K}'$ inertial frame in the lab. frame $\mathscr{K}$ (see text).

by a specific observer, and it is described by a sequential path in space-time. For example, the world line of an inertial observer at rest in the spatial origin $\mathscr{O}$ of $\mathscr{K}$ coincides by definition with the time axis, ct, of $\mathscr{K}$. In an analogous way, the world line of another inertial observer at rest in the spatial origin $\mathscr{O}'$ of $\mathscr{K}'$ coincides by definition with the time axis, ct', of $\mathscr{K}'$. It follows that Eq (2.12) with $x' = y' = z' = 0$ provides the position of the ct' axis in the $x - ct$ plane:

$$\frac{ct}{x} = \frac{1}{\beta} \tag{2.18}$$

This equation describes a straight line with angular coefficient $1/\beta$ (see Figure 2.2a). Following a similar line of reasoning, we conclude that the x-axis of $\mathscr{K}$ is constituted by all the events that in $\mathscr{K}$ are simultaneous with $\mathscr{O}$, that is, those with $t = 0$. Likewise, the x'-axis is constituted by all the events that in $\mathscr{K}'$ are simultaneous with $\mathscr{O}'$, that is, those with $t' = 0$. Thus, we can use Eq (2.17) with $t' = 0$ to identify the position of the x'-axis in the $x - ct$ plane.

$$\frac{ct}{x} = \beta \tag{2.19}$$

This equation still describes a straight line in the $x - ct$ plane, but now with an angular coefficient β (see Figure 2.2b). In conclusion, in a Minkowski space-time, a hyperbolic rotation "squeezes" the $\mathscr{K}'$ coordinate axes, ct' and x', toward the bisector of the first (and third) quadrants of the $x - ct$ plane (see Figure 2.2c). If $\beta \to -\beta$, $\mathscr{K}'$ moves in the opposite direction to that of the x-axis: the ct'- and x'-axes are squeezed now toward the bisector of the second (and fourth) quadrants of the $x - ct$ plane.

Let's conclude this section with few consideration that immediately follows from what we have just discussed. First, the two bisectors ($ct = x$ or $ct = -x$) are defined by light-like events with $ds^2 = 0$ and constitute the so-called *light cone*. Light signal propagates along these bisectors in every reference frame, as a consequence of the invariance of the speed of light. Secondly, the portion of space-time inside the light cone is constituted by time-like events, with $ds^2 > 0$. It is *always possible* to find an inertial frame in which a time-like event occurs at the spatial origin of the chosen reference frame (see Figures 2.3a and 2.3b). It is however *impossible* to find

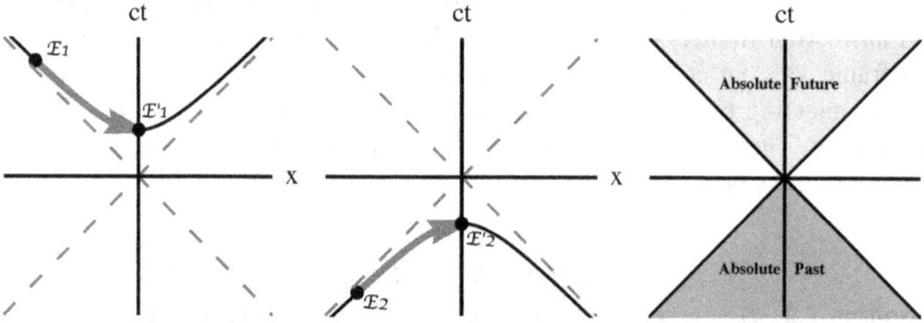

Figure 2.3 The event E_1 (E_2) in the lab. frame $\mathscr{K}$ is moved by the hyperbolic rotation of Eq.(2.14). Thus, we can tune the rotation angle (or the drag velocity) to find a new frame, $\mathscr{K}'$, where this event, E'_1 (E'_2), is seen to occur at the origin $\mathscr{O}'$ but not a $t' = 0$, see text.

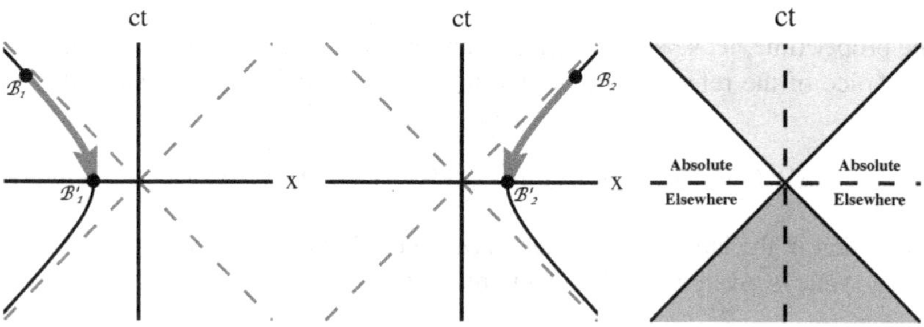

Figure 2.4 The event $\mathscr{B}_1$ ($\mathscr{B}_2$) in the lab. frame $\mathscr{K}$ is moved by the hyperbolic rotation of Eq.(2.14). Thus, we can tune the rotation angle (or the drag velocity) to find a new frame, $\mathscr{K}'$, where this event, $\mathscr{B}'_1$ ($\mathscr{B}'_2$), is seen to occur at $t' = 0$, but not at the origin $\mathscr{O}'$, see text.

a reference frame where time-like events are simultaneous to the one occurring in $\mathscr{O}$ ($\mathscr{O}'$) at $t = 0$ ($t' = 0$). This is why the upper (lower) part of the light cone is called the *absolute future* (*past*) of an observer at the origin $\mathscr{O}$ of $\mathscr{K}$ (see Figure 2.3c). Finally, the portion of space-time outside the light cone contains space-like events, with $ds^2 < 0$. It is *always possible* to find an inertial frame in which a space-like event is simultaneous to the one occurring in $\mathscr{O}$ ($\mathscr{O}'$) at $t = 0$ ($t' = 0$) (see Figures 2.4a and 2.4b). It is however *impossible* to find a reference frame where space-like events are seen to occur at the same position in space. This is why the portion of the space-time outside the light-cone is often called the *absolute elsewhere* (see Figure 2.4c).

2.7 PROPER TIME AND PROPER LENGTH

Proper time

As discussed in Section 2.2, the *when* and the *where* of an event are defined by its space-time coordinates, ξ^α ($\alpha = 0,3$). As such, they depend upon the chosen reference frame [*c.f.* Eq.(2.13)]. In Section 2.6, we have seen that it is always possible to find an inertial frame, $\mathscr{K}'$, where a time-like event occurs at the spatial origin of that frame, $\mathscr{O}'$, but at $t' \neq 0$ [see Figures 2.3a and 2.3b]. Thus, in $\mathscr{K}'$, the line element connecting two time-like, copunctual events reduces to $ds = cdt'$. But ds is an invariant. Then, also this partiuclar dt' must be so. This allows us to define a particular time, the *proper time*.

PROPER TIME

Definition 2.2. *The proper time* τ *is defined as the time elapsed between two time-like events that occur at the* same position *in space*

$$\boxed{d\tau = \frac{ds}{c}} \tag{2.20}$$

The proper time, let's stress it again, is an invariant and, as such, is independent of the choice of the reference frame. On the basis of Eq.(2.20), we can write [*c.f.* Eq.(2.2)]

$$c^2 d\tau^2 = c^2 dt^2 + \eta_{ij} d\xi^i d\xi^j = c^2 dt^2 \left(1 - \delta_{ij} \frac{d\xi^i}{cdt} \frac{d\xi^j}{cdt}\right) \tag{2.21}$$

It follows that in the lab. reference frame, $\mathscr{K}$, the relation between proper time and coordinate time is given by the following relation:

$$d\tau = dt\sqrt{1 - \beta^2} \tag{2.22}$$

where $\beta = V/c$ and $V = \sqrt{\delta_{ij}\dot{\xi}^i\dot{\xi}^j}$ is the magnitude of the dragging speed of $\mathscr{K}'$ *w.r.t.* $\mathscr{K}$[7]. In Special Relativity, Eq.(2.22) describes the well-known *time dilation phenomenon*. In fact, since $0 \leq \beta \leq 1$, then $d\tau \leq dt$: the time of a moving observer, its proper time, flows more slowly than inferred by an observer at rest *w.r.t.* the lab.

Proper length

Since time and space coordinates are alike in Minkowski space-time, we do expect that Lorenz transformations affect not only time—but also space—intervals, that is lengths. Just on the basis of the Lorentz transformation, it is possible to define a particular length, the *proper length*.

PROPER LENGTH

Definition 2.3. *The proper length is the distance between two space-like events that occur at the* same time.

[7] Eq.(2.22) can also be derived by the Lorentz transformations [*c.f.* Eq.(2.17)]. In fact, for an observer in the spatial origin $\mathscr{O}'$ of $\mathscr{K}'$, we have $cdt = \gamma cdt' = \gamma cd\tau$.

Consider once again the inertial reference frame $\mathscr{K}'$ uniformly sliding along the x-axis of the lab. frame $\mathscr{K}$. Consider a ruler at rest *w.r.t.* $\mathscr{K}'$ and aligned with the x' axis. The proper length of the ruler is then given by $L_p \equiv x'_2 - x'_1$. An observer at rest *w.r.t.* $\mathscr{K}$ measures the coordinates of the ruler's extrema (*i.e.*, x_1 and x_2) at the same time t. By using Eq.(2.17), one immediately gets

$$L_p \equiv x'_2 - x'_1 = \gamma(x_2 - x_1) = \gamma L \tag{2.23}$$

implying

$$L = L_p\sqrt{1-\beta^2} \tag{2.24}$$

Thus, the maximum length of a ruler is L_p, the one measured in the reference frame where the ruler is at rest. The length of the same ruler will be shorter than L_p in those reference frames *w.r.t.* which the ruler is moving, *e.g.*, in the lab. frame $\mathscr{K}$. This is the *length contraction* phenomenon of Special Relativity.

2.8 MOTION OF A FREE TEST-PARTICLE

In Section 1.5 we have discussed the variational approach to the study of a free test-particle in Classical Mechanics. Let's extend a similar approach to the case of Special Relativity. Since for the Relativity Principle the motion of a free-particle has to be the same for all the inertial observers, it makes sense to define the action of a free test-particle in terms of an invariant. The obvious invariant to take in consideration is the line element, ds. Remember that for a massive particle, ds is, by definition, time-like and positively defined in *any* reference frame. Then, let's define the action of a free-particle as follows:

$$S = -mc\int_{\mathscr{E}_1}^{\mathscr{E}_2} ds \tag{2.25}$$

Here m is the rest mass of the test-particle, c is the speed of light, while $\mathscr{E}_1$ and $\mathscr{E}_2$ identify the initial and final "positions" of the free-particle in space-time. Because of Eq.(2.22), the action can always be written as follows:

$$S = -mc\int_{\mathscr{E}_1}^{\mathscr{E}_2} cdt\sqrt{1-\beta^2} = \int_{\mathscr{E}_1}^{\mathscr{E}_2} L(\mathrm{v}^2)dt \tag{2.26}$$

where $\beta^2 = \mathrm{v}^2/c^2$ and $\mathrm{v}^2 = \delta_{ij}\dot{\xi}^i\dot{\xi}^j$ is the square magnitude of the particle 3D velocity in the chosen inertial reference frame. It follows that for a free test-particle the Lagrangian is given by

$$L = -mc^2\sqrt{1-\mathrm{v}^2/c^2} \tag{2.27}$$

Note that this Lagrangian depends only on the magnitude of the particle velocity, as in Eq.(1.26), because of the *homogeneity* and the *isotropy* of the Minkowski space-time. These properties are embedded in the functional form of the line element [see Section 2.3]. By imposing that the action is stationary, we arrive to write the Euler-Lagrangian equations [*c.f.* Eq.(1.25)]

$$\frac{d}{dt}\frac{\partial L}{\partial\dot{\xi}^k} = \frac{\partial L}{\partial\xi^k} \tag{2.28}$$

leading to [*c.f.* Eq.(1.27)]

$$\frac{d^2\xi^i}{dt^2} = 0 \quad \Rightarrow \quad \frac{d\xi^i}{dt} = c^i|\vec{v}| \tag{2.29}$$

where the integration constants have again been expressed in terms of the direction cosines, c^i, and of the constant magnitude of the test-particle 3D velocity, $|\vec{v}|$.

2.9 FOUR-VELOCITY AND FOUR-ACCELERATION VECTORS

It is convenient to fully exploit the four-dimensional formalism of the Minkowski space-time to define the contravariant components of the test-particle *four-velocity*

$$u^\alpha \equiv \frac{d\xi^\alpha}{ds} \tag{2.30}$$

and *four-acceleration.*

$$a^\alpha \equiv \frac{d^2\xi^\alpha}{ds^2} \tag{2.31}$$

These are clearly contravariant components of four-vectors. In fact, under a *linear* Lorentz transformation [*c.f.* Eq.(2.7)],

$$\frac{d\xi^\alpha}{ds} = L^\alpha{}_\mu \frac{d\xi'^\mu}{ds}; \qquad \frac{d^2\xi^\alpha}{ds^2} = L^\alpha{}_\mu \frac{d^2\xi'^\mu}{ds^2} \tag{2.32}$$

consistently with the definition given in Eq.(2.8). Being $ds = cdt/\gamma_p$, the test-particle four-velocity has the following contravariant and covariant components:

$$u^\alpha = \gamma_p \left\{1, \frac{\vec{v}_p}{c}\right\}; \qquad u_\beta = \eta_{\beta\mu} u^\mu = \gamma_p \left\{1, -\frac{\vec{v}_p}{c}\right\} \tag{2.33}$$

where the 3D velocity vector, $\vec{v}_p$, has contravariant components $v_p^{(k)} \equiv \dot{\xi}^k$, and magnitude $|\vec{v}_p| = \delta_{ij}\dot{\xi}^i\dot{\xi}^j$. The Lorentz factor writes in the usual form: $1/\gamma_p = \sqrt{1-|\vec{v}_p|^2/c^2}$. Note that, by construction, the components of the particle four-velocity are *adimensional*. Note also that they are not independent. In fact[8],

$$u_\alpha u^\alpha = 1 \tag{2.34}$$

Being an inner product [*c.f.* Eq.(B2.1.f)], Eq.(2.34) is valid in any reference frame. Let's conclude this section by noting that the four-acceleration is always orthogonal to the four-velocity. In fact, by deriving Eq.(2.34) *w.r.t.* ds, one gets

$$2u_\alpha \frac{du^\alpha}{ds} = 0 \tag{2.35}$$

implying that the inner product of four-velocity and four-acceleration vanishes in *any* reference frame.

---

[8] Eq.(2.34) can be verified either by remembering that $ds^2 = d\xi_\alpha d\xi^\alpha$ [*c.f.* Eq.(B2.1.c) and Eq.(2.30)] or by direct substitution: $u'_\alpha u'^\alpha = \eta_{\alpha\beta} u'^\alpha u'^\beta$ [*c.f.* Eq.(B2.1.b) and Eq.(2.33)].

2.10 GEODESICS IN MINKOWSKI SPACE-TIME

As discussed in Box 1.1, a free-particle moves along the geodesic that connects the initial with the final particle positions. So, let's substitute the line element $d\ell$ of Box 1.1 with the line element ds, proper of a Minkowski space-time[9].

$$\delta \int_{\mathscr{E}_1}^{\mathscr{E}_2} ds = \delta \int_{\mathscr{E}_1}^{\mathscr{E}_2} \frac{ds}{d\lambda} d\lambda = 0 \tag{2.36}$$

As in Box 1.1, λ is a parameter defined along the trajectory of the particle, whereas, as usual, $\delta\xi^\alpha(\mathscr{E}_1) = \delta\xi^\alpha(\mathscr{E}_2) = 0$. Eq.(2.2) allows us to write

$$\frac{ds}{d\lambda} = \sqrt{\eta_{\alpha\beta} \frac{d\xi^\alpha}{d\lambda} \frac{d\xi^\beta}{d\lambda}} \tag{2.37}$$

Then, Eq.(2.36) provides

$$\int_{\mathscr{E}_1}^{\mathscr{E}_2} \frac{d\lambda}{ds} \eta_{\alpha\beta} \frac{d\xi^\alpha}{d\lambda} \frac{d\delta\xi^\beta}{d\lambda} d\lambda = \int_{\mathscr{E}_1}^{\mathscr{E}_2} \eta_{\alpha\beta} \frac{d\xi^\alpha}{ds} \frac{d\delta\xi^\beta}{ds} ds = -\int_{\mathscr{E}_1}^{\mathscr{E}_2} \eta_{\alpha\beta} \frac{d^2\xi^\alpha}{ds^2} \delta\xi^\beta ds = 0 \tag{2.38}$$

Because of the arbitrariness of $\delta\xi_\alpha = \eta_{\alpha\beta}\delta\xi^\beta$, we have to conclude that the four-acceleration of a free test-particle must vanish

$$\frac{d^2\xi^\alpha}{ds^2} = 0 \tag{2.39}$$

The four-acceleration is a vector [*c.f.* Eq.(2.32)]. Because of the *linearity* of the Lorentz transformations [*c.f.* Eq.(2.7)], the vanishing of the four-acceleration must be verified in *any* inertial frame [*c.f.* Eq.(2.32)]. Eq.(2.39) clearly implies that

$$\frac{d\xi^\alpha}{ds} = const \tag{2.40}$$

Therefore, in geometrical terms, the vector tangent to the particle trajectory, $d\xi^\alpha/ds$, identifies always the same direction. So, we can state the following:

THE MOTION OF A FREE TEST-PARTICLE

A free test-particle moves along a geodesic, a curve of stationary length [*c.f.* Eq.(2.36)], the straightest line one can draw between two events in the Minkowski space-time [*c.f.* Eq.(2.40)].

Let's conclude this section with two comments. First, note that the straightness of the Minkowski geodesics reflects the flatness of the space-time [*c.f.* Eq.(2.4)]. Secondly, remember that $ds = cdt\sqrt{1-(|\vec{v}_p|/c)^2}$ and that $|\vec{v}_p| = const$: it follows that the spatial components of Eq. (2.39) and Eq.(2.40) are completely consistent with those given in Eq.(2.29).

9 Remember that for a massive test-particle the line element ds is time-like and positive defined.

2.11 FOUR-MOMENTUM

Let's use Eq.(2.28) to define the covariant components of the 3D test-particle momentum [*c.f.* Eq.(1.36)]

$$\boxed{p_k} \equiv \frac{\partial L}{\partial \dot{\xi}^k} = \boxed{-\gamma m \dot{\xi}_k} \tag{2.41}$$

where L is given by Eq.(2.27). On a similar line, we can write the energy of the test-particle as follows [*c.f.* Eq.(1.39) and Eq.(2.27)]:

$$\boxed{E} = \frac{\partial L}{\partial \dot{\xi}^k}\dot{\xi}^k - L = mv^2\gamma - \left(-mc^2\sqrt{1-\beta^2}\right) = \boxed{\frac{mc^2}{\sqrt{1-\beta^2}}} \tag{2.42}$$

where, again, $\beta^2 = \delta_{ij}\dot{\xi}^i\dot{\xi}^j/c^2$. This is a well-known result of Special Relativity, stating the following.

THE REST-MASS ENERGY

When a particle is at rest, its energy does not vanish, but it is instead equal to its *rest-mass energy*: $E = mc^2$. As a consequence, mass can be converted into energy and energy can be converted into mass.

The energy and the momentum of a test-particle can be conveniently written as the components of a single four-vector, the test-particle *four-momentum*

$$p^\alpha \equiv mc\,u^\alpha = \left\{\frac{E}{c}, \vec{p}\right\} \tag{2.43}$$

The squared magnitude of the four-momentum is given by

$$p^\alpha p_\alpha = p^0 p_0 + p^k p_k = \frac{E^2}{c^2} - |\vec{p}|^2 = m^2c^2 \tag{2.44}$$

Thus, the relativistic relation between energy and momentum of a free test-particle can be written as follows:

$$\boxed{\frac{E^2}{c^2} = m^2c^2 + |\vec{p}|^2} \tag{2.45}$$

For ultra-relativistic particles, with either $m = 0$ or $m \ll E/c^2$, Eq.(2.45) provides the well-known result for relativistic particles: $|\vec{p}| = E/c$.

Let's use Eq.(B2.1.a) to find how energy and momentum transform from one inertial to another (still inertial) frame: $p^\alpha = L^\alpha{}_\beta p'^\beta$. By using Eq.(2.13), we find $p^0 = \gamma\left(p'^0 + \beta p'^1\right)$, that is,

$$E = \frac{E' + Vp'^1}{\sqrt{1 - V^2/c^2}} \tag{2.46}$$

where V is the magnitude of the drag velocity of $\mathscr{K}'$, oriented along the x-axis of $\mathscr{K}$. In a similar way, using again Eq.(2.13), we find $p^k = \gamma\left(\beta p'^0 + p'^1\right)$, leading to

$$p^1 = \frac{E'V/c^2 + p'^1}{\sqrt{1 - V^2/c^2}}; \qquad p^2 = p'^2; \qquad p^3 = p'^3 \tag{2.47}$$

2.12 RELATIVISTIC VELOCITY COMPOSITION LAW

Let's conclude this chapter by discussing the relativistic velocity composition law. To do so, consider an inertial frame, $\mathscr{K}'$, uniformly sliding along the x-axis of the lab. frame, $\mathscr{K}$. The corresponding Lorentz transformations are described by the hyperbolic rotation matrix of Eq.(2.12):

$$L^{\alpha}{}_{\beta} = \begin{pmatrix} \gamma_{\mathscr{K}'} & \gamma_{\mathscr{K}'}\beta_{\mathscr{K}'} & 0 & 0 \\ \gamma_{\mathscr{K}'}\beta_{\mathscr{K}'} & \beta_{\mathscr{K}'} & 0 & 0 \\ 0 & 0 & 1 & 0 \\ 0 & 0 & 0 & 1 \end{pmatrix} \tag{2.48}$$

where the subscripts "$\mathscr{K}'$" indicate that both the Lorentz, $\gamma_{\mathscr{K}'}$, and the $\beta_{\mathscr{K}'}$ factors depend on the dragging velocity V of $\mathscr{K}'$ *w.r.t.* $\mathscr{K}$. The contravariant components of the particle four-velocity in $\mathscr{K}$ are then given by $u^{\alpha} = L^{\alpha}{}_{\beta}u'^{\beta}$, where $u'^{\beta} = \gamma_p(1, \vec{\mathrm{v}}'_p/c)$ [*c.f.* Eq.(2.33)] and $\vec{\mathrm{v}}'_p \equiv \{\mathrm{v}'^{(x)}_p, \{\mathrm{v}'^{(y)}_p, \{\mathrm{v}'^{(z)}_p\}$.

$$u^{\alpha} \equiv \gamma_p \left\{ \gamma_{\mathscr{K}'} \left(1 + \frac{\mathrm{v}'^{(x)}_p V}{c^2} \right), \gamma_{\mathscr{K}'} \frac{\mathrm{v}'^{(x)}_p + V}{c}, \frac{\mathrm{v}'^{(y)}_p}{c}, \frac{\mathrm{v}'^{(z)}_p}{c} \right\} \tag{2.49}$$

To have a more direct link with the Galilean velocity composition law, let's derive the transformation law of the spatial components of the particle four-velocity. First, note that

$$u^i = \frac{d\xi^i}{ds} = \frac{d\xi^0}{ds}\frac{d\xi^i}{d\xi^0} = u^0 \frac{d\xi^i}{cdt} = u^0 \frac{\mathrm{v}^{(i)}_p}{c} \tag{2.50}$$

that is $\mathrm{v}^{(i)}_p = cu^i/u^0$. Let's then use Eq.(2.49) to derive the following expressions:

$$\mathrm{v}^{(x)}_p = \frac{\mathrm{v}'^{(x)}_p + V}{1 + \mathrm{v}'^{(x)}_p V/c^2}; \quad \mathrm{v}^{(y)}_p = \frac{\mathrm{v}'^{(y)}_p \sqrt{1 - V^2/c^2}}{1 + \mathrm{v}'^{(x)}_p V/c^2}; \quad \mathrm{v}^{(z)}_p = \frac{\mathrm{v}'^{(z)}_p \sqrt{1 - V^2/c^2}}{1 + \mathrm{v}'^{(x)}_p V/c^2}; \tag{2.51}$$

Note that for $c \to \infty$, Eq.(2.51) is perfectly consistent with Eq.(1.28), the Galileo's velocity composition law. If the test-particle moves also along the x'-axis of $\mathscr{K}'$ with velocity $\mathrm{v}'^{(x)}_p = \mathrm{v}'_p$, then the only non-vanishing component is given by

$$\mathrm{v}^{(x)}_p \equiv \boxed{\mathrm{v}_p = \frac{\mathrm{v}'_p + V}{1 + V\mathrm{v}'_p/c^2}} \tag{2.52}$$

Note that for a massive test-particle $\mathrm{v}'_p < c$. Also, the dragging velocity of $\mathscr{K}'$ *w.r.t.* to $\mathscr{K}$ has to be less than the speed of light. Then, Eq.(2.52) implies the wanted condition for a massive test-particle in $\mathscr{K}$: $\mathrm{v}_p < c$. In fact, $\lim_{V \to c} \mathrm{v}_p = c$ and $\lim_{\mathrm{v}'_p \to c} \mathrm{v}_p = c$. This (re-)states the invariance of the velocity of light in vacuum.

3 From Inertial to Non-Inertial Reference Frames

3.1 INTRODUCTION

Inertial reference frames are key elements of the logical structure of Classical Mechanics and Special Relativity. However, non-inertial reference frames play a key role in the path from Special to General Relativity. The purpose of this chapter is to use these frames to further discuss the physical meaning of the metric of the space-time and, most of all, to further highlight the connection between geometry and dynamics that we started to discuss both in Chapters 1 and 2.

3.2 LINEAR VS. NON-LINEAR COORDINATE TRANSFORMATIONS

Inertial reference frames share the properties of being in uniform motion one *w.r.t.* all the others. The Minkowski space-time observed from any of these frames appears to be always homogeneous and isotropic. This is enforced by the Lorentz transformations [*c.f.* Eq.(2.17)] that connect them all. Let's stress again that these transformations are *linear*, as they depend only on the constant drag velocity of one inertial frame *w.r.t.* another (still inertial) one.

For the sake of generality, let's ask what happens if we consider a *non-linear* coordinate transformation. The line of reasoning is as follows. If linear coordinate transformations connect among themselves inertial frames, it is conceivable to presume that a *non-linear* coordinate transformation should allow one to move from an inertial to a *non-inertial* frame. To discuss a concrete example, consider the (inertial) lab. reference frame $\mathscr{K}$, of coordinates ξ^{α}, where the metric has its standard Minkowski form [*c.f.* Eq.(2.2)]. Consider also a new reference frame $\mathscr{X}$, of coordinates $x^{\tau} \equiv \{x^0, x, y, z\}$, rotating counterclockwise around the ξ^3-axis of $\mathscr{K}$ with a constant angular velocity ω. Thus, we can write the coordinate transformation as follows:

$$\xi^0 = x^0 \tag{3.1a}$$

$$\xi^1 = x\cos(\tilde{\omega}x^0) - y\sin(\tilde{\omega}x^0) \tag{3.1b}$$

$$\xi^2 = x\sin(\tilde{\omega}x^0) + y\cos(\tilde{\omega}x^0) \tag{3.1c}$$

$$\xi^3 = z \tag{3.1d}$$

where $x^0 = ct$ and $\tilde{\omega} = \omega/c$. This is very similar to what was done in Chapter 1: Eq.(3.1b-d) are exactly those given in Eq.(1.53). Working within a space-time, we

DOI: 10.1201/9781003141259-3

have also to prescribe how the time coordinate changes under a given transformation. Note that Eq.(3.1a) leaves unchanged the time coordinate. At first sight, this seems odd given the discussion we had in the previous chapter [see, *e.g.*, Eq.(2.17)]. However, since time is just one of the four coordinates, it is up to us to decide what kind of coordinate transformation to consider. In the case of Eq.(3.1a), we choose as a new time coordinate, x^0, the *central time*: this is the time measured by an observer on the rotation axis z, at rest *w.r.t.* the lab. The transformation given by Eq.(3.1) is clearly *non-linear*, with the Jacobian matrix given by

$$\frac{\partial \xi^\alpha}{\partial x^\mu}(x^\tau) = \begin{pmatrix} 1 & 0 & 0 & 0 \\ -\tilde{\omega}x\sin\left(\tilde{\omega}x^0\right) - \omega y\cos\left(\tilde{\omega}x^0\right) & \cos\left(\tilde{\omega}x^0\right) & -\sin\left(\tilde{\omega}x^0\right) & 0 \\ \tilde{\omega}x\cos\left(\tilde{\omega}x^0\right) - \tilde{\omega}y\sin\left(\tilde{\omega}x^0\right) & \sin\left(\tilde{\omega}x^0\right) & \cos\left(\tilde{\omega}x^0\right) & 0 \\ 0 & 0 & 0 & 1 \end{pmatrix} \tag{3.2}$$

Note that this matrix depends explicitly on the coordinates, x^τ, unlike the case of the *linear* Lorentz transformations of Eq.(2.12). Given this matrix, we can then evaluate the metric in the rotating frame $\mathscr{X}$

$$ds^2 = \eta_{\alpha\beta}d\xi^\alpha d\xi^\beta = \eta_{\alpha\beta}\frac{\partial \xi^\alpha}{\partial x^\mu}dx^\mu \frac{\partial \xi^\beta}{\partial x^\nu}dx^\nu = g_{\mu\nu}(x^\tau)dx^\mu dx^\nu \tag{3.3}$$

where the metric coefficients are now given by

$$g_{\mu\nu}(x^\tau) \equiv \eta_{\alpha\beta}\frac{\partial \xi^\alpha}{\partial x^\mu}(x^\tau)\frac{\partial \xi^\beta}{\partial x^\nu}(x^\tau) \tag{3.4}$$

Note the similarity of the transformation law of the metric coefficients [*c.f.* Eq.(1.12) and Eq.(2.10) with Eq.(3.4)]. We will discuss this point in the next chapter from a more formal point of view. Note also that now the metric coefficients, $g_{\mu\nu}$, depend on the position in space-time because the Jacobian matrix does. Then, by using Eq.(3.2) and Eq.(3.4), we find [see Exercise A.13]

$$\begin{aligned} g_{00} &= 1 - \omega^2(x^2+y^2)/c^2 \\ g_{11} &= g_{22} = g_{33} = -1 \\ g_{01} &= \omega y/c \\ g_{02} &= -\omega x/c \end{aligned} \tag{3.5}$$

Thus, in the rotating frame $\mathscr{X}$, the metric of the space-time becomes

$$\boxed{ds^2 = \left[1 - \frac{\omega^2(x^2+y^2)}{c^2}\right]dx^{0 2} - dx^2 - dy^2 - dz^2 + 2\frac{\omega y}{c}dx^0 dx - 2\frac{\omega x}{c}dx^0 dy} \tag{3.6}$$

This expression appears to be more complicated than the Minkowski one [*c.f.* Eq.(2.2)]: i) the metric coefficients are functions of the coordinates; ii) the matrix of the metric coefficients is not diagonal anymore. This is not unexpected, as the space-time observed from $\mathscr{X}$ is neither homogeneous (being on the rotation axes or not does indeed matter) nor isotropic (there is a preferred direction given by the rotation axis z).

The transformation given in Eq.(3.1), as the one given in Eq.(1.53) in the framework of Classical Mechanics, describes what happens when an observer jumps from the lab. on a merry-go-round. Clearly, we must always be able to jump off from the merry-go-round to go back to the lab. So, we want that the transformation of Eq.(3.1) admits its inverse. This is the case, as the Jacobian of Eq.(3.2) is not vanishing.

$$J \equiv \left| \frac{\partial \xi^\alpha}{\partial x^\mu} \right| = 1 \tag{3.7}$$

Thus, the inverse transformation exists and it is described by the following Jacobian matrix:

$$\frac{\partial x^\tau}{\partial \xi^\alpha}[\xi^\beta(x^\tau)] = \begin{pmatrix} 1 & 0 & 0 & 0 \\ \tilde{\omega} y & \cos\left(\tilde{\omega} x^0\right) & \sin\left(\tilde{\omega} x^0\right) & 0 \\ -\tilde{\omega} x & -\sin\left(\tilde{\omega} x^0\right) & \cos\left(\tilde{\omega} x^0\right) & 0 \\ 0 & 0 & 0 & 1 \end{pmatrix} \tag{3.8}$$

By construction, the Jacobian matrixes given in Eq.(3.2) and Eq.(3.8) are one the inverse of the other. Then,

$$\frac{\partial \xi^\alpha}{\partial x^\tau}\frac{\partial x^\tau}{\partial \xi^\beta} = \delta^\alpha_\beta; \qquad \frac{\partial x^\tau}{\partial \xi^\alpha}\frac{\partial \xi^\alpha}{\partial x^\sigma} = \delta^\tau_\sigma \tag{3.9}$$

To verify that we actually can jump off the merry-go-round, let's evaluate how the metric coefficients transform when the observer moves from the rotating to the lab.frame. Now we have to use the Jacobian matrix given in Eq.(3.8)

$$ds^2 = g_{\mu\nu}(x^\tau) dx^\mu dx^\nu = g_{\mu\nu}\frac{\partial x^\mu}{\partial \xi^\alpha} d\xi^\alpha \frac{\partial x^\nu}{\partial \xi^\beta} d\xi^\beta = \tilde{g}_{\alpha\beta} d\xi^\alpha d\xi^\beta \tag{3.10}$$

where

$$\tilde{g}_{\alpha\beta} = g_{\mu\nu}(x^\tau)\frac{\partial x^\mu}{\partial \xi^\alpha}(\xi^\gamma)\frac{\partial x^\nu}{\partial \xi^\beta}(\xi^\gamma) \tag{3.11}$$

By using Eq.(3.5) and Eq.(3.8), it is easy to verify that

$$\tilde{g}_{\alpha\beta} = \eta_{\alpha\beta} = \text{diag}\{1,-1,-1,-1\} \tag{3.12}$$

as it should be in the inertial reference frame of the lab [see Exercise A.14].

3.3 MOTION OF A FREE TEST-PARTICLE IN A ROTATING FRAME

We have seen that in an inertial reference frame the four-acceleration of a free test-particle vanishes [*c.f.* Section 2.8].

$$\frac{d^2 \xi^\alpha}{ds^2} = 0 \tag{3.13}$$

To find the equation of motion of that particle in the rotating reference frame, we can use the Jacobian matrix of Eq.(3.2) and write

$$\frac{d^2\xi^\alpha}{ds^2} = \frac{d}{ds}\left(\frac{dx^\beta}{ds}\frac{\partial\xi^\alpha}{\partial x^\beta}\right) = \frac{d^2x^\beta}{ds^2}\frac{\partial\xi^\alpha}{\partial x^\beta} + \frac{dx^\beta}{ds}\frac{dx^\gamma}{ds}\frac{\partial^2\xi^\alpha}{\partial x^\beta\partial x^\gamma} = 0 \tag{3.14}$$

After multiplying the last equality by $\partial x^\tau/\partial\xi^\alpha$ and renaming $\beta\rightarrow\mu$ and $\gamma\rightarrow\nu$, we get

$$\boxed{\frac{d^2x^\tau}{ds^2} + \Gamma^\tau{}_{\mu\nu}\frac{dx^\mu}{ds}\frac{dx^\nu}{ds} = 0} \tag{3.15}$$

where, as usual, we sum over the dummy indexes μ and ν. The quantities

$$\boxed{\Gamma^\tau_{\mu\nu} = \frac{\partial x^\tau}{\partial\xi^\alpha}\frac{\partial^2\xi^\alpha}{\partial x^\mu\partial x^\nu}} \tag{3.16}$$

are the so-called *Christoffel symbols* that we have already encountered in Box 1.2 [*c.f.* Eq.(B1.2.f]. Note that the Christoffel symbols are symmetric under inversion of the lower indexes: $\Gamma^\lambda_{\mu\nu} = \Gamma^\lambda_{\nu\mu}$. Note also that Eq.(3.15) provides second-order differential equations. Their solutions are uniquely determined by the boundary conditions on the position, x^τ, and the direction of the test-particle (*i.e.*, its four-velocity dx^τ/ds) at some initial time. As it is obvious from Eq.(3.15), the Christoffel symbols contribute to define the four-acceleration of a free-particle in the rotating frame.

Finding the Christoffel symbols by using Eq.(3.16) implies knowing where we came from (the lab. reference frame) and where we ended up (the merry-go-round). This is clearly not very satisfactory. Fortunately, once we have chosen a reference frame, it is possible to express the Christoffel symbols directly in terms of the metric coefficients and of their derivatives in that frame. This can be done by exploiting the transformation law of the metric coefficients. In fact, deriving Eq.(3.4) yields

$$g_{\mu\nu,\lambda} = \eta_{\alpha\beta}\frac{\partial^2\xi^\alpha}{\partial x^\mu\partial x^\lambda}\frac{\partial\xi^\beta}{\partial x^\nu} + \eta_{\alpha\beta}\frac{\partial\xi^\alpha}{\partial x^\mu}\frac{\partial^2\xi^\beta}{\partial x^\nu\partial x^\lambda} \tag{3.17}$$

whereas multiplying Eq.(3.16) by $\partial\xi^\sigma/\partial x^\tau$ yields

$$\frac{\partial^2\xi^\sigma}{\partial x^\mu\partial x^\nu} = \frac{\partial\xi^\sigma}{\partial x^\tau}\Gamma^\tau_{\mu\nu} \tag{3.18}$$

Thus, given Eq.(3.4) and Eq.(3.18), we can rewrite Eq.(3.17) in a more compact way.

$$g_{\mu\nu,\lambda} = g_{\tau\nu}\Gamma^\tau_{\mu\lambda} + g_{\tau\mu}\Gamma^\tau_{\nu\lambda} \tag{3.19}$$

Now let's perform a cyclic permutation of the three free indexes (the last becomes the first and pushes the others on the right). We can then write

$$\begin{aligned} g_{\mu\nu,\lambda} &= g_{\tau\nu}\Gamma^\tau_{\mu\lambda} + g_{\tau\mu}\Gamma^\tau_{\nu\lambda} \\ g_{\lambda\mu,\nu} &= g_{\tau\mu}\Gamma^\tau_{\lambda\nu} + g_{\tau\lambda}\Gamma^\tau_{\mu\nu} \\ g_{\nu\lambda,\mu} &= g_{\tau\lambda}\Gamma^\tau_{\nu\mu} + g_{\tau\nu}\Gamma^\tau_{\lambda\mu} \end{aligned} \tag{3.20}$$

After algebraically summing these equation (the first with a minus sign and the last two with a plus sign), we get

$$-g_{\mu\nu,\lambda}+g_{\lambda\mu,\nu}+g_{\nu\lambda,\mu}=2g_{\tau\lambda}\Gamma^{\tau}_{\nu\mu} \tag{3.21}$$

We can extend the operation of lowering a contravariant index given in Eq.(1.17) and Eq.(B2.1.b) by writing

$$dx_{\alpha}=g_{\alpha\beta}(x^{\tau})dx^{\beta} \tag{3.22}$$

The identity matrix, $\delta_{\mu}{}^{\nu}$, can be formally given by the following expression: $\delta_{\mu}{}^{\nu}=g_{\mu\sigma}\delta^{\sigma\nu}$. The matrix $\delta^{\sigma\nu}$ must necessarily be the inverse of $g_{\mu\sigma}$, and it is indicated with the symbol $g^{\sigma\nu}$, where both the indexes are contravariant. Thus, we can write that

$$g^{\mu\sigma}g_{\sigma\nu}=g_{\nu\rho}g^{\rho\mu}=\delta^{\mu}{}_{\nu} \tag{3.23}$$

Having this in mind, we can multiply Eq.(3.21) by $g^{\alpha\lambda}$ to obtain a self-consistent expression. We can then state the following.

THE CHRISTOFFEL SYMBOLS

Definition 3.1. *Given a reference frame and a coordinate system, the Christoffel symbols can be written in terms of the metric coefficients and of their derivatives consistently evaluated in that frame*

$$\Gamma^{\alpha}_{\mu\nu}=\frac{1}{2}g^{\alpha\lambda}\left(-g_{\mu\nu,\lambda}+g_{\lambda\mu,\nu}+g_{\nu\lambda,\mu}\right) \tag{3.24}$$

We leave as an exercise to use this definition to derive the Christoffel symbols given in Eq.(B1.2.e) [see Exercise A.15]. Then, Eq.(3.15) together with Eq.(3.24) fully defines the trajectory followed by a free-particle in the rotating frame of Section 3.2, without knowing anything about the lab. reference frame from where we actually came. Let's note that the metric coefficients, $g_{\mu\nu}$, describe the geometry of the space-time *as observed* from the chosen reference frame. On the other hand, the Christoffel symbols are defined in terms of the metric coefficients and of their derivatives. Thus, also the Christoffel symbols know the geometry of the space-time *seen from a given reference frame*. So, we can state the following:

THE FOUR-ACCELERATION OF A FREE TEST-PARTICLE

The geometry of space-time, as observed in a given reference frame, defines the four-acceleration of a free-particle in that reference frame [*c.f.* Eq.(3.15)].

3.4 GEODESICS IN A GENERIC SPACE-TIME

The trajectory followed by a free test-particle has a very precise geometrical meaning: it is a geodesic of the space [*c.f.* Box 2.1] or space-time [*c.f.* Section 2.10] within which the particle moves. A geodesic is a line of stationary length. Then, in analogy with Eq.(2.36), we can write

$$\delta \int_{\mathscr{E}_1}^{\mathscr{E}_2} ds = \delta \int L ds = 0 \tag{3.25}$$

where the quadratic form

$$L = g_{\mu\nu}(x^\gamma)\dot{x}^\mu \dot{x}^\nu \tag{3.26}$$

is by construction equal to unity and formally plays the role of a Lagrangian. Here and below, the dot indicates a derivative *w.r.t.* ds. Note that for sake of generality, we assume that the metric coefficients could depend on the position in space-time. The corresponding Euler-Lagrange equations

$$\frac{d}{ds}\left(\frac{\partial L}{\partial \dot{x}^\alpha}\right) = \frac{\partial L}{\partial x^\alpha} \tag{3.27}$$

provide $2d\left(g_{\alpha\nu}\dot{x}^\nu\right)/ds = g_{\mu\nu,\alpha}\dot{x}^\mu\dot{x}^\nu$, that is,

$$g_{\alpha\nu}\frac{d^2x^\nu}{ds^2} + g_{\alpha\nu,\mu}\frac{dx^\mu}{ds}\frac{dx^\nu}{ds} - \frac{1}{2}g_{\mu\nu,\alpha}\frac{dx^\mu}{ds}\frac{dx^\nu}{ds} = 0 \tag{3.28}$$

Since both μ and ν are dummy indexes, we can write $g_{\alpha\nu,\mu}\dot{x}^\mu\dot{x}^\nu = g_{\alpha\nu,\mu}\dot{x}^\mu\dot{x}^\nu/2 + g_{\alpha\mu,\nu}\dot{x}^\mu\dot{x}^\nu/2$. Then Eq.(3.28) becomes

$$g_{\alpha\nu}\frac{d^2x^\nu}{ds^2} + \frac{1}{2}\left(-g_{\mu\nu,\alpha} + g_{\alpha\mu,\nu} + g_{\nu\alpha,\mu}\right)\frac{dx^\mu}{ds}\frac{dx^\nu}{ds} = 0 \tag{3.29}$$

After multiplying Eq.(3.29) by $g^{\tau\alpha}$ and making use of the Christoffel symbols' definition [*c.f.* Eq.(3.24)], we recover Eq.(3.15). Then, in line with the discussion done in Section 2.10, we can state the following:

THE MOTION OF A FREE TEST-PARTICLE

A free-particle subjected only to inertial forces moves along the geodesics of the space-time.

$$\boxed{\frac{d^2x^\tau}{ds^2} + \Gamma^\tau{}_{\mu\nu}\frac{dx^\mu}{ds}\frac{dx^\nu}{ds} = 0} \tag{3.30}$$

3.5 GEODESIC MOTION IN THE ROTATING REFERENCE FRAME

We are now in the position of evaluating the trajectory of a free test-particle in the rotating frame $\mathscr{X}$. Given the metric of Eq.(3.6), the Lagrangian of the problem is given by Eq.(3.26):

$$L = \left[1 - \frac{\omega^2(x^2+y^2)}{c^2}\right](\dot{x}^0)^2 - \dot{x}^2 - \dot{y}^2 - \dot{z}^2 + 2\frac{\omega y}{c}\dot{x}^0\dot{x} - 2\frac{\omega x}{c}\dot{x}^0\dot{y} \tag{3.31}$$

The Euler-Lagrange equations [*c.f.* Eq.(3.27)] provide [see Exercise A.16]

$$\ddot{x}^0 = 0 \tag{3.32a}$$

$$\ddot{x} - \frac{\omega^2}{c^2}x(\dot{x}^0)^2 - 2\frac{\omega}{c}\dot{x}^0\dot{y} = 0 \tag{3.32b}$$

$$\ddot{y} - \frac{\omega^2}{c^2}y(\dot{x}^0)^2 + 2\frac{\omega}{c}\dot{x}^0\dot{x} = 0 \tag{3.32c}$$

$$\ddot{z}^0 = 0 \tag{3.32d}$$

We can easily identify the non-vanishing Christoffel symbols by comparing Eq.(3.32) with Eq.(3.30) . The prefactors of $(\dot{x}^0)^2$ in Eq.(3.32b) and Eq.(3.32c) then provide

$$\Gamma^1_{00} = -\frac{\omega^2}{c^2}x \qquad \Gamma^2_{00} = -\frac{\omega^2}{c^2}y \tag{3.33}$$

The prefactors of $\dot{x}^0\dot{y}$ in Eq.(3.32b) and of $\dot{x}^0\dot{x}$ in Eq.(3.32c) provide $(\Gamma^1_{02} + \Gamma^1_{20})$ and $(\Gamma^2_{01} + \Gamma^2_{10})$. Given the symmetry of the Christoffel symbols, it follows that

$$\Gamma^1_{02} = -\frac{\omega}{c} \qquad \Gamma^2_{01} = \frac{\omega}{c} \tag{3.34}$$

In the non-relativistic limit ($ds \simeq c\,dt$, $\dot{x}^0 \simeq 1$), Eq.(3.32) reduces to

$$\begin{aligned} \frac{d^2x}{dt^2} &= \omega^2 x + 2\omega\frac{dy}{dt} \\ \frac{d^2y}{dt^2} &= \omega^2 y - 2\omega\frac{dx}{dt} \end{aligned} \tag{3.35}$$

These are the classical equations of motion of a particle subjected only to inertial forces [*c.f.* Eq.(1.56)].

At this point, we can draw two important conclusions. First, as expected, a non-linear transformation can allow one to pass from an inertial to a non-inertial reference frame. Secondly, the formalism developed up to now proves to correctly contain the non-relativistic limit. Note that the classical inertial forces, centrifugal and Coriolis forces, are now described in geometrical terms as they are embedded in the Christoffel symbols. So, let's stress again what already stated at the end of Section 3.3.

GEOMETRY AND DYNAMICS

It is the geometry of the space-time as seen from the rotating reference frame that defines the four-acceleration of a free-particle observed in that frame.

3.6 SOMETHING MORE ON PROPER TIME

We can generalize the definition of *proper time* taking into account that the metric coefficients can depend on the coordinates. According to its definition [*c.f.* Eq.(2.15)], we can then state the following.

THE PROPER TIME

Definition 3.2. *The proper time τ is defined as the time elapsed between two time-like events that occur at the* same position *in space. The proper time of an observe in x^α writes then as follows*

$$\boxed{d\tau \equiv \frac{ds}{c} = \sqrt{g_{00}(x^\alpha)}dt} \tag{3.36}$$

In the case of the rotating frame, we can write the metric given in Eq.(3.6) using cylindrical (rather than Cartesian) coordinates to get [see Exercise A.17]

$$ds^2 = \left[1 - \frac{\omega^2 r^2}{c^2}\right] dx^{0^2} - dr^2 - dz^2 - r^2 d\phi^2 - 2\frac{\omega r^2}{c} dx^0 d\phi \tag{3.37}$$

Then, in the rotating frame, the proper and coordinate times are related by the following relation:

$$d\tau = \sqrt{1 - \frac{\omega^2 r^2}{c^2}} dt \tag{3.38}$$

Since ωr is the rotational velocity of a point at distance r from the rotational axis, Eq.(3.38) is consistent with the time dilation formula of Special Relativity [*c.f.* Eq.(2.22); see also Exercise A.18]. Note that if $r = 0$, then $d\tau = dt$: proper time and central time coincide, as expected, given that an observer at $r = 0$ is at rest *w.r.t.* the lab. On the contrary, if $r = c/\omega$, then $g_{00} = 0$ and $d\tau = 0$. This singularity in the time-time metric coefficient reflects the impossibility of building a physical reference frame, realized with clocks and rods, when rotation velocities become larger than the speed of light. So, the chosen set of coordinates cannot cover all the space-time, but only the region with $r < c/\omega$.

Let's conclude this section by noting that it is always possible to write a finite proper time interval by integrating Eq. (2.20):

$$\tau = \frac{1}{c} \int_{\mathscr{E}_1}^{\mathscr{E}_2} ds \tag{3.39}$$

However, it must be stressed that this integral clearly depends on the integration path. Therefore, it is not possible to univocally define a finite proper time interval: only infinitesimal proper time intervals, $d\tau = ds/c$, have the property of being invariants.

3.7 CLOCK SYNCHRONIZATION

Let's consider two observers, A and B, with spatial coordinates x^k and $x^k + dx^k$ in a generic reference frame. The observer A sends a light signal to B, who receives it at time x^0 and, with a mirror, immediately reflects it to A. According to A, the coordinate time interval elapsed from the emission and the receiving of the light signal can be calculated as follows. Write the metric of Eq.(3.3) separating the time-time and space-time from the space-space components.

$$ds^2 = g_{00} dx^{0^2} + 2g_{0i} dx^0 dx^i + g_{ik} dx^i dx^k \tag{3.40}$$

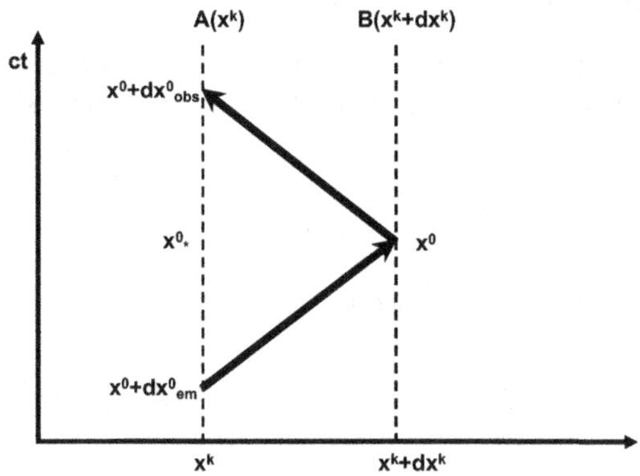

Figure 3.1 The world lines of observer A at position x^k and observer B at position $x^k + dx^k$ is shown for the sake of simplicity in a Minkowski space-time.

For light signals, $ds^2 = 0$. Thus, given for known the metric coefficients, Eq.(3.40) becomes a second-order algebraic equation for dx^0. The two solutions identify the times of the signal emission $x^0 + dx^0_{em}$ (with dx^0_{em} negatively defined), and of the signal observation, $x^0 + dx^0_{obs}$:

$$\begin{pmatrix} dx^0_{obs} \\ \\ dx^0_{em} \end{pmatrix} = -\frac{g_{0i}}{g_{00}} dx^i \begin{pmatrix} + \\ \\ - \end{pmatrix} \sqrt{\frac{1}{g_{00}} \left(-g_{ik} + \frac{g_{0i} g_{0k}}{g_{00}} \right) dx^i dx^k} \tag{3.41}$$

Now, according to A (see Figure 3.1), the light signal was received by B not at x^0, but rather at

$$x^0_\star = \frac{(x_0 + dx^0_{obs}) + (x_0 + dx^0_{em})}{2} = x_0 - \frac{g_{0i}}{g_{00}} dx^i \tag{3.42}$$

that is, at half of the way between the signal emission and receiving times. So, in general, $x^0_* \neq x^0$, unless $g_{0i} = 0$. However, even if this is not the case, we can use the previous expression to synchronize the clocks along any open path. Unfortunately, if $g_{0i} \neq 0$, the synchronization around a closed path is in general impossible, unless

$$\oint \frac{g_{0i}}{g_{00}} dx^i = 0 \tag{3.43}$$

Let's see this point by using as an example the rotating reference frame $\mathscr{X}$ of the merry-go-round. The metric of Eq.(3.37) yields

$$\oint \frac{g_{03}}{g_{00}} d\phi = -\oint \frac{r^2\omega/c}{1-\omega^2 r^2/c^2} d\phi = -2\pi \frac{r^2\omega/c}{1-\omega^2 r^2/c^2} \neq 0 \tag{3.44}$$

Thus, at the end of a closed path, the A's watch can't be synchronized with itself.

3.8 PROPER SPATIAL DISTANCES

Consider again the observers A and B of Section 3.7. According to A, the coordinate time interval between the emission and the receiving of the light signal is given by

$$\Delta x^0 = (x^0 + dx^0_{obs}) - (x^0 + dx^0_{em}) = \frac{2}{\sqrt{g_{00}}}\sqrt{\left(-g_{ik} + \frac{g_{0i}g_{0k}}{g_{00}}\right) dx^i dx^k} \tag{3.45}$$

The proper time interval associated to this coordinate time interval is [*c.f.* Eq.(3.31)]:

$$\Delta\tau = \sqrt{g_{00}}\frac{\Delta x^0}{c} = \frac{2}{c}\sqrt{\left(-g_{ik} + \frac{g_{0i}g_{0k}}{g_{00}}\right) dx^i dx^k} \tag{3.46}$$

Then, it is natural for A to define its proper spatial distance from B as follows:

$$d\ell = \frac{c\Delta\tau}{2} \tag{3.47}$$

Squaring this quantity provides the three-dimensional metric of the purely spatial sector (or spatial hypersurface) of the space-time:

$$d\ell^2 = \gamma_{ik}(x^\tau) dx^i dx^k \tag{3.48}$$

where the 3D metric coefficients

$$\gamma_{ik}(x^\tau) = -g_{ik}(x^\tau) + \frac{g_{0i}(x^\tau)g_{0k}(x^\tau)}{g_{00}(x^\tau)} \tag{3.49}$$

depend on the space-time coodinates, x^τ. If $g_{0k} = 0$, then the γ_{ik} is the space-space minor of the four dimensional metric, with an obvious change of sign. For the Minkowski metric, Eq.(3.49) provides $\gamma_{ik} = diag(+1, +1, +1)$, implying $d\ell^2 = dx^2 + dy^2 + dz^2$. So, we have now formally derived something we have anticipated in Section 2.2: the spatial, 3D hypersurface of a Minkowski space-time has an Euclidean geometry. When using cylindrical coordinate,

$$ds^2 = c^2 d\bar{t}^2 - d\bar{r}^2 - d\bar{z}^2 - \bar{r}^2 d\bar{\varphi}^2 \tag{3.50}$$

and

$$d\bar{\ell}^2 = d\bar{r}^2 + d\bar{z}^2 + \bar{r}^2 d\bar{\varphi}^2 \tag{3.51}$$

are the Minkowski and 3D spatial metrics. As expected in an Euclidean geometry, the length of a circumference of radius $\bar{r} = R$ at $\bar{z} = const$ is

$$\bar{\ell} = \int_0^{2\pi} R d\bar{\varphi} = 2\pi R \tag{3.52}$$

On the other hand, in the rotating reference frame [*c.f.* Eq.(3.37)], the spatial metric writes

$$d\ell^2 = dr^2 + dz^2 + \frac{r^2 d\phi^2}{1 - \omega^2 r^2/c^2} \tag{3.53}$$

and the proper length of a circumference of radius R at $z = const$ will be

$$\ell = \int_0^{2\pi} \frac{R\,d\phi}{\sqrt{1 - \omega^2 R^2/c^2}} = \frac{2\pi R}{\sqrt{1 - \omega^2 R^2/c^2}} > 2\pi R \qquad (3.54)$$

The interpretation of this result is simple, and it is basically due to the Lorentz contraction phenomenon. Consider a measuring rod at rest in lab. reference frame and tangent to the circumference whose length we want to measure. In the lab, we have the result given in Eq.(3.52). In the rotating frame, the same rod will be shorter, and this justifies the result of Eq.(3.54). Remember that verifying only one of the Euclidean relations is a necessary *but not* sufficient condition to state that the geometry is Euclidean. On the contrary, if only one of the Euclidean relations is violated, then this is a necessary *and* sufficient condition for stating that the geometry is non-Euclidean. We can then conclude that a spatial 3D hypersurface as observed from the rotating frame is not-Euclidean.

These results are clearly unsatisfactory. The *apparent* geometry of the 3D space clearly depends upon the chosen reference frame [*c.f.* Eq.(2.2) with Eq.(3.6)]. However, the space-time is the space-time: can we assess the *intrinsic* geometrical properties of our space-time *independently* of the chosen frame and coordinate system? To properly answer to this question, we have first to discuss in a more systematic way some elements of tensor calculus that we have anticipated here and there in the previous chapters. This is the goal of the next two chapters.

4 Pseudo-Riemannian Spaces

4.1 INTRODUCTION

When the dimensionality of a space exceeds the familiar three dimensions of the Euclidean space, a proper visualization is far from being either obvious or intuitive. Nonetheless, it is still possible to work by abstractions and analogies, using in fact a unified geometrical language. The tool necessary for reaching this goal is the tensor calculus[1]. It turns out that the formalism necessary to study the geometry of a space is indeed independent of the space dimensionality. Also, tensor calculus allows us to formulate relations that are indeed independent of the choice of the coordinate system. The goal of this chapter is to present the basics of tensor calculus by reviewing in a more systematic way some of the concepts already introduced here and there in the previous chapters.

4.2 MANIFOLDS

In Chapter 3, we extended the concepts of proper time and proper length of Special Relativity to the case of space-times where, unlike the Minkowski case, the metric coefficients do depend on the space-time coordinates:

$$ds^2 = g_{\mu\nu}(x^\tau)dx^\mu dx^\nu \tag{4.1}$$

When evaluated at a specific point $P = P(x^\tau)$, Eq.(4.1) can be seen as a quadratic form with constant coefficients. This form can always be diagonalized with a suitable coordinate transformation. This implies that the matrix of the metric coefficients can be written in a diagonal form, with its elements given by the eigenvalues of $g_{\mu\nu}$ evaluated at P. These eigenvalues define the signature of the metric. We are of course interested in working with metrics that have the same signature as Minkowski's metric: $\text{sign}(\eta_{\alpha\beta}) = \{0,1,3\}$[2]. This suggests to extend our investigation to a wider class of space-times described by pseudo-Riemannian geometries.

PSEUDO-RIEMANNIAN SPACE-TIMES

Definition 4.1. *A space-time is said to be pseudo-Riemannian if its metric is described by a second-order, symmetric metric tensor, $g_{\mu\nu}$, that* locally *(that is, around but very near to any of its points) can be expressed in terms of the Minkowski metric tensor, $\eta_{\alpha\beta}$.*

---

[1] Tensor calculus was developed by Gregorio Ricci-Curbastro and his student Tullio Levi-Civita [74] at the very beginning of the last century.

[2] The signature of a quadratic form is defined in terms of the numbers of vanishing (N_0), positive (N_+) and negative (N_-) eigenvalues of the quadratic form. Thus, in general, $\text{sign}(g_{\mu\nu}) = \{N_0, N_+, N_-\}$.

DOI: 10.1201/9781003141259-4

Thus, a pseudo-Riemannian space-time has in general a non-vanishing curvature. However, *locally*, it can be very well approximated by a flat, Minkowski space-time.

Curved spaces

The simpler way of visualizing a curved N-dimensional space is to embed it in a flat space of dimension $N+1$. For example, the 2D curved surface of a sphere can be easily visualized by embedding it in a 3D flat Euclidean space. Consider a point P on the surface of the sphere, for example, the North Pole. In a sufficiently small neighborhood of P—that is, *locally*—the surface of the sphere can be well approximated by the flat plane tangent in P [see Figure 4.1a]. This is why we are able to *locally* use a very accurate flat map: let's call it a *chart*. We can then *locally* use Cartesian coordinates, such that the distance between two points on the chart, P and Q say, is correctly described by the Euclidean relation: $\Delta\ell = \sqrt{\Delta x^2 + \Delta y^2}$. Clearly, there is a chart for each point of the sphere [see Figure 4.1b for a point at latitude 45° and Figure 4.1c for a point on the equator]. As it is well known, it is *not* possible to cover *all* the surface of the 2D sphere with just a single (flat) chart. So, we are forced to "tesselate" the surface of the sphere with a large number of flat charts [see Figure 4.2]. Since different charts have their own independent coordinate system, we have to find a way to *connect* (infinitesimally) nearby ones. This can be done in different ways. For example, one can imagine to *rigidly* slide one chart into the nearby one.

We can generalize these considerations to an N-dimensional surface, which is *locally* very similar to a flat N-dimensional hyperplane, but *globally* very different from it. The mathematical object that formalizes this intuitive picture is the manifold.

MANIFOLDS AND CHARTS

Definition 4.2. *An N-dimensional manifold is constituted by a set of points, each identified by a N-tuple of numbers (i.e., their "coordinates"). The points of an N-dimensional manifold can be* locally *mapped* one-to-one *with the point belonging to a portion of a flat N-dimensional space, where a* local *coordinate system has been used. This* one-to-one *mapping is called a chart.*

Pseudo-Riemannian manifolds

By construction, a manifold can be "visualized" as a smooth surface without any discontinuity. On the basis of the Definition 4.1, a pseudo-Riemannian space-time is a (curved) manifold. Given a point P of this manifold, it is possible to find a chart—that is, a flat Minkowski space-time tangent to the manifold at that point. Because of this, we can easily extend to the manifolds concepts like time-, light- and space-like intervals discussed in the framework of Special Relativity. In particular, the condition $ds^2 = 0$ enforces also in pseudo-Riemannian space-times the invariance of the speed of light.

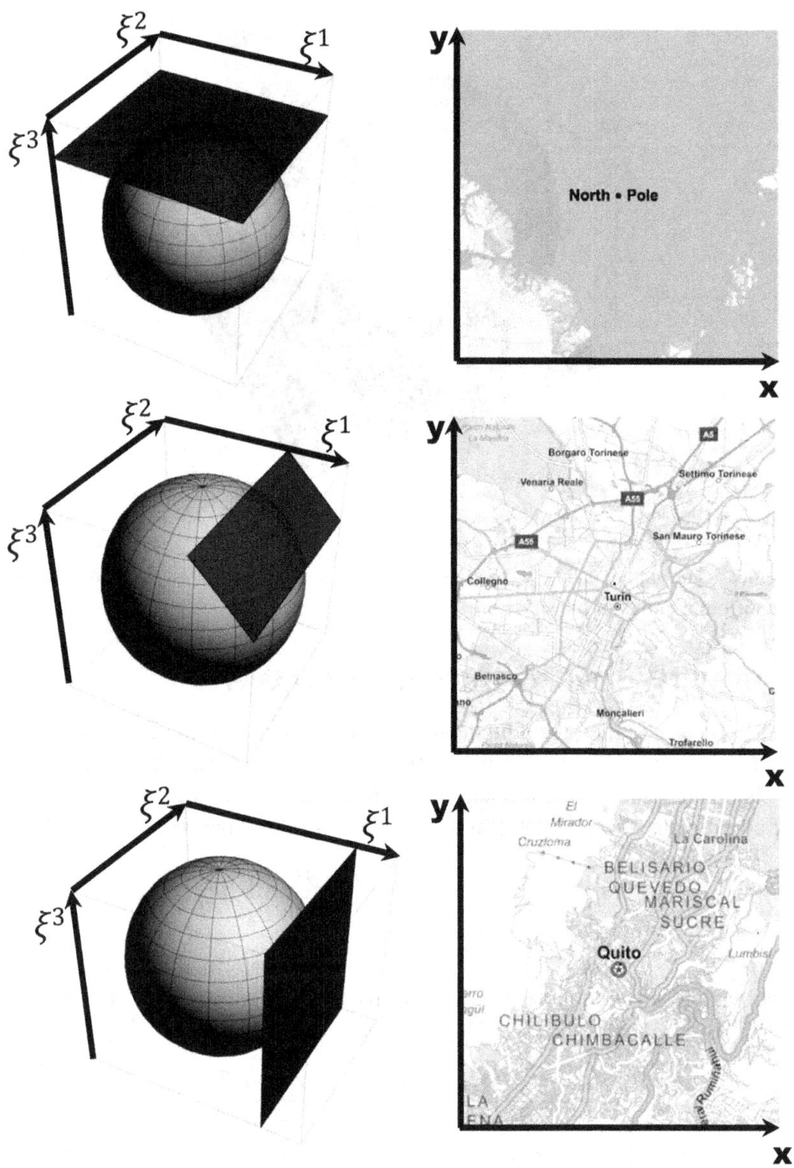

Figure 4.1 A 2D spherical surface (a *manifold*) is embedded in a flat 3D Euclidean space. The chosen coordinate system is Cartesian, with axes ξ^1, ξ^2 and ξ^3. The flat (dark gray) 2D plane is tangent to the sphere at the North Pole (top-left panel) at a point of latitude $\theta = 45°$ (center-left panel), and at a point on the equator (bottom-left panel). *Locally*, it is possible to perform a "one-to-one" mapping of the points of the sphere into the points of the tangent flat planes, obtaining *charts* of the polar cap (top-right panel), of the center of Turin, Italy (middle-right panel), or Quito, Ecuador (bottom-right panel).

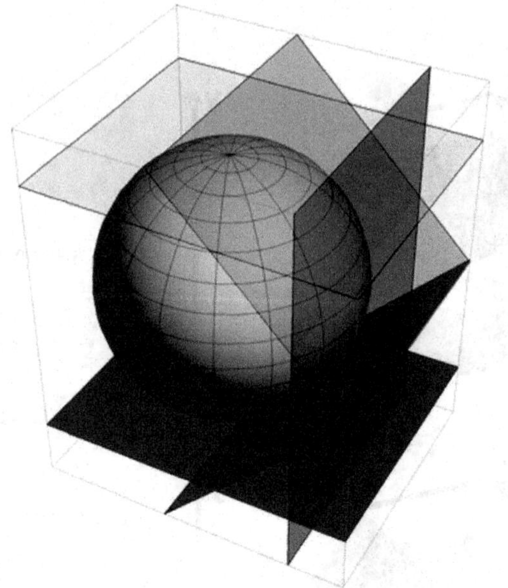

Figure 4.2 The sphere can be "tessellated" with an infinite number of flat planes, each tangent to different points of the sphere.

4.3 HOW TO MOVE TO THE TANGENT PLANE?

In Eq.(3.16), we defined the Christoffel symbols in terms of the ξ^α and x^μ coordinates of an inertial ($\mathscr{K}$) and a generic ($\mathscr{X}$) frame. Let's now ask how the Christoffel symbols change under a coordinate transformation from frame $\mathscr{X}$ to a new frame $\overline{\mathscr{X}}$ of coordinates $\bar{x}^\tau = \bar{x}^\tau(x^\mu)$. It is straightforward to derive [see Exercise A.19]

$$\Gamma^\lambda_{\mu\nu} = \frac{\partial x^\lambda}{\partial \bar{x}^\tau}\frac{\partial^2 \bar{x}^\tau}{\partial x^\mu \partial x^\nu} + \frac{\partial x^\lambda}{\partial \bar{x}^\rho}\frac{\partial \bar{x}^\tau}{\partial x^\mu}\frac{\partial \bar{x}^\sigma}{\partial x^\nu}\overline{\Gamma}^\rho_{\tau\sigma} \tag{4.2}$$

where $\overline{\Gamma}^\lambda_{\mu\nu} = (\partial^2\xi^\alpha/\partial\bar{x}^\mu\partial\bar{x}^\nu)(\partial\bar{x}^\lambda/\partial\xi^\alpha)$. If the transformation is single valued, continuous and differentiable, it will have its inverse: $x^\mu = x^\mu(\bar{x}^\tau)$. It follows that $(\partial x^\lambda/\partial\bar{x}^\sigma)\times(\partial\bar{x}^\sigma/\partial x^\nu) = \delta^\lambda{}_\nu$. Deriving this relation yields

$$\frac{\partial x^\lambda}{\partial \bar{x}^\sigma}\frac{\partial^2 \bar{x}^\sigma}{\partial x^\mu \partial x^\nu} = -\frac{\partial^2 x^\lambda}{\partial \bar{x}^\sigma \partial \bar{x}^\tau}\frac{\partial \bar{x}^\tau}{\partial x^\mu}\frac{\partial \bar{x}^\sigma}{\partial x^\nu} \tag{4.3}$$

Then, Eq.(4.2) can be rewritten as follows

$$\boxed{\Gamma^\lambda_{\mu\nu} = \frac{\partial x^\lambda}{\partial \bar{x}^\rho}\frac{\partial \bar{x}^\tau}{\partial x^\mu}\frac{\partial \bar{x}^\sigma}{\partial x^\nu}\overline{\Gamma}^\rho_{\tau\sigma} - \frac{\partial^2 x^\lambda}{\partial \bar{x}^\sigma \partial \bar{x}^\tau}\frac{\partial \bar{x}^\tau}{\partial x^\mu}\frac{\partial \bar{x}^\sigma}{\partial x^\nu}} \tag{4.4}$$

Let's assume, for the sake of simplicity, that the two frames $-\overline{\mathscr{X}}$ and $\mathscr{X}-$ share the same origins, $\overline{\mathscr{O}} \equiv \mathscr{O}$. Then, let's consider a *non-linear* transformation to move from

$\overline{\mathscr{X}}$ to $\mathscr{X}$ and

$$\boxed{x^\alpha = \overline{x}^\alpha + \frac{1}{2}\overline{\Gamma}^\alpha_{\sigma\tau}(\overline{\mathscr{O}})\overline{x}^\sigma\overline{x}^\tau} \tag{4.5}$$

where

$$\left.\frac{\partial x^\alpha}{\partial \overline{x}^\tau}\right|_{\overline{\mathscr{O}}} = \delta^\alpha{}_\tau; \qquad \left.\frac{\partial \overline{x}^\mu}{\partial x^\beta}\right|_{\mathscr{O}\equiv\overline{\mathscr{O}}} = \delta^\mu{}_\beta; \qquad \left.\frac{\partial^2 x^\alpha}{\partial \overline{x}^\mu \partial \overline{x}^\nu}\right|_{\overline{\mathscr{O}}} = \overline{\Gamma}^\alpha_{\mu\nu}(\overline{\mathscr{O}}) \tag{4.6}$$

We want to claim that the transformation given in Eq.(4.5) is the one needed to move from a point $\overline{\mathscr{O}}$ of a pseudo-Riemannian space-time to the point $\mathscr{O}$ of the tangent Minkowski space-time. If this is the case, we should find that the Christoffel symbols evaluated at the origin $\mathscr{O}$ of the $\mathscr{X}$ frame are identically equal to zero. This would be consistent with Eq.(3.24): if the metric is Minkowskian, the Christoffel symbols vanish. It would also be consistent with the geodesic equation [*c.f.* Eq.(3.30)]: in a Minkowskian frame, the four-acceleration of a free-particle vanishes [*c.f.* Eq.(2.39)]. To verify the statement, let's replace Eq.(4.6) in Eq.(4.4) to get

$$\left.\Gamma^\lambda_{\mu\nu}\right|_{\mathscr{O}} = \delta^\lambda{}_\rho\delta^\tau{}_\mu\delta^\sigma{}_\nu\overline{\Gamma}^\rho_{\tau\sigma}(\overline{\mathscr{O}}) - \overline{\Gamma}^\lambda_{\tau\sigma}(\overline{\mathscr{O}})\delta^\tau{}_\mu\delta^\sigma{}_\nu = 0 \tag{4.7}$$

as indeed expected in a tangent Minkowski space-time. We will be back on this important point in Chapter 6.

4.4 VECTOR FIELDS

Let's consider a pseudo-Riemannian space-time, and let's choose two different frames, $\mathscr{X}$ and $\mathscr{X}'$ say. These frames are completely arbitrary, in no way connected neither with the lab. nor with the merry-go-round frame of Chapter 3. A generic event in space-time is identified by a set of four coordinates: x^τ in $\mathscr{X}$ and x'^τ in $\mathscr{X}'$. The two sets of coordinates are linked together by a single-valued, continuous and differentiable function: $x'^\mu(x^\tau)$. When the Jacobian of transformation, $J = |\partial x'^\mu/\partial x^\tau|$, does not vanish, it is possible to find the inverse transformation, mapping the new coordinates to the old ones: $x^\alpha(x'^\nu)$. Both these transformations can in principle be *non-linear* and arbitrarily complex. However, if we limit ourselves to the coordinate differentials, these transformations simplify in a very significant way.

$$dx'^\mu = J^\mu{}_\nu(x^\beta)dx^\nu; \qquad dx^\mu = K^\mu{}_\nu(x'^\gamma)dx'^\nu \tag{4.8}$$

Here $J^\mu{}_\nu(x^\beta) \equiv \partial x'^\mu/\partial x^\nu|_{x^\beta}$ and $K^\mu{}_\nu \equiv \partial x^\mu/\partial x'^\nu|_{x'^\gamma}$ are the corresponding Jacobian matrixes, satisfying the obvious conditions $K^\mu{}_\alpha J^\alpha{}_\nu = \delta^\mu{}_\nu$ and $J^\alpha{}_\mu K^\mu{}_\beta = \delta^\alpha{}_\beta$. Eq.(4.8) highlights an important point: the *differentials* of the coordinates undergo *linear* and *homogeneous* transformations, whose coefficients [either $J^\alpha{}_\mu(x^\beta)$ or $K^\alpha{}_\mu(x'^\gamma)$] are evaluated at a specific position in space-time [either x^β or x'^γ]. On the other hand, the coordinate differentials can be interpreted as the contravariant components of an infinitesimal displacement vector [*c.f.* Eq.(1.2) and Eq.(2.2)]. Then, we can extend the definitions already given in Eq.(1.13) and Eq.(B2.1.a) to the case of an arbitrary coordinate transformation [see Exercise A.20].

CONTRAVARIANT COMPONENTS OF A FOUR-VECTOR

Definition 4.3. *A set of four quantities V^α ($\alpha = 0,3$) are the contravariant components of a vector if they transform, on a change of coordinates, as in Eq.(4.8).*

$$V'^\mu = \frac{\partial x'^\mu}{\partial x^\alpha} V^\alpha; \qquad V^\alpha = \frac{\partial x^\alpha}{\partial x'^\nu} V'^\nu; \tag{4.9}$$

Consider a scalar, invariant quantity ϕ. Its gradients in $\mathscr{X}$ and in $\mathscr{X}'$ are related.

$$\frac{\partial \phi}{\partial x'^\mu} = K^\nu{}_\mu \frac{\partial \phi}{\partial x^\alpha}; \qquad \frac{\partial \phi}{\partial x^\mu} = J^\nu{}_\mu \frac{\partial \phi}{\partial x'^\nu} \tag{4.10}$$

Note that in Eq.(4.9) we used the Jacobian matrixes $\mathbf{J}$ to pass from $\mathscr{X}$ to $\mathscr{X}'$, whereas in Eq.(4.11) we used its inverse, $\mathbf{K} = \mathbf{J}^{-1}$. The components of the gradient of an invariant are an example of *covariant* components of a vector. So, we can generalize the definitions given in in Eq.(1.16) and Eq.(B2.1.e) to the case of an arbitrary coordinate transformation.

COVARIANT COMPONENTS OF A FOUR-VECTOR

Definition 4.4. *A set of four quantities, V_α ($\alpha = 0,3$), are the covariant components of a vector if they transform, on a change of coordinates, as in Eq.(4.10)*

$$V'_\mu = \frac{\partial x^\alpha}{\partial x'^\mu} V_\alpha; \qquad V_\alpha = \frac{\partial x'^\nu}{\partial x^\alpha} V'_\nu \tag{4.11}$$

As discussed in the previous chapters, covariant and contravariant components of a vector can be different, even if they refer to the same geometrical object. As already seen [*c.f.* Eq.(1.17) and Eq.(B2.1.b)], they are connected by the operation of lowering and raising of indices. Thus, we can write

$$V_\alpha = g_{\alpha\beta}(x^\tau) V^\beta; \qquad V^\alpha = g^{\alpha\beta}(x^\tau) V_\beta; \tag{4.12}$$

These relations allow us to extend the definitions given in Eq.(1.22) and Eq.(B2.1.f) for the inner product of vectors [see Exercise A.21].

INNER PRODUCT OF FOUR-VECTORS

Definition 4.5. *The inner product of two vectors is an invariant, obtained from the saturated product of the contravariant components of one vector with the covariant components of the other one (or vice versa):*

$$A'^\mu B'_\mu = A^\alpha B_\alpha \tag{4.13}$$

Note that because of Eq.(4.12), $A'^{\mu}B'_{\mu} = A'_{\sigma}B'^{\sigma}$ and $A^{\alpha}B_{\alpha} = A_{\tau}B^{\tau}$. Another interesting operation among vectors is the *outer product*. It is obtained by writing one vector after the other, while keeping different indices. Thus, the outer product of two four-vectors, A^{μ} and B^{ν} say, will be given by $A^{\mu}B^{\nu}$, a 4×4 matrix. More in general, we can state the following:

OUTER PRODUCT OF FOUR-VECTORS

Definition 4.6. *The outer product of N vectors will be a multidimensional matrix with $4\times N$ components: $A^{\mu}B^{\nu}..C_{\tau}D_{\sigma}$. On a change of coordinates, the outer product of vectors transforms as follows:*

$$A'^{\mu}B'^{\nu}\times..\times C'_{\tau}D'_{\sigma} = \frac{\partial x'^{\mu}}{\partial x^{\alpha}}\frac{\partial x'^{\nu}}{\partial x^{\beta}}\times..\times\frac{\partial x^{\gamma}}{\partial x'^{\tau}}\frac{\partial x^{\delta}}{\partial x'^{\sigma}}A^{\alpha}B^{\beta}C_{\gamma}D_{\delta} \tag{4.14}$$

4.5 TENSORS

Eq.(4.14) allows us to generalize the previous considerations to a wider class of geometrical objects, the *tensors*. In very general terms, a tensor is an object with m indices that transforms as suggested by Eq.(4.14): each index (either covariant or contravariant) follows the same transformation rule of the corresponding (either covariant or contravariant) vector component [*c.f.* Eq. (4.9) and Eq.(4.11)]. The number of tensor indices defines its *order* or *rank*. We can then state the following:

TENSORS

Definition 4.7. *A set of quantities $T'^{\mu\nu\dots}{}_{\dots\rho\sigma}$ are the components of a tensor if under a change of coordinates they transform as follows:*

$$T'^{\mu\nu\dots}{}_{\dots\rho\sigma} = \frac{\partial x'^{\mu}}{\partial x^{\alpha}}\frac{\partial x'^{\nu}}{\partial x^{\beta}}\times\dots\dots\times\frac{\partial x^{\gamma}}{\partial x'^{\rho}}\frac{\partial x^{\delta}}{\partial x'^{\sigma}}T^{\alpha\beta\dots}{}_{\dots\gamma\delta} \tag{4.15}$$

Clearly a vector is a tensor of rank 1, whereas an invariant is a tensor of rank 0. Note that the outer product of two or more vectors is a tensor [*c.f.* Eq.(4.14)], while the reverse is not necessarily true [see Exercise A.22]. On the basis of Eq.(4.15), it is immediate to demonstrate that the sum of two tensors of the same order and type is a tensor $C^{\alpha\beta}{}_{\mu\nu} = A^{\alpha\beta}{}_{\mu\nu} + B^{\alpha\beta}{}_{\mu\nu}$. A tensor is said to be symmetrical *w.r.t.* a pair of indices (both contravariant or both covariant) if the value of its components does not depend on the order of the indices in the pair. Vice versa, the tensor is skew symmetric or antisymmetric *w.r.t.* a pair of indices (again, both contravariant or both covariant) if the sign of its components depends on the order of the indices in the pair. The properties of symmetry or skew symmetry are clearly conserved under coordinate transformations [see Exercise A.23]. As a consequence, a second-order tensor can always be expressed as the sum of its symmetric and skew-symmetric components:

$$A^{\alpha\beta} = \frac{1}{2}\left(A^{\alpha\beta} + A^{\beta\alpha}\right) + \frac{1}{2}\left(A^{\alpha\beta} - A^{\beta\alpha}\right) \tag{4.16}$$

As an extension of Eq.(4.14), we can define the outer product of tensors.

OUTER PRODUCT OF TENSORS

Definition 4.8. *The outer product of tensors is obtained simply by writing the tensors one after the other, while keeping different their indexes.*

$$A^{\alpha\beta}B_{\mu\nu} = C^{\alpha\beta}{}_{\mu\nu} \qquad U^{\rho}{}_{\tau}V^{\sigma}{}_{\gamma} = Z^{\rho\sigma}{}_{\tau\gamma} \tag{4.17}$$

The outer product of tensors is a tensor of higher order.

There is another important operation with tensor that needs to be mentioned, the *tensor contraction*. Lets' consider the fourth-order tensor $A^{\mu\beta}{}_{\gamma\lambda}$. The contraction of a tensor is obtained in two steps: first, by setting equal to each other a covariant index and a contravariant index; then, by summing over those indices, according to Einstein's summation convention. Thus, writing $A^{\alpha\nu}{}_{\nu\lambda}$ implies summing over the dummy index ν that appears both in a contravariant and in a covariant position. The contraction reduces the tensor rank by 2. In fact, under a coordinate transformation, we have

$$A'^{\mu\nu}{}_{\nu\lambda} = \frac{\partial x'^{\mu}}{\partial x^{\alpha}}\frac{\partial x'^{\nu}}{\partial x^{\beta}}\frac{\partial x^{\gamma}}{\partial x'^{\nu}}\frac{\partial x^{\delta}}{\partial x'^{\lambda}}A^{\alpha\beta}{}_{\gamma\delta} = \frac{\partial x'^{\mu}}{\partial x^{\alpha}}\frac{\partial x^{\delta}}{\partial x'^{\lambda}}A^{\alpha\beta}{}_{\beta\delta} \tag{4.18}$$

Indeed, we started with a fourth-order tensor, and, after contracting over the two dummy indices, we end with a transformation law proper of a tensor of order two [*c.f.* Eq.(4.15]. This leads to the following definition [see Exercise A.24]

INNER PRODUCT OF TENSORS

Definition 4.9. *The inner product of two tensors is the saturated product obtained by summing over two dummy indices, one covariant and the other contravariant.*

$$A^{\alpha\mu}B_{\mu\nu} = C^{\alpha}{}_{\nu}; \qquad V^{\rho}{}_{\sigma}Z^{\sigma}{}_{\gamma} = Z^{\rho}{}_{\gamma}; \tag{4.19}$$

4.6 HOW TO RECOGNIZE TENSORS

The tensorial properties of a multi-index object can be verified by explicitly studying its transformation under a change of coordinates [*c.f.* Eq.(4.15)]. Consider, for example, the Christoffel symbols. We have already seen how they transform from the $\overline{\mathscr{X}}$ to the $\mathscr{X}$ reference frame. Looking at Eq.(4.4), it is immediate to note that the first term transforms in line with the Definition 4.7. On the contrary, the second term doesn't. Note that this second term would vanish *if and only if* the considered transformation $x^{\mu}(\overline{x}^{\tau})$ were linear. This is not what happens in the more general case of *non-linear* transformations. Thus, we have to conclude that the Christoffel symbols are *not* tensors. However, it is considerably easier (and faster) to test the tensorial properties of a multi-indexed object by exploiting the following theorem:

Theorem 4.1
If the inner product of a multi-index object of order n with n arbitrary vectors is an invariant, then the multi-index object is a tensor. ■

Proof. Consider the simple case with $n = 3$. By hypothesis,

$$T'^{\mu\nu}{}_{\lambda} D'_{\mu} E'_{\nu} F'^{\lambda} = T^{\alpha\beta}{}_{\gamma} D_{\alpha} E_{\beta} F^{\gamma} \tag{4.20}$$

The vectors D_α, E_β and F^γ transform according to Eq.(4.9) and Eq.(4.11). Then, Eq.(4.20) can be written as follows:

$$\left(T'^{\mu\nu}{}_{\lambda} - T^{\alpha\beta}{}_{\gamma} \frac{\partial x'^{\mu}}{\partial x^{\alpha}} \frac{\partial x'^{\nu}}{\partial x^{\beta}} \frac{\partial x^{\gamma}}{\partial x'^{\lambda}} \right) D'_{\mu} E'_{\nu} F'^{\lambda} = 0 \tag{4.21}$$

The three vectors are, by hypothesis, arbitrary ones. Then, Eq.(4.21) is satisfied *if and only if*

$$T'^{\mu\nu}{}_{\lambda} = \frac{\partial x'^{\mu}}{\partial x^{\alpha}} \frac{\partial x'^{\nu}}{\partial x^{\beta}} \frac{\partial x^{\gamma}}{\partial x'^{\lambda}} T^{\alpha\beta}{}_{\gamma} \tag{4.22}$$

QED [*c.f.* Eq.(4.15)]. □

On the basis of the Theorem 4.1 [see Exercise A.25], we can immediately conclude that the metric coefficients are the components of a tensor, the *metric tensor*. In fact, if we multiply the second-order tensor, $g_{\mu\nu}$, by two (arbitrary) displacement vectors, dx^μ and dx^ν, we obtain an invariant, ds^2 [*c.f.* Eq.(4.1)].

It is also useful, and sometimes even faster, to use a corollary of the Theorem 4.1

COROLLARY 4.1
If the inner product of a multi-index object of order n with a tensor of order $m < n$ is a tensor, then the multi-index object is a tensor.

Proof. Consider the inner product of a multi-index object, $T'^{\mu\nu}{}_{\rho\sigma}$, with an arbitrary vector D'_μ. Assume that $S'^{\nu}{}_{\rho\sigma} = T'^{\mu\nu}{}_{\rho\sigma} D'_{\mu}$ is a tensor. Then, the quantity

$$S'^{\nu}{}_{\rho\sigma} A'_{\nu} B'^{\rho} C'^{\sigma} = S^{\alpha}{}_{\beta\gamma} A_{\alpha} B^{\beta} C^{\gamma} \tag{4.23}$$

is an invariant [*c.f.* Eq.(4.20)], QED. □

On the basis of the Corollary 4.1, we can immediately conclude that the Kronecker symbol is a tensor: multiplying $\delta^\alpha{}_\beta$ by a generic vector, V^β say, generates a vector: $V^\alpha = \delta^\alpha{}_\beta V^\beta$. Also, the inverse of the metric tensor is a tensor. In fact, when multiplying it with the metric tensor, we still have a tensor: $g^{\alpha\mu} g_{\mu\beta} = \delta^\alpha{}_\beta$.

4.7 GENERAL COVARIANCE

It must be stressed that the importance of tensors rests on their transformations [*c.f.* Eq.(4.15)]. If a tensor vanishes in one reference frame, it will vanish in all the other ones. If two tensors are equal in one reference frame, they will be equal in all the other ones. Thus, if we write equations in tensorial form, we can use these equations in any reference frame, without being restricted to the inertial ones. This is clearly important, but also quite reasonable. In fact, coordinates do not exist by themselves. It is up to us to choose one reference frame, rather than another one. It follows that coordinates and frames cannot explicitly enter into the definition of physical laws. This leads us to state the following principle.

PRINCIPLE OF GENERAL COVARIANCE

Physical laws must be invariant under any arbitrary, invertible and differentiable coordinate transformation. So, they must be written in terms of tensor fields.

5 The Riemann-Christoffel Curvature Tensor

5.1 INTRODUCTION

In the last chapter, we introduced the concepts of curved manifolds and charts. Curvature is an intrinsic property of a manifold, and any observer should be able to measure it, independently of the chosen coordinate system. This implies the existence of a tensor able to describe a manifold curvature. This tensor is the *Riemann-Christoffel curvature tensor.*[1] The goal of the present chapter is to discuss the main properties of the derivatives of a vector field in curved space-times to arrive at a formal definition of the Riemann-Christoffel curvature tensor.

5.2 PARALLEL TRANSPORT IN CURVED SPACE: THE 2D CASE

Calculating the derivative of a vector is an operation to be treated with great care. We already discussed some of these issues in Section 1.7 and in Box 1.2. Let's extend that discussion to the case of a 3D Euclidean space. An infinitesimal 3D displacement vector can then be written in terms of its components in either the Cartesian or the polar basis vectors.

$$\begin{aligned} d\vec{\ell} =& dx\,,\widehat{e}_x + dy\widehat{e}_y + dz\widehat{e}_z = \left(\frac{\partial x}{\partial r}dr + \frac{\partial x}{\partial \theta}d\theta + \frac{\partial x}{\partial \phi}d\phi\right)\widehat{e}_x \\ &+ \left(\frac{\partial y}{\partial r}dr + \frac{\partial y}{\partial \theta}d\theta + \frac{\partial y}{\partial \phi}d\phi\right)\widehat{e}_y + \left(\frac{\partial z}{\partial r}dr + \frac{\partial z}{\partial \theta}d\theta + \frac{\partial z}{\partial \phi}d\phi\right)\widehat{e}_z \\ =& dr\ \widehat{e}_r + d\theta\ \widehat{e}_\theta + d\phi\ \widehat{e}_\phi \end{aligned} \tag{5.1}$$

By considering the terms proportional to dr, $d\theta$ and $d\phi$, we can write the polar basis vectors in terms of the Cartesian ones

$$\widehat{e}_r = +\sin\theta\cos\phi\widehat{e}_x + \sin\theta\sin\phi\widehat{e}_y + \cos\theta\widehat{e}_z \tag{5.2a}$$

$$\widehat{e}_\theta = +r\cos\theta\cos\phi\widehat{e}_x + r\cos\theta\sin\phi\widehat{e}_y - r\sin\theta\widehat{e}_z \tag{5.2b}$$

$$\widehat{e}_\phi = -r\sin\theta\sin\phi\widehat{e}_x + r\sin\theta\cos\phi\widehat{e}_y \tag{5.2c}$$

since $x = r\cos\phi\sin\theta$, $y = r\sin\phi\sin\theta$ and $z = r\cos\theta$. As already mentioned in Box 1.2, $\widehat{e}_r$ is a unit vector, whereas $\widehat{e}_\theta$ and $\widehat{e}_\phi$ are not.

Consider the case of a (unitary radius) 2D sphere as an example of curved space. Points on the surface of the sphere are naturally identified by their spherical coordinates. So, the position vector of a point P is given by the curvilinear coordinate

[1] In literature and books, this tensor is often called either *Riemann tensor* or *curvature tensor*.

DOI: 10.1201/9781003141259-5

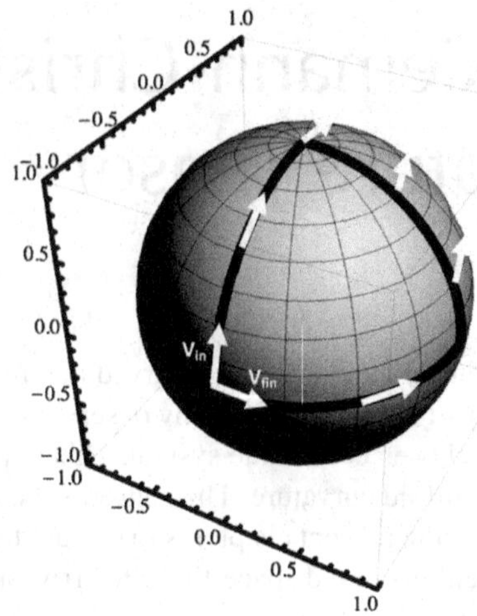

Figure 5.1 Parallel transport in a 2D curved space. The vector $\vec{V}_{in}$ is parallel transported along an equilateral spherical triangle: first, from the Equator to the North Pole; then, from the North Pole back to the Equator; finally, along the Equator, back to its initial position. The two meridians are 90° away. So, at the end of its journey, the vector rotates by 90° *w.r.t.* to initial orientation.

$\vec{\rho} = \theta \widehat{e}_\theta + \phi \widehat{e}_\phi$. In an analogous way, we can write a 2D vector field on the sphere in terms of its components in the chosen vector basis.

$$\vec{V} = V^\theta(\theta,\phi)\,\widehat{e}_\theta + V^\phi(\theta,\phi)\,\widehat{e}_\phi \tag{5.3}$$

Let's evaluate the change of these components when the vector is moved from one point of application to another on the 2D spherical surface. This is done, as in Section 1.7, by keeping constant the angle between the vector and the tangent to the arc of geodesic connecting the two points. In our case, we deal with arcs of great circles (the shortest and straightest path connecting two points on a sphere). In particular, let's *parallel transport* a (constant) vector along a closed path taken, for the sake of simplicity, to be an equilateral, spherical triangle. It is easy to see that, in this case, the *parallel transport* introduces a rotation of 90° of the vector, in spite of the fact that the vector is intrinsically constant [see Figure 5.1 and Exercise A.26].

From a more formal point of view, the change of a vector $\vec{V}(\theta,\phi)$ when moving from $P(\theta,\phi)$ to $Q(\theta+d\theta,\phi+d\phi)$ is given by

$$\begin{aligned} D\vec{V} = &\left(\frac{\partial V^\theta}{\partial\theta}d\theta + \frac{\partial V^\theta}{\partial\phi}d\phi\right)\widehat{e}_\theta + V^\theta\left(\frac{\partial\widehat{e}_\theta}{\partial\theta}d\theta + \frac{\partial\widehat{e}_\theta}{\partial\phi}d\phi\right) + \\ &\left(\frac{\partial V^\phi}{\partial\theta}d\theta + \frac{\partial V^\phi}{\partial\phi}d\phi\right)\widehat{e}_\phi + V^\phi\left(\frac{\partial\widehat{e}_\phi}{\partial\theta}d\theta + \frac{\partial\widehat{e}_\phi}{\partial\phi}d\phi\right) \end{aligned} \tag{5.4}$$

The variation of the basis vectors can be derived from Eq.(5.2).

$$\frac{\partial \hat{e}_r}{\partial r} = 0; \qquad \frac{\partial \hat{e}_r}{\partial \theta} = \frac{\partial \hat{e}_\theta}{\partial r} = \frac{\hat{e}_\theta}{r}; \qquad \frac{\partial \hat{e}_r}{\partial \phi} = \frac{\partial \hat{e}_\phi}{\partial r} = \frac{\hat{e}_\phi}{r};$$

$$\frac{\partial \hat{e}_\theta}{\partial \theta} = -r\hat{e}_r; \quad \frac{\partial \hat{e}_\theta}{\partial \phi} = \frac{\partial \hat{e}_\phi}{\partial \theta} = \cot\theta\, \hat{e}_\phi; \quad \frac{\partial \hat{e}_\phi}{\partial \phi} = -\sin\theta\cos\theta\, \hat{e}_\theta - r\sin\theta^2\, \hat{e}_r \tag{5.5}$$

We can now use these relations to re-express the derivative of some polar basis vectors in terms of other polar basis vectors. We will of course neglect any term aligned with the radial direction, as it is orthogonal to the surface of the sphere where we are bound to be. Regrouping the various terms, we finally get

$$D\vec{V} = \left[\frac{\partial V^\theta}{\partial \rho^k} d\rho^k - V^\phi \sin\theta\cos\theta d\phi\right]\hat{e}_\theta + \left[\frac{\partial V^\phi}{\partial \rho^k} d\rho^k + \cot\theta\left(V^\theta d\phi + V^\phi d\theta\right)\right]\hat{e}_\phi \tag{5.6}$$

5.3 MORE ABOUT THE CHRISTOFFEL SYMBOLS

The metric of our 2D (unitary radius) spherical surface is given by $d\ell^2 = d\theta^2 + \sin^2\theta d\phi^2$. The Lagrangian associated to this metric is $L = \dot{\theta}^2 + \sin^2\theta\dot{\phi}^2$ [*c.f.* Eq.(3.26)]. The corresponding Euler-Lagrangian equations [*c.f.* Eq.(3.27)] provide

$$\begin{aligned} \ddot{\theta} - \sin\theta\cos\theta\dot{\phi}^2 &= 0 \\ \ddot{\phi} + 2\cot\theta\dot{\theta}\dot{\phi} &= 0 \end{aligned} \tag{5.7}$$

Note that here the dot symbol denotes a derivative *w.r.t.* ℓ. It follows that the only non-vanishing Christoffel symbols are[2] $\Gamma^2{}_{33} = -\sin\theta\cos\theta$; $\Gamma^3{}_{23} = \Gamma^3{}_{32} = \cot\theta$, where the indexes 2 and 3 correspond to θ and ϕ, respectively. Now, having in mind Eq.(5.5), we can write

$$\begin{aligned} \frac{\partial \hat{e}_\phi}{\partial \phi} &= -\sin\theta\cos\theta\hat{e}_\theta = \Gamma^2{}_{33}\hat{e}_\theta \\ \frac{\partial \hat{e}_\theta}{\partial \phi} &= \cot\theta\hat{e}_\phi = \Gamma^3{}_{23}\hat{e}_\phi \\ \frac{\partial \hat{e}_\phi}{\partial \theta} &= \cot\theta\hat{e}_\phi = \Gamma^3{}_{32}\hat{e}_\phi \end{aligned} \tag{5.8}$$

These relations can be summarized in a single expression

$$\boxed{\frac{\partial \hat{e}_i}{\partial \rho^j} = \Gamma^k{}_{ij}\hat{e}_k} \tag{5.9}$$

Consider now the vector field defined in Eq.(5.3) and write it in a more compact form: $\vec{V} = V^m\hat{e}_m$. Here $m = 1,2$, $\hat{e}_1 = \hat{e}_\theta$ and $\hat{e}_2 = \hat{e}_\phi$. Eq.(5.4) can then be written

[2] See what was done in Section 3.5 to derive, for example, Eq.(3.33)

as follows:

$$D\vec{V} = \frac{\partial V^m}{\partial \rho^k} d\rho^k \hat{e}_m + V^m \frac{\partial \hat{e}_m}{\partial \rho^k} d\rho^k = \left(\frac{\partial V^m}{\partial \rho^k} + V^j \Gamma^m_{jk} \right) d\rho^k \hat{e}_m = DV^m \hat{e}_m \qquad (5.10)$$

So, we can discuss the derivative of a vector in terms of the derivatives of its components, without explicitly referring to the chosen basis vector.

Let's write the components of $D\vec{V}$ in the following form:

$$DV^m = dV^m - \delta V^m \qquad (5.11)$$

Thus, we can state that the *intrinsic variation* of the k-th component of a vector, DV^k, is obtained by subtracting to its *total variation*, $dV^k = (\partial V^k / \partial \rho^j) d\rho^j$, the *spurious variation*, $\delta V^k = -\Gamma^k_{ij} V^j d\rho^i$, the latter being introduced by the "*parallel transport*." The *spurious variation* clearly depends on the chosen coordinate system, as the Christoffel symbols do.

5.4 PARALLEL TRANSPORT IN CURVED MANIFOLDS

The extension of the previous discussion to curved, pseudo- Riemannian manifold is quite straightforward. Let's consider a vector field **V** and let's express it in terms of its components in the chosen coordinate system: $\mathbf{V} = V^\alpha \hat{e}_\alpha$, where the Greek indexes are now running from 0 to 3. The differential of the vector is then given by [*c.f.* Eq.(5.10)]

$$D\vec{V} = \left(\frac{\partial V^\alpha}{\partial x^\beta} dx^\beta \right) \hat{e}_\alpha + V^\alpha \left(\frac{\partial \hat{e}_\alpha}{\partial x^\beta} dx^\beta \right) \qquad (5.12)$$

The derivative of one basis vector can be expressed in terms of the other basis vectors [*c.f.* Eq.(5.9)].

$$\boxed{\frac{\partial \hat{e}_\alpha}{\partial x^\beta} = \Gamma^\gamma_{\alpha\beta} \hat{e}_\gamma} \qquad (5.13)$$

After substituting this in Eq.(5.12), it is immediate to write [*c.f.* Eq.(5.10)]

$$DV^\alpha = \left(\frac{\partial V^\alpha}{\partial x^\beta} + \Gamma^\alpha_{\tau\beta} V^\tau \right) dx^\beta \qquad (5.14)$$

Remembering the definitions of manifolds and charts given in Section 4.2, it should be clear that the extra-term proportional to the Christoffel symbols is due to the change of the basis vectors when passing from one chart, tangent in P, to the nearby one tangent in Q: the *parallel transport* keeps the vector in the tangent planes, but it changes the vector components. By writing Eq.(5.14) as follows:

$$DV^\alpha = dV^\alpha - \delta V^\alpha \qquad (5.15)$$

we can generalize the statement: the *intrinsic variation* of the α-component of a four-vector, DV^α, is obtained by subtracting to its *total variation*, $dV^\alpha = (\partial V^\alpha / \partial x^\beta) dx^\beta$, the *spurious variation*,

$$\boxed{\delta V^\alpha = -\Gamma^\alpha_{\tau\beta} V^\tau dx^\beta} \qquad (5.16)$$

introduced by the "*parallel transport*."

5.5 COVARIANT DERIVATIVE AND COVARIANT DIFFERENTIAL

Let's consider a vector of contravariant components V'^{α} in a given inertial frame, $\mathscr{K}'$ say, of coordinates ξ^{β}. The same vector has contravariant components V^{μ} in another arbitrary (not necessarily inertial) reference frame, $\mathscr{X}$, of coordinates x^{ν}. So, let's ask how the ordinary derivative of a vector transforms in passing from $\mathscr{K}'$ to $\mathscr{X}$. By exploiting how vectors transform under a change of coordinates [*c.f.* Eq.(4.11)], it is immediate to verify the following[3]:

$$V^{\mu}{}_{,\nu} = \frac{\partial}{\partial x^{\nu}}\left[\frac{\partial x^{\mu}}{\partial \xi^{\alpha}} V'^{\alpha}\right] = \frac{\partial \xi^{\beta}}{\partial x^{\nu}} \frac{\partial^2 x^{\mu}}{\partial \xi^{\alpha}\partial \xi^{\beta}} V'^{\alpha} + \frac{\partial x^{\mu}}{\partial \xi^{\alpha}} \frac{\partial \xi^{\gamma}}{\partial x^{\nu}} V'^{\alpha}{}_{,\gamma} \tag{5.17}$$

The previous equation clearly shows that in general the ordinary derivative of a covariant vector is not a tensor of higher order [*c.f.* Eq.(4.15)]. It would be so *if and only if* $\partial^2 x^{\mu}/\partial \xi^{\alpha}\partial \xi^{\beta} = 0$—that is, only in the case of linear transformations. Now, let's assume that the magnitude of V'^{α} is intrinsically constant, so that in the $\mathscr{K}'$ reference frame $V'^{\alpha}{}_{,\gamma} = 0$. Because of Eq.(5.17), the vanishing quantity in $\mathscr{X}$ is not the ordinary derivative, but rather the combination.

$$V^{\mu}{}_{,\nu} - \frac{\partial \xi^{\beta}}{\partial x^{\nu}} \frac{\partial^2 x^{\mu}}{\partial \xi^{\alpha} \xi^{\beta}} \frac{\partial \xi^{\alpha}}{\partial x^{\tau}} V^{\tau} = 0 \tag{5.18}$$

where we used again Eq.(4.11). On the other hand [*c.f.* Eq.(4.3) and Eq.(3.16)],

$$\frac{\partial^2 x^{\mu}}{\partial \xi^{\alpha}\partial \xi^{\beta}} \frac{\partial \xi^{\beta}}{\partial x^{\nu}} \frac{\partial \xi^{\alpha}}{\partial x^{\tau}} = -\frac{\partial x^{\mu}}{\partial \xi^{\alpha}} \frac{\partial^2 \xi^{\alpha}}{\partial x^{\nu}\partial x^{\tau}} = -\Gamma^{\mu}_{\nu\tau} \tag{5.19}$$

So, we have to conclude that the vanishing of $V'^{\alpha}{}_{,\gamma}$ in $\mathscr{K}'$ corresponds to the vanishing of the so-called *covariant derivative.*[4]

COVARIANT DERIVATIVE OF CONTRAVARIANT VECTOR COMPONENTS

Definition 5.1. *The covariant derivative of the contravariant components of a vector field w.r.t. to the coordinate x^{ν} is defined as follows:*

$$\boxed{V^{\mu}{}_{;\nu} \equiv V^{\mu}{}_{,\nu} + \Gamma^{\mu}_{\nu\tau} V^{\tau}} \tag{5.20}$$

5.6 COVARIANT DERIVATIVES ARE TENSORS

The importance of the covariant derivative stands on its behavior under a transformation of coordinates. Let's move from the $\mathscr{X}$ reference frame of coordinate x^{μ} to

[3] Hereafter we will indicate the partial derivative *w.r.t.* the coordinate x^{τ} with a comma followed by the index τ: $V^{\mu}{}_{,\tau} \equiv \partial V^{\mu}/\partial x^{\tau}$.

[4] Here and in the following, we will indicate the covariant derivative *w.r.t.* the coordinate x^{ν} with a semi-colon followed by the index ν: $V^{\mu}{}_{;\nu}$.

a new reference frame $\overline{\mathscr{X}}$ of coordinates $\bar{x}^\tau(x^\mu)$. After using Eq.(4.9) together with Eq.(4.4), we obtain [see Exercise A.27].

$$V^\mu{}_{;\nu} = \frac{\partial x^\mu}{\partial \bar{x}^\alpha}\frac{\partial \bar{x}^\beta}{\partial x^\nu}\overline{V}^\alpha{}_{,\beta} + \frac{\partial x^\mu}{\partial \bar{x}^\sigma}\frac{\partial \bar{x}^\beta}{\partial x^\nu}\overline{\Gamma}^\sigma{}_{\alpha\beta}\overline{V}^\alpha \tag{5.21}$$

This clearly shows the tensorial properties of the covariant derivative. In fact, we can write Eq.(5.21) in a more compact way:

$$\boxed{V^\mu{}_{;\nu} = \frac{\partial \bar{x}^\beta}{\partial x^\nu}\frac{\partial x^\mu}{\partial \bar{x}^\alpha}\overline{V}^\alpha{}_{;\beta}} \tag{5.22}$$

This is indeed in line with Eq.(4.15). Thus, if the covariant derivative of a contravariant vector vanishes in one reference frame, it must vanish in any other reference frame. If two covariant derivatives are equal in one reference frame, they will be equal in all frames. After comparing Eq.(5.14) and Eq.(5.20), we can arrive at the following definition:

COVARIANT (OR ABSOLUTE) DIFFERENTIAL OPERATOR

Definition 5.2. *The covariant (or absolute) differential of the contravariant components of a vector is defined as follows:*

$$\boxed{DV^\alpha = V^\alpha{}_{;\beta}dx^\beta} \tag{5.23}$$

The absolute differential of a vector is obtained as the saturated product of the covariant derivative of a vector (which is a tensor) with the differential of the coordinates (the components of an infinitesimal displacement vector). It is indeed a first-order tensor because of the Corollary 4.1.

5.7 MORE ON GEODESICS

We have derived the geodesic equations in Section 3.3 (by performing a non-linear coordinate transformation) and in Section 3.4 (by using a variational principle). The question that we can now ask is the following: why do we have the Christoffel symbols in Eq.(3.30), if they are related to the variation of the basis vectors [*c.f.* Eq.(5.13)]? We already implicitly discussed the point, but let's render it more explicit here. Consider a curve $x^\mu = x^\mu(s)$, where s is a parameter along the curve. The unit tangent vector to this curve is $V^\alpha \equiv dx^\alpha/ds$. By construction, the magnitude of this vector is constant in every point of the curve: $V_\alpha V^\alpha = 1$. So, if we "*parallel transport*" the tangent vector along the curve (this means parallel transporting the vector "parallel to itself"), we do expect its intrinsic variation to be null. This implies writing [*c.f.* Eq.(5.23)]

$$\frac{DV^\alpha}{ds} = V^\alpha{}_{;\beta}\frac{dx^\beta}{ds} = 0 \tag{5.24}$$

that is

$$V^{\alpha}{}_{,\beta}\frac{dx^{\beta}}{ds}+\Gamma^{\alpha}_{\mu\nu}V^{\mu}\frac{dx^{\nu}}{ds}=0 \tag{5.25}$$

If $V^{\alpha}=dx^{\alpha}/ds$, then Eq.(5.25) yields

$$\boxed{\frac{d^2x^{\alpha}}{ds^2}+\Gamma^{\alpha}_{\mu\nu}\frac{dx^{\mu}}{ds}\frac{dx^{\nu}}{ds}=0} \tag{5.26}$$

This is indeed the geodesic equation already derived in Eq.(3.30). We can now give a formal definition of a geodesic in an arbitrarily curved manifold.

GEODESICS IN CURVED MANIFOLDS

Definition 5.3. *A geodesic is a line connecting two points on a manifold. It has two properties: i) it is a line of stationary length [c.f. Eq.(3.25)]; ii) it is the straightest line one can draw in a manifold of arbitrarily complex geometry [c.f. Eq.(5.26)].*

5.8 MORE ON COVARIANT DERIVATIVE

In Section 5.5, we have discussed the covariant derivative of the contravariant components of a vector. We clearly have to extend our discussion to consider also the covariant derivative of the covariant components of a vector. To see this, in a slight different and faster way, consider an intrinsically constant vector, initially bound to x^{β} and then transported in $x^{\beta}+dx^{\beta}$. Because of its scalar nature, the squared magnitude of the vector is unaffected by the "*parallel transport*". So, we can write

$$\delta(V^{\alpha}V_{\alpha})=\delta V^{\tau}V_{\tau}+V^{\alpha}\delta V_{\alpha}=-\Gamma^{\tau}_{\alpha\beta}V^{\alpha}V_{\tau}dx^{\beta}+V^{\alpha}\delta V_{\alpha}=0 \tag{5.27}$$

Due to the arbitrariness of V^{α}, we can conclude that the *spurious variation* of the covariant components of a vector at the end of its "*parallel transport*" is given by

$$\boxed{\delta V_{\alpha}\equiv\Gamma^{\tau}_{\alpha\beta}V_{\tau}dx^{\beta}} \tag{5.28}$$

After lowering its contravariant indices, Eq.(5.15) yields $DV_{\alpha}=dV_{\alpha}-\delta V_{\alpha}$. We can then conclude the following:

COVARIANT DERIVATIVE OF COVARIANT VECTOR COMPONENTS

Definition 5.4. *The covariant derivative of the covariant components of a vector field, V_{α}, w.r.t. to the coordinate x^{β} is defined as follows:*

$$\boxed{V_{\alpha;\beta}=V_{\alpha,\beta}-\Gamma^{\tau}_{\alpha\beta}V_{\tau}} \tag{5.29}$$

Note that after lowering its contravariant indices, Eq.(5.23) yields $DV_\alpha = V_{\alpha;\beta}dx^\beta$. This defines the covariant (or absolute) differential of the covariant components of a vector.

5.9 COVARIANT DERIVATIVE OF A TENSOR

Let's extend the concept of covariant derivative to tensors. Consider first two arbitrary vectors, A^α and B_β say. We want to define the covariant derivative of the outer product, $(A^\alpha B_\beta)_{;\gamma} = D(A^\alpha B_\beta)/dx^\gamma$, by enforcing the requirement that in the tangent hyperplane $(A^\alpha B_\beta)_{;\gamma} \to (A^\alpha B_\beta)_{,\gamma} = A^\alpha{}_{,\gamma} B_\beta + A^\alpha B_{\beta,\gamma}$. On the other hand, the metric of the tangent hyperplane is Minkowskian and the Christoffel symbols vanish [*c.f.* Section 4.3]. Then, in the tangent plane, $A^\alpha{}_{,\gamma} = A^\alpha{}_{;\gamma}$ and $B_{\beta,\gamma} = B_{\beta;\gamma}$. So, we can write the following tensorial equation:

$$(A^\alpha B_\beta)_{;\gamma} = A^\alpha{}_{;\gamma} B_\beta + A^\alpha B_{\beta;\gamma} \tag{5.30}$$

On the basis of the definitions given in Eq.(5.20) and Eq.(5.29), we can expand Eq.(5.30) and write

$$\begin{aligned}(A^\alpha B_\beta)_{;\gamma} &= \left(A^\alpha{}_{,\gamma} + \Gamma^\alpha_{\tau\gamma} A^\tau\right) B_\beta + A^\alpha \left(B_{\beta,\gamma} - \Gamma^\tau_{\beta\gamma} B_\tau\right) \\ &= (A^\alpha B_\beta)_{,\gamma} + \Gamma^\alpha_{\tau\gamma} A^\tau B_\beta - \Gamma^\tau_{\beta\gamma} A^\alpha B_\tau\end{aligned} \tag{5.31}$$

As already discussed [see Section 4.5], the outer product of m vectors can always be written as a tensor of rank m. This allows us to arrive at the following definition:

COVARIANT DERIVATIVE OF A TENSOR

Definition 5.5. *The covariant derivative of a tensor w.r.t. to the coordinate x^μ is obtained by adding to the ordinary derivative a number of terms equal to its rank, proportional to the Christoffel symbol and consistent with either Eq.(5.20) or Eq.(5.29), depending on whether the considered tensor index is contravariant or covariant.*

$$\begin{aligned}\left(T^{\alpha\beta\ldots}{}_{\sigma\tau\ldots}\right)_{;\mu} &= \left(T^{\alpha\beta\ldots}{}_{\sigma\tau\ldots}\right)_{,\mu} + \Gamma^\alpha_{\rho\mu} T^{\rho\beta\ldots}{}_{\sigma\tau\ldots} + \Gamma^\beta_{\rho\mu} T^{\alpha\rho\ldots}{}_{\sigma\tau\ldots} + \cdots \\ &\quad - \Gamma^\lambda_{\sigma\mu} T^{\alpha\beta\ldots}{}_{\lambda\tau\ldots} - \Gamma^\lambda_{\tau\mu} T^{\alpha\beta\ldots}{}_{\sigma\lambda\ldots} - \cdots\end{aligned} \tag{5.32}$$

It is easy to show that the covariant derivative of the metric tensor vanishes. In other words, the metric tensor is *covariantly constant*. In fact, following Eq.(5.32), one has $g_{\alpha\beta;\gamma} = g_{\alpha\beta,\gamma} - \Gamma^\tau{}_{\alpha\gamma} g_{\tau\beta} - \Gamma^\tau{}_{\beta\gamma} g_{\alpha\tau}$. Remembering Eq.(3.19), it is immediate to show that

$$g_{\alpha\beta;\gamma} = 0 \tag{5.33}$$

This is not surprising. In fact, in the tangent (Minkowskian) plane [*c.f.* Definition 4.1], the first derivatives of the metric tensor *locally* vanish, and with them, all the Christoffel symbols [*c.f.* Eq.(3.24)]. On the other hand, the covariant derivative of a

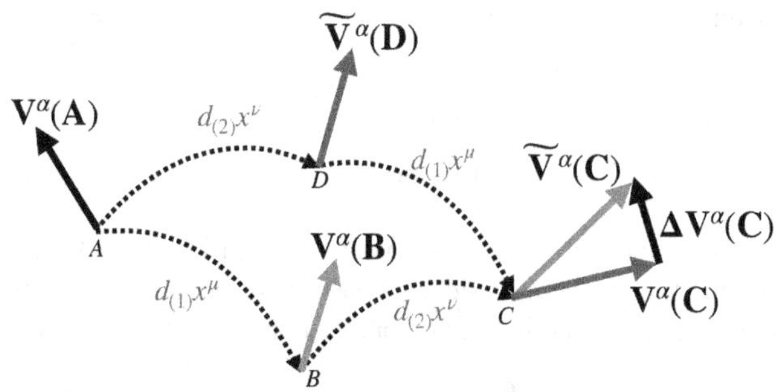

Figure 5.2 The vector V^α is parallel transported from A to C, first passing through B and then through D. If the space is curved, the vectors in C obtained by parallel transporting the *same* vector through two different paths are not the same, because the spurious variation introduced by the operation of parallel transport depends on the followed path.

tensor is a tensor. So, if the covariant derivative of the metric tensor vanishes in the tangent plane, it must vanish in all the other reference frames. This is the important meaning of Eq.(5.33).

5.10 THE RIEMANN-CHRISTOFFEL CURVATURE TENSOR

Consider an arbitrary manifold and choose a reference frame $\mathscr{X}$ of coordinates x^τ. In this frame, the metric tensor will be a function of the position in space-time: $g_{\mu\nu}(x^\tau)$. Now, let's address a very specific question: is this manifold *intrinsically* flat or curved? To answer this question, consider the vector $\vec{V}$, originally bound to point A, and "*parallel transport*" it first in B, along the geodesic arc $\widehat{AB}$, and then in C, along the geodesic arc $\widehat{BC}$ (see Figure 5.2). The result of this operation provides a vector of contravariant components $V^\alpha(C)$. Alternatively, "*parallel transport*" $\vec{V}$ first in D, along the geodesic arc $\widehat{AD}$, and then in C, along the geodesic arc $\widehat{DC}$. The result of this second operation still provides a vector, also bounded in C, but of contravariant components $\tilde{V}^\alpha(C)$ (see Figure 5.2). For the sake of simplicity, let's assume the following symmetry: $\widehat{AB} = \widehat{DC} = d_{(1)}x^\mu$ and $\widehat{AD} = \widehat{BC} = d_{(2)}x^\nu$. Here the subscripts (1) and (2) are not covariant indexes; they are meant to identify the different geodesic arcs of Figure 5.2.

On the basis of the considerations given in Section 5.4, the vector $V^\alpha(C)$—applied in C—is the vector $V^\alpha(B)$—applied in B—plus its spurious variation $\delta_{(2)}V^\alpha(B)$ due to its "*parallel transport*" from B to C. Similarly, the vector $V^\alpha(B)$—applied in B—is the vector $V^\alpha(A)$—applied in A—plus its spurious variation $\delta_{(1)}V^\alpha(A)$ due to its "*parallel transport*" from A to B. We can therefore write, with an obvious meaning of the symbols:

$$\begin{aligned} V^\alpha(C) &= V^\alpha(B) + \delta_{(2)}V^\alpha(B) \\ &= \left[V^\alpha(A) + \delta_{(1)}V^\alpha(A)\right] + \delta_{(2)}\left[V^\alpha(A) + \delta_{(1)}V^\alpha(A)\right] \end{aligned} \tag{5.34}$$

A similar relation can be written for the "*parallel transport*" of V^α from A to C passing through D:

$$\begin{aligned}\tilde{V}^\alpha(C) &= V^\alpha(D) + \delta_{(1)}V^\alpha(D)\\ &= \left[V^\alpha(A) + \delta_{(2)}V^\alpha(A)\right] + \delta_{(1)}\left[V^\alpha(A) + \delta_{(2)}V^\alpha(A)\right]\end{aligned} \tag{5.35}$$

At this point, $V^\alpha(C)$ and $\tilde{V}^\alpha(C)$ are two vectors, both bounded to C, obtained by parallel transporting the *same*, intrinsically constant vector along two different paths. Their difference is, by construction, a vector. By using Eq.(5.16), we first get

$$\begin{aligned}\Delta V^\alpha \equiv V^\alpha(C) - \tilde{V}^\alpha(C) &= \delta_2\left[\delta_1 V^\alpha(A)\right] - \delta_1\left[\delta_2 V^\alpha(A)\right]\\ &= \delta_2\left[-\Gamma^\alpha_{\nu\mu}V^\mu dx_1^\nu\right] - \delta_1\left[-\Gamma^\alpha_{\mu\rho}V^\mu dx_2^\rho\right]\end{aligned} \tag{5.36}$$

and then

$$\begin{aligned}\Delta V^\alpha = &\left(-\Gamma^\alpha_{\nu\mu,\rho}dx_2^\rho\right)V^\mu dx_1^\nu - \Gamma^\alpha_{\nu\mu}\left(-\Gamma^\mu_{\sigma\rho}V^\sigma dx_2^\rho\right)dx_1^\nu\\ &+\left(\Gamma^\alpha_{\mu\rho,\nu}dx_1^\nu\right)V^\mu dx_2^\rho + \Gamma^\alpha_{\mu\rho}\left(-\Gamma^\mu_{\sigma\nu}V^\sigma dx_1^\nu\right)dx_2^\rho\end{aligned} \tag{5.37}$$

After renaming some dummy indexes, we can finally write

$$\boxed{\Delta V^\alpha = \left(-\Gamma^\alpha_{\mu\nu,\rho} + \Gamma^\alpha_{\mu\rho,\nu} + \Gamma^\alpha_{\sigma\nu}\Gamma^\sigma_{\mu\rho} - \Gamma^\alpha_{\sigma\rho}\Gamma^\sigma_{\mu\nu}\right)V^\mu dx_1^\nu dx_2^\rho} \tag{5.38}$$

Note that the round parenthesis, an object of four indexes, has indeed tensorial properties. Infact, its saturated product with three arbitrary vectors, $V^\mu dx_1^\nu dx_2^\rho$, provides a vector, ΔV^α (*c.f.* Corollary 4.1). It follows that the squared parenthesis defines a tensor, the so-called *Riemann-Christoffel curvature tensor.*

THE RIEMANN-CHRISTOFFEL CURVATURE TENSOR

Definition 5.6. *The Riemann-Christoffel curvature tensor can be conveniently expressed as the sum of two determinants.*

$$\boxed{R^\alpha{}_{\mu\nu\rho} \equiv \begin{vmatrix} {}_{,\nu} & {}_{,\rho} \\ \Gamma^\alpha_{\mu\nu} & \Gamma^\alpha_{\mu\rho}\end{vmatrix} + \begin{vmatrix} \Gamma^\alpha_{\sigma\nu} & \Gamma^\alpha_{\sigma\rho} \\ \Gamma^\sigma_{\mu\nu} & \Gamma^\sigma_{\mu\rho}\end{vmatrix}} \tag{5.39}$$

where $_{,\nu} \equiv \partial/\partial x^\nu$ *and* $_{,\rho} \equiv \partial/\partial x^\rho$.

Note that the Riemann-Christoffel curvature tensor depends on the Christoffel symbols and on their derivatives. This of course implies a dependence on the metric tensor and in particular: i) a linear dependence on the second derivatives (first determinant); ii) a non-linear dependence on the first derivatives (second determinant).

So, to recap, we considered an arbitrary, intrinsically constant, vector and "*parallel transport*" it along two different paths, from point A to point C. Then, we evaluate the difference ΔV^α between the two resulting vectors, both bounded in C. Now, if $\Delta V^\alpha = 0$, then the manifold is flat: the "*parallel transport*" does *not* introduce any spurious variation. But if $\Delta V^\alpha = 0$, also $R^\alpha{}_{\beta\mu\nu} = 0$. Vice versa, if $\Delta V^\alpha \neq 0$, then the manifold is intrinsically curved: the "*parallel transport*" does indeed introduce spurious variations. But if $\Delta V^\alpha \neq 0$, also $R^\alpha{}_{\beta\mu\nu} \neq 0$. It follows that the Riemann-Christoffel curvature tensor indeed enables us to assess the *intrinsic* curvature of space-time in a covariant way, that is, independently of the chosen coordinate system.

5.11 SECOND COVARIANT DERIVATIVES

Consider the second covariant derivative of the vector V^α. From Eq.(5.32), we have

$$V^\alpha{}_{;\mu\nu} = \frac{\partial(V^\alpha{}_{;\mu})}{\partial x^\nu} + \Gamma^\alpha_{\tau\nu} V^\tau{}_{;\mu} - \Gamma^\tau_{\mu\nu} V^\alpha{}_{;\tau} \tag{5.40}$$

Thus, by further using Eq.(5.20), it is immediate to derive the following expressions:

$$\begin{aligned} V^\alpha{}_{;\mu\nu} &= \left(V^\alpha_{,\mu} + \Gamma^\alpha_{\tau\mu} V^\tau\right)_{,\nu} + \Gamma^\alpha_{\tau\nu}\left(V^\tau_{,\mu} + \Gamma^\tau_{\sigma\mu} V^\sigma\right) - \Gamma^\tau_{\mu\nu}\left(V^\alpha_{,\tau} + \Gamma^\alpha_{\tau\sigma} V^\sigma\right) \\ V^\alpha_{;\nu\mu} &= \left(V^\alpha_{,\nu} + \Gamma^\alpha_{\tau\nu} V^\tau\right)_{,\mu} + \Gamma^\alpha_{\tau\mu}\left(V^\tau_{,\nu} + \Gamma^\tau_{\sigma\nu} V^\sigma\right) - \Gamma^\tau_{\nu\mu}\left(V^\alpha_{,\tau} + \Gamma^\alpha_{\tau\sigma} V^\sigma\right) \end{aligned} \tag{5.41}$$

We can then conclude that

$$\boxed{V^\alpha_{;\mu\nu} - V^\alpha_{;\nu\mu} = R^\alpha{}_{\tau\nu\mu} V^\tau} \tag{5.42}$$

So, in curved manifolds, covariant derivatives do not commute. This happens only in flat space-times—that is, for $R^\alpha{}_{\tau\nu\mu} = 0$. In these cases, we find the familiar rule: $\partial^2 V^\alpha / \partial x^\mu \partial x^\nu = \partial^2 V^\alpha / \partial x^\nu \partial x^\mu$.

5.12 SYMMETRIES OF THE RIEMANN-CHRISTOFFEL TENSOR

In a four-dimensional space-time, the Riemann-Christoffel curvature tensor has 256 components. Fortunately, due to its numerous symmetries, the number of independent components decreases by a bit more than an order of magnitude. Let's see why. From Eq.(5.39), it is immediate to show that the Riemann-Christoffel curvature tensor is antisymmetric in the second pair of (covariant) indices.

$$\boxed{R^\alpha{}_{\beta\mu\nu} = -R^\alpha{}_{\beta\nu\mu}} \tag{5.43}$$

It is quite convenient to evaluate the curvature tensor in the flat tangent hyperplane, where the Christoffel symbols vanish whereas their derivatives do *not*. So, the writing of the Riemann-Christoffel curvature tensor in this (hat) frame reduces to the first determinant of Eq.(5.39).

$$\begin{aligned} \widehat{R}^\alpha{}_{\beta\mu\nu} &= \widehat{\Gamma}^\alpha_{\beta\nu,\mu} - \widehat{\Gamma}^\alpha_{\beta\mu,\nu} \\ \widehat{R}^\alpha{}_{\nu\beta\mu} &= \widehat{\Gamma}^\alpha_{\nu\mu,\beta} - \widehat{\Gamma}^\alpha_{\nu\beta,\mu} \\ \widehat{R}^\alpha{}_{\mu\nu\beta} &= \widehat{\Gamma}^\alpha_{\mu\beta,\nu} - \widehat{\Gamma}^\alpha_{\mu\nu,\beta} \end{aligned} \tag{5.44}$$

It is immediate to verify that, thanks to the symmetries of the Christoffel symbols ($\Gamma^{\rho}_{\sigma\tau} = \Gamma^{\rho}_{\tau\sigma}$), the sum of these equations provides a null result.

$$\widehat{R}^{\alpha}{}_{\beta\mu\nu} + \widehat{R}^{\alpha}{}_{\nu\beta\mu} + \widehat{R}^{\alpha}{}_{\mu\nu\beta} = 0 \tag{5.45}$$

Although evaluated in the tangent hyperplane, this is a tensorial equation and, as such, it must be valid in any reference frame. So, let's rewrite it ignoring the hat symbol.

$$\boxed{R^{\alpha}{}_{\beta\mu\nu} + R^{\alpha}{}_{\nu\beta\mu} + R^{\alpha}{}_{\mu\nu\beta} = 0} \tag{5.46}$$

This equation provides a relation among the components of the curvature tensor: we will be back to the implications of this equation just below.

The Riemann-Christoffel curvature tensor can also be written in a fully covariant form by lowering the first contravariant index: $R_{\alpha\beta\mu\nu} = g_{\alpha\sigma} R^{\sigma}{}_{\beta\mu\nu}$. Let's work again in the tangent plane. Then,

$$\begin{aligned} \widehat{R}_{\alpha\beta\mu\nu} &= \widehat{g}_{\alpha\sigma}\left[-\widehat{\Gamma}^{\sigma}_{\beta\nu,\mu} + \widehat{\Gamma}^{\sigma}_{\beta\mu,\nu}\right] = \left(\widehat{g}_{\alpha\sigma}\widehat{\Gamma}^{\sigma}_{\beta\nu}\right)_{,\mu} - \left(\widehat{g}_{\alpha\sigma}\widehat{\Gamma}^{\sigma}_{\beta\mu}\right)_{,\nu} \\ &= \frac{1}{2}\left(-\widehat{g}_{\beta\nu,\alpha\mu} + \widehat{g}_{\nu\alpha,\beta\mu} + \widehat{g}_{\beta\mu,\alpha\nu} - \widehat{g}_{\mu\alpha,\beta\nu}\right) \end{aligned} \tag{5.47}$$

where $\widehat{g}_{\beta\nu,\alpha\mu} \equiv \partial^2 \widehat{g}_{\beta\nu}/(\partial x^{\alpha} \partial x^{\nu})$. Eq.(5.47) shows that the Riemann-Christoffel curvature tensor is antisymmetric in the first pair of (covariant) indices, whereas it is symmetric if we invert the first with the second pairs of indexes. Eq.(5.47) is a tensorial equation, and it is valid in any reference frame. Then, we can re-write it ignoring the hat symbol.

$$\boxed{R_{\beta\alpha\mu\nu} = -R_{\alpha\beta\mu\nu}} \tag{5.48}$$

$$\boxed{R_{(\mu\nu)(\alpha\beta)} = R_{(\alpha\beta)(\mu\nu)}} \tag{5.49}$$

Thus, the number of independent pairs is $M(M+1)/2$ [because of the symmetry shown in Eq.(5.49)], where M is the number of independent combination of indexes in a pair. Because of the asymmetry shown in Eq.(5.43) and Eq.(5.48), $M = N(N-1)/2$, where $N = 4$ is the dimension of our space-time. Moreover, Eq.(5.46) provides an independent constraint on the Riemann-Christoffel curvature tensor components *if and only if* the four indexes are *all* different. This means that Eq.(5.46) provides only one independent relation, the others being trivial identities, as it easy to verify. So, in conclusion, the number of independent components of the Riemann-Christoffel curvature tensor is

$$\boxed{\frac{1}{12}N^2\left(N^2 - 1\right) \underset{N=4}{\Longrightarrow} 20} \tag{5.50}$$

Let's conclude this section by considering the covariant derivative of the Riemann-Christoffel curvature tensor. In the hat reference frame:

$$\begin{aligned} \widehat{R}^{\alpha}{}_{\beta\mu\nu,\lambda} &= \widehat{\Gamma}^{\alpha}_{\beta\nu,\mu\lambda} - \widehat{\Gamma}^{\alpha}_{\beta\mu,\nu\lambda} \\ \widehat{R}^{\alpha}{}_{\beta\lambda\mu,\nu} &= \Gamma^{\alpha}_{\beta\mu,\lambda\nu} - \Gamma^{\alpha}_{\beta\lambda,\mu\nu} \\ \widehat{R}^{\alpha}{}_{\beta\nu\lambda,\mu} &= \Gamma^{\alpha}_{\beta\lambda,\nu\mu} - \Gamma^{\alpha}_{\beta\nu,\lambda\mu} \end{aligned} \tag{5.51}$$

By summing the previous relation, one finds the so-called *Bianchi identities*.

THE BIANCHI IDENTITIES

Definition 5.7. *The Bianchi identities provide a linear combination of covariant derivatives of the Riemann-Christoffel curvature tensor, obtained by cyclic permuting the last three covariant indexes.*

$$\boxed{R^{\alpha}{}_{\beta\mu\nu;\lambda} + R^{\alpha}{}_{\beta\lambda\mu;\nu} + R^{\alpha}{}_{\beta\nu\lambda;\mu} = 0} \tag{5.52}$$

6 From Non-inertial Frames to Gravity: The Equivalence Principle

6.1 INTRODUCTION

The geometry of pseudo-Riemannian space-times and the physics of gravitation are linked together by the *local equivalence* between inertial forces and gravitation fields. It is this equivalence, known as the *Equivalence Principle*, that allows to exploit all the formalism developed in Chapter 3 to describe also gravitational phenomena. The goal of this chapter is to discuss this equivalence together with its theoretical consequences.

6.2 THE EQUIVALENCE PRINCIPLE

The classical description of gravitational phenomena rests on the equality of inertial and gravitational masses. The inertial mass describes the inertia of a test-particle to change its status of motion. The gravitational mass describes the response of the same particle to an external gravitational field. The equality of these two masses explains why in a uniform gravitational field any test-particle falls with the same acceleration. This equality has one important consequence: it is always possible to *locally* cancel the effects of a non-uniform, gravitational field by choosing a suitable non-inertial reference frame. Let us consider the reference frame $\mathscr{X}$ bound to the International Space Station. $\mathscr{X}$ is clearly non-inertial, as it is centripetally accelerated by the Earth's gravitational field. Nonetheless, astronauts and objects move *w.r.t.* $\mathscr{X}$ as they are expected to move *w.r.t.* an inertial frame in the absence of gravity [see Figure 6.1]. At the same time, it is possible to *locally* mimic a uniform gravitational field by choosing a suitable non-inertial frame. This is the case of centrifuges used to separate particles of different masses in suspension. The separating power of a centrifuge is indeed expressed by the ratio of the centrifugal force to the gravitational one. Thus, we can state the following:

THE EQUIVALENCE PRINCIPLE

"*The effects of inertial forces and of gravitational fields are* locally *indistinguishable.*"

or

"*A frame located in a uniform gravitational field is mechanically equivalent to a frame which is uniformly accelerating w.r.t. inertial ones.*"

DOI: 10.1201/9781003141259-6

Figure 6.1 NASA astronaut Jessica Meir in the weightless environment of the International Space Station. See `www.nasa.gov/image-feature/expedition-62-flight-engineer-jessica-meir-hovers-for-a-portrait`.

6.3 SWITCHING OFF GRAVITY: THE FREE-FALL

From a geometrical point of view, we have seen that the Minkowski space-time describes the geometry of a hyperplane tangent to any of the points of a curved four-dimensional manifold [*c.f.* Definition 4.1]. Moreover, we have shown that it is possible to move to this tangent hyperplane by means of a *local*, non-linear coordinate transformation [*c.f.* Eq.(4.5)]:

$$x^{\alpha} = \overline{x}^{\alpha} + \frac{1}{2}\overline{\Gamma}^{\alpha}_{\sigma\tau}(\overline{\mathscr{O}})\overline{x}^{\sigma}\overline{x}^{\tau} \tag{6.1}$$

Let's remind here that $\overline{\mathscr{X}}$ (with coordinates $\overline{x}^{\tau}$) and $\mathscr{X}$ (with coordinates x^{ρ}) are two reference frames that share the same origin, $\overline{\mathscr{O}} = \mathscr{O}$, assumed to be the point of tangency between the curved manifold and the Minkowski space-time. The transformation of Eq.(6.1) allows to *locally* set to zero the Christoffel symbols, so that in $\mathscr{O}$ the metric is *locally* Minkowskian.

It could be not unreasonable to presume that, in agreement with the Principle of Equivalence, the same transformation of Eq.(6.1) allows us to move from the reference frame $\overline{\mathscr{X}}$—bound to the source of the gravitational field—to the frame $\mathscr{X}$—in free-fall *w.r.t.* $\overline{\mathscr{X}}$. To see if this is actually the case, remember that a test-particle moves along geodesics of the space-time. So, in the $\overline{\mathscr{X}}$ frame, we have

$$\frac{d^2\overline{x}^{\alpha}}{ds^2} + \overline{\Gamma}^{\alpha}_{\mu\nu}\frac{d\overline{x}^{\mu}}{ds}\frac{d\overline{x}^{\nu}}{ds} = 0 \tag{6.2}$$

On the other hand, Eq.(6.1) provides

$$\left.\frac{d^2x^{\alpha}}{ds^2}\right|_{\mathscr{O}} = \left.\left(\frac{d^2\overline{x}^{\alpha}}{ds^2} + \overline{\Gamma}^{\alpha}_{\mu\nu}\frac{d\overline{x}^{\mu}}{ds}\frac{d\overline{x}^{\nu}}{ds} + \overline{\Gamma}^{\alpha}_{\mu\nu}\frac{d^2\overline{x}^{\mu}}{ds^2}\overline{x}^{\nu}\right)\right|_{\overline{\mathscr{O}}} = 0 \tag{6.3}$$

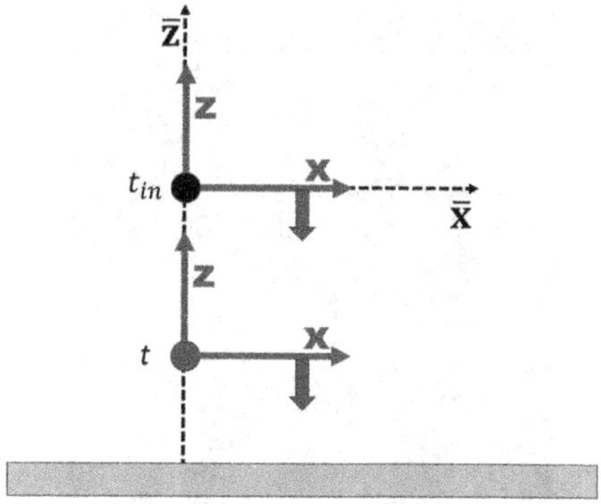

Figure 6.2 The frame $\overline{\mathscr{X}}$ (dashed black lines) is bound to the source M of a uniform gravitational field. Its $\bar{z}$-axis defines the vertical directions. The frame $\mathscr{X}$ (gray lines) is bound to a test-particle in free-fall *w.r.t.* M. The origins, $\overline{\mathscr{O}}$ and $\mathscr{O}$, coincide with the position of the test-particle at a given initial time, t_{in}. See text.

In fact, the round parenthesis vanishes: the first two terms because of Eq.(6.2); the last term because it is evaluated at the origin $\overline{\mathscr{O}}$. It follows that in $\mathscr{O} = \overline{\mathscr{O}}$ the four-acceleration of the test-particle actually vanishes as expected in an inertial frame.

To see the problem from another perspective, let's refer to Figure 6.2, where the $\bar{z}$ axis identifies the vertical *w.r.t.* the mass generating the gravitational field. The origins, $\overline{\mathscr{O}}$ and $\mathscr{O}$, are chosen to coincide with the position of the test-particle at a given initial time, t_{in}. In the weak field approximation, $\bar{t} \simeq t$, $ds \simeq c d\bar{t}$, $\dot{\bar{x}}^0 \simeq 1$ and $\dot{\bar{x}}^k \simeq \bar{v}^k/c \ll 1$. In these limits, Eq.(6.2) provides $\ddot{\bar{z}} + c^2\bar{\Gamma}^3_{00} = 0$. On the other hand, in Classical Mechanics we have $\ddot{\bar{z}} + g = 0$, with solution $\bar{z} = -g\bar{t}^2/2$. By comparing the previous two equations for $\ddot{z}$, we conclude that in the weak field limit $\bar{\Gamma}^3_{00} = g/c^2$. Thus, the coordination transformation given in Eq.(6.1) yields

$$z = \bar{z} + \frac{1}{2}\bar{\Gamma}^3_{00}\bar{x}^0\bar{x}^0 \simeq -\frac{1}{2}gt^2 + \frac{1}{2}gt^2 = 0 \tag{6.4}$$

The test-particle, initially at the origin of $\mathscr{X}$, remains in $\mathscr{O}$ even at later times, showing that $\mathscr{X}$ is indeed in free-fall with the test-particle [*c.f.* Figure 6.2].

To conclude, from a geometrical point of view, Eq.(6.1) allows to approximate a pseudo-Riemannian manifold with its tangent (in $\overline{\mathscr{O}}$) Minkowski space-time [*c.f.* Section 4.3]. From a physical point of view, Eq.(6.1) allows to identify a *local* inertial frame, $\mathscr{X}$, in free-fall in a given gravitational field.

6.4 "CREATING GRAVITY": NON-INERTIAL FRAMES

On the basis of the Equivalence Principle, we can also "create" gravity by using a non-inertial frame, for example, in rotation *w.r.t.* inertial ones. This is why in famous science fiction movies spaceships are often represented as "big wheels." The most striking example is that of the Stanley Kubrick's movie 2001: A Space Odyssey [see

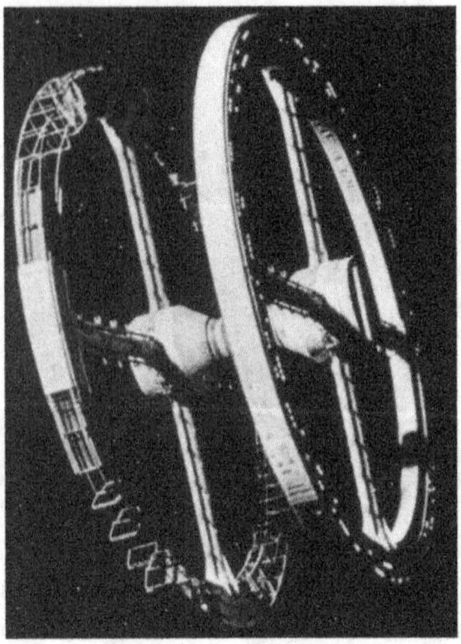

Figure 6.3 The spaceship in the movie "2001: A Space Odyssey" by Stanley Kubrick in 1968. In the movie, it was 300 *m* in diameter and orbited at 300 *km* above Earth. See `https://www.esa.int/ESA_Multimedia/Images/2012/01/Space_station_from_2001_A_Space_Odyssey`.

Figure 6.3]. Because of their slow rotation, these large structures can simulate the effects of gravity. It is easy to see that this is indeed the case by considering again the rotating frame of Section 3.2, with the metric written in cylindrical coordinates [*c.f.* Eq.(3.37)]. The corresponding Lagrangian [*c.f.* Eq.(3.26)] writes

$$L = \left(1 - \frac{\omega^2 r^2}{c^2}\right)(\dot{x}^0)^2 - \dot{r}^2 - r^2\dot{\phi}^2 - d\dot{z}^2 - 2\frac{\omega r^2}{c}\dot{x}^0\dot{\phi} \tag{6.5}$$

Here the dot, as usual, means derivative *w.r.t.* the line element, *ds*. The corresponding Euler-Lagrangian equations [*c.f.* Eq.(3.27)] are given by the following expressions:

$$\left(1 - \frac{\omega^2 r^2}{c^2}\right)\ddot{x}^0 - 2\frac{\omega^2 r}{c^2}\dot{x}^0\dot{r} - 2\frac{\omega r}{c}\dot{r}\dot{\phi} - \frac{\omega r^2\ddot{\phi}}{c} = 0 \tag{6.6a}$$

$$\ddot{r} - \frac{\omega^2 r}{c^2}(\dot{x}^0)^2 - 2\frac{\omega r}{c}\dot{x}^0\dot{\phi} - r\dot{\phi}^2 = 0 \tag{6.6b}$$

$$\ddot{\phi} + 2\frac{\omega}{cr}\dot{x}^0\dot{r} + \frac{\omega}{c}\ddot{x}^0 + \frac{2}{r}\dot{r}\dot{\phi} = 0 \tag{6.6c}$$

$$\ddot{z} = 0 \tag{6.6d}$$

After some obvious substitutions, we are left with $\ddot{x}^0 = 0$, $\ddot{x}^3 = 0$ and

$$\ddot{r} = \frac{\omega^2 r}{c^2}(\dot{x}^0)^2 + 2\frac{\omega r}{c}\dot{x}^0\dot{\phi} + r\dot{\phi}^2 \tag{6.7a}$$

$$\ddot{\phi} = 2\frac{\omega}{cr}\dot{r}\dot{x}^0 - \frac{2}{r}\dot{r}\dot{\phi} \tag{6.7b}$$

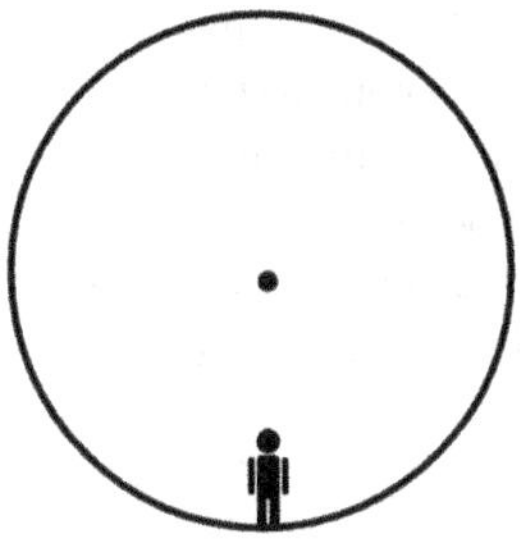

"big-wheel" spaceship in slow counterclockwise rotation around the central l to the page. Consider the non-inertial frame bound to the spaceship. The as- cted to centrifugal and Coriolis inertial forces. See text.

ativistic limit ($ds \simeq cdt$ and $\dot{x}^0 \simeq 1$), Eq.(6.7a) provides

$$\frac{d^2r}{dt^2} = \omega^2 r \left(1 + \frac{d\phi/dt}{\omega}\right)^2 \tag{6.8}$$

6.7b) yields

$$\frac{d^2\phi}{dt^2} = 2\frac{\omega}{r}\frac{dr}{dt}\left(1 - \frac{d\phi/dt}{\omega}\right) \tag{6.9}$$

astronaut standing and still ($dr/dt = 0$, $d\phi/dt = 0$) as in Figure 6.4. a G-force[1] equal to $\omega^2 R$, where R is the radius of the spaceship. By gular rotation velocity ω, it is possible to have a G-force equal to 1— ssible to *locally* reproduce the Earth's gravity. However, be aware of the es. They deflect any object that varies its distance from the rotation axis. od for the astronaut. In fact, when he/she stood up or lowered inside a eship, the inner ear fluids would be deflected *w.r.t.* the local "vertical," inuous contradictory signals to the brain. However, this shouldn't be a ong as $d\phi/dt << \omega$. In Stanley Kubrick's movie, the spaceship had a ut 150 m and the rotation period is of about 60 s, to simulate the Moon these numbers $\omega \simeq 0.1 s^{-1}$, while a normal walking velocity would $\simeq 7 \times 10^{-3} s^{-1}$.

RAVITY CASE: A METRIC SPACE

of the Equivalence Principle, inertial forces and gravitational fields can onfused." Then, it is conceivable to describe gravitational phenomena all the formalism used, *e.g.*, in Section 3.5 to study inertial forces. eed as follows. Consider a Minkowski space-time. Imagine to put by

---

G-force is a measure of an acceleration in units of the gravitational acceleration on Earth. e equal to 1 corresponds to an acceleration of $\simeq 981\ cm/s^2$.

at the origin of a chosen spatial reference frame $\mathscr{K}$ a point particle of mass iven the symmetry of the problem, let's choose a spherical coordinate system: $\{x^0, r, \theta, \phi\}$. It is reasonable to assume that asymptotically—that is, infinitely far M the space-time preserves its Minkowski geometry. This will be progressively s we are getting nearer and nearer to M. However, if we are sufficiently far from e should be allowed to write the metric tensor in a perturbative way.

$$g_{\mu\nu}(r) = \eta_{\mu\nu} + h_{\mu\nu}(r) \tag{6.10}$$

$\eta_{\mu\nu}$ is the Minkowski metric tensor, $h_{\mu\nu}$ are its first-order perturbations due , $|h_{\mu\nu}| \ll |\eta_{\mu\nu}|$, and $g_{\mu\nu}$ is then assumed to be diagonal. Also, a point mass is cted to produce a static, $h_{\mu\nu,0} = 0$, and isotropic, $h_{\mu\nu} = h_{\mu\nu}(r)$, metric perturns. Since the metric tensor given in Eq.(6.10) is assumed to be diagonal, we can the metric in the following form:

$$ds^2 = g_{00}(r)dx^{0^2} + g_{11}(r)dr^2 + g_{22}(r)d\theta^2 + g_{33}(r)\sin^2\theta d\phi^2 \tag{6.11}$$

ider now another frame, $\mathscr{K}'$, bound to a test-particle free-falling in the gravital field generated by M. Let's also assume, for the sake of simplicity, that the lian and equatorial planes of $\mathscr{K}$ and $\mathscr{K}'$ coincide. On the basis of the Equiva- Principle and of the discussion of Section 6.3, the metric seen by $\mathscr{K}'$ must be owskian

$$ds^2 = c^2dt'^2 - d\ell_{\parallel}^2 - d\ell_{\perp,\phi}^2 - d\ell_{\perp,\theta}^2 \tag{6.12}$$

e the symbols $d\ell_{\parallel}$, $d\ell_{\perp,\phi}$ and $d\ell_{\perp,\theta}$ refer to the proper lengths along the radial tion, along meridians ($\phi = const$) and along the equator ($\theta = \pi/2$), respectively. to Eq.(3.48), we can write

$$d\ell_{\parallel} = \sqrt{-g_{11}(r)}dr \tag{6.13a}$$

$$d\ell_{\perp,\phi} = \sqrt{-g_{22}(r)}d\theta \tag{6.13b}$$

$$d\ell_{\perp,\theta} = \sqrt{-g_{33}(r,\theta)}d\phi \tag{6.13c}$$

wing the intuition of Special Relativity, we don't expect changes in the direction ndicular to the radial motion of the test-particle. So, we can conclude that in ne angular part of the metric remains Minkowskian.

$$\boxed{g_{22} = -r^2} \tag{6.14a}$$

$$\boxed{g_{33} = -r^2\sin^2\theta} \tag{6.14b}$$

ne contrary, because of the length contraction [*c.f.* Eq.(2.24)], we expect that $\mathscr{K}$ the proper radial distances writes

$$d\ell_{\parallel} = \frac{dr}{\sqrt{1 - \mathrm{v}^2(r)/c^2}} \simeq dr\sqrt{1 + \frac{\mathrm{v}^2(r)}{c^2}} \tag{6.15}$$

of simplicity, let's assume that the test-particle started its motion toward itially vanishing velocity: $v(\infty) \simeq 0$. Then, $v^2(r) = 2GM/r$. It follows 5) that we can write

$$\boxed{g_{11} \simeq -\left(1+\frac{2GM}{c^2 r}\right)} \tag{6.16}$$

ct a time dilation effect [*c.f.* Eq.(2.22)]

$$dt' = \sqrt{1-\frac{v^2(r)}{c^2}}dt = \sqrt{1-\frac{2GM}{c^2 r}}dt \tag{6.17}$$

s [*c.f.* Eq.(3.31)]

$$\boxed{g_{00}(r) = 1-\frac{2GM}{c^2 r}} \tag{6.18}$$

t the non-vanishing components of the perturbed metric tensor are

$$\boxed{h_{00} = h_{11} = -\frac{2GM}{c^2 r}} \tag{6.19}$$

, by exploiting only the Equivalence Principle, we were able to write—–the metric of a pseudo-Riemannian space-time with a point particle of e origin.

ST-ORDER METRIC IN THE PRESENCE OF A POINT MASS

$$\left(1-\frac{2GM}{c^2 r}\right)dx^{0\,2} - dr^2\left(1+\frac{2GM}{c^2 r}\right) - r^2\left[d\theta^2+\sin^2\theta d\phi^2\right] \tag{6.20}$$

gain that Eq.(6.17) rests only on the assumed validity of the Equivalence on the assumption of Eq.(6.10): $|h_{\mu\nu}| \ll |\eta_{\mu\nu}|$, that is $2GM/c^2 \ll r$. A derivation of Eq.(6.20) will be given in Section 8.8.

MOTION OF A TEST-PARTICLE IN A GRAVITATIONAL FIELD

article subjected *only to inertial forces* follows a geodesic of a pseudo-space-time [*c.f.* Section 3.5]. Because of the Equivalence Principle, let's a test-particle subjected *only to gravitational fields* should also move ics of the pseudo- Riemannian space-time described by the metric given Let's see if this makes sense at all. To evaluate the radial trajectory of a subjected only to the gravitational field of M, let's write the Lagrangian m [*c.f.* Eq.(3.25)]

$$L = \left(1-\frac{2GM}{c^2 r}\right)(\dot{x}^0)^2 - \left(1+\frac{2GM}{c^2 r}\right)\dot{r}^2 \tag{6.21}$$

e we have assumed a radial infall—that is, $\theta = const$ and $\phi = const$. The Euler-
ngian equations provide

$$\ddot{x}^0 \left(1 - \frac{2GM}{c^2 r}\right) + \frac{2GM}{c^2 r^2} \dot{r} \dot{x}^0 = 0 \tag{6.22a}$$

$$\ddot{r} \left(1 + \frac{2GM}{c^2 r}\right) - \frac{GM}{c^2 r^2} \dot{r}^2 + \frac{GM}{c^2 r^2} (\dot{x}^0)^2 = 0 \tag{6.22b}$$

weak field limit ($ds \simeq cdt$; $\dot{x}^0 \simeq 1$; $r \gg 2GM/c^2$; $\dot{r} \ll 1$), Eq.(6.22b) provides

$$\frac{d^2 r}{dt^2} \simeq -\frac{GM}{r^2} \tag{6.23}$$

pected in Newtonian mechanics. Since the Euler-Lagrangian equation do pro-
the geodesic equations [*c.f.* Eq.(3.30)], we can conclude the following:

GEODESIC MOTION

est-particle *subjected only to gravitational fields* moves along the geodesics of a
udo-Riemannian space-time.

is an important result, as it shows how the Equivalence Principle links together
eometry of pseudo-Riemannian space-times with the physics of gravity.

GEODESIC DEVIATION

consider two non-interacting test-particles, A and B say, subjected only to an
nal gravitational field. Each of them is in free-fall, and each of them follows its
geodesic [*c.f.* Section 6.4]. To simplify the calculations, let's study the motion of
particles in the local inertial reference frame, $\mathscr{X}'$ say, bound to the A particle.
is frame, the coordinates of A and B are x'^α and $x'^\alpha + \delta x'^\alpha$, $\delta x'^\alpha$ being the
acement vector which takes us from A to B. The geodesic equations for the A
particles can then be written as follows:

$$\frac{x'^\alpha}{s^2} = 0 \tag{6.24a}$$

$$\frac{}{2} (x'^\alpha + \delta x'^\alpha) + \Gamma'^\alpha_{\mu\nu} (x'^\tau + \delta x'^\tau) \frac{d}{ds} (x'^\mu + \delta x'^\mu) \frac{d}{ds} (x'^\nu + \delta x'^\nu) = 0 \tag{6.24b}$$

.(6.24a), we set to zero the terms proportional to the Christoffel symbols, as in
cal inertial reference frame they are all vanishing [*c.f.* Eq.(4.7)]. This is not the
of Eq.(6.24b), where we have to consider the displacement four-vector $\delta x'^\alpha$ of B
A. However, if we consider $\delta x'^\alpha$ as a first-order quantity, we can Taylor expand
hristoffel symbols around the A position: $\Gamma'^\alpha_{\mu\nu} (x'^\tau + \delta x'^\tau) = \Gamma'^\alpha_{\mu\nu,\rho} (x'^\tau) \delta x'^\rho$.
, after eliminating the zeroth order solution, Eq.(6.24b) becomes

$$\boxed{\frac{d^2}{ds^2} \delta x'^\alpha + \Gamma'^\alpha_{\mu\nu,\rho} \delta x'^\rho \frac{dx'^\mu}{ds} \frac{dx'^\nu}{ds} = 0} \tag{6.25}$$

first term of Eq.(6.25) is not a tensor. In fact, the second term of the n contains the ordinary derivative of the Christoffel symbols—that do orial properties. So, in order to write Eq.(6.25) in a covariant form, let's tion of covariant differential operator we introduced in Eq.(5.23). In the free-fall with A, we have

$$\frac{D}{ds}\delta x'^{\alpha} \equiv \delta x'^{\alpha}{}_{;\mu}\frac{dx'^{\mu}}{ds} = \frac{d}{ds}\delta x'^{\alpha} + \Gamma'^{\alpha}_{\mu\rho}\,\delta x'^{\rho}\frac{dx'^{\mu}}{ds} \tag{6.26}$$

t the second absolute derivative writes as follows:

$$\frac{D}{ds}\left[\frac{d}{ds}\delta x'^{\alpha} + \Gamma'^{\alpha}_{\mu\rho}\,\delta x'^{\rho}\frac{dx'^{\mu}}{ds}\right]$$

$$\frac{d}{ds}\left[\frac{d}{ds}\delta x'^{\alpha} + \Gamma'^{\alpha}_{\mu\rho}\,\delta x'^{\rho}\frac{dx'^{\mu}}{ds}\right] + \Gamma'^{\alpha}_{\sigma\tau}\left[\frac{d}{ds}\delta x'^{\sigma} + \Gamma'^{\sigma}_{\mu\rho}\,\delta x'^{\rho}\frac{dx'^{\mu}}{ds}\right]\frac{dx'^{\tau}}{ds} \tag{6.27}$$

$= 0$: we can then ignore all the terms proportional to the Christoffel et

$$\frac{D^2}{ds^2}\delta x'^{\alpha} = \frac{d^2}{ds^2}\delta x'^{\alpha} + \Gamma'^{\alpha}_{\mu\rho,\lambda}\,\delta x'^{\rho}\frac{dx'^{\lambda}}{ds}\frac{dx'^{\mu}}{ds} \tag{6.28}$$

use Eq.(6.25) to substitute $d^2\delta x^{\alpha}/ds^2$ in Eq.(6.28). We then obtain

$$\boxed{\frac{D^2}{ds^2}\delta x'^{\alpha} = \left(\Gamma'^{\alpha}_{\mu\rho,\nu} - \Gamma'^{\alpha}_{\mu\nu,\rho}\right)\delta x'^{\rho}\frac{dx'^{\nu}}{ds}\frac{dx'^{\mu}}{ds}} \tag{6.29}$$

of the Corollary 4.1 of Section 4.6, we can conclude that the quantity s has tensorial properties: the saturated product with three arbitrary vec- hs provides a the vector on the *lhs* . Indeed, by comparing it with the ristoffel tensor definition, we can verify that the round parenthesis in given by the first of the two determinants of Eq.(5.39), the second being Eq.(6.29) is a tensorial equation that can be written in a generic frame e Riemann-Christoffel tensor.

GEODESIC DEVIATION EQUATION

$$\boxed{\frac{D^2}{ds^2}\delta x^{\alpha} = -R^{\alpha}{}_{\mu\rho\nu}\frac{dx^{\mu}}{ds}\delta x^{\rho}\frac{dx^{\nu}}{ds}} \tag{6.30}$$

BETWEEN GEOMETRY AND DYNAMICS

ection 6.3, gravity can be *locally* eliminated in the (non-inertial) frame st-particle free-falling in a given gravitational field. The adverb *locally* s here. But what does it mean *locally*? How much *local* is *locally*? The iation [*c.f.* Eq.(6.30)] allows us to answer this question.

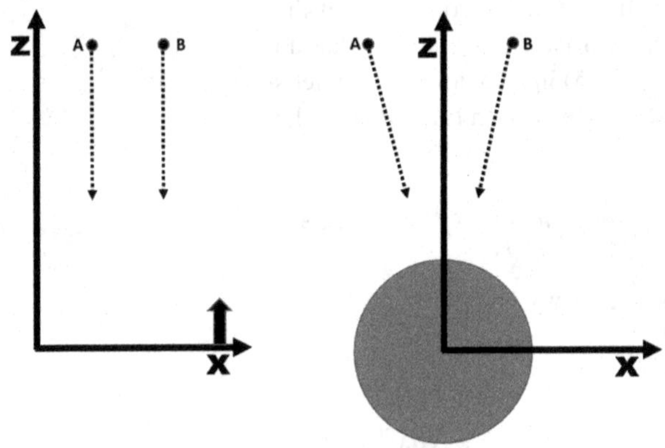

e 6.5 The motion of two, non-interacting particles in two different frames: a non-
al one (left panel) with a constant acceleration along the z axis; a frame bound to the
M (right panel). The trajectories of the two test-particles are parallel in the first case, but
the second one.

t's consider the case of the A and B particles that at some initial time have a linate distance δx^α and zero relative velocity. If δx^α remains constant at later —that is, $D^2\delta x^\alpha/ds^2 = 0$—then the Riemann-Christoffel tensor has to be iden- y equal to zero. But if the Riemann-Christoffel tensor vanishes, the space-time t, as both the Christoffel symbols and their derivatives would be zero. On the of the Definition 4.1, this means that it is possible to find an inertial reference $\mathscr{K}$ of coordinates ξ^α covering the *entire* space-time with a Minkowski metric q.(2.2)]. In this reference frame the non-interacting A and B particles are either t one *w.r.t.* the other, or in a uniform motion along straight, *parallel* geodesics. e other hand, the relation $D^2\delta x^\alpha/ds^2 = 0$ has covariant properties, so it has to rified in all frames, in particular in the non-inertial ones. For the sake of sim- y, consider the A and B particles at rest in the lab. frame. In a non-inertial frame, rmly accelerating in one direction, the two particles would be seen accelerat- n the opposite direction, but always under the condition $D^2\delta x^\alpha/ds^2 = 0$. This es that they still move along straight, *parallel* geodesics (*see* Fg. 6.5a). The con- on of all this reasoning is that in flat space-time the Riemann-Christoffel tensor be always identically equal to zero, no matter what the frame of reference is ow complicated the space-time metric in that frame is.

onsider now a different situation. At some initial time, two non-interacting par- A and B, subjected *only* to an external gravitational field, are at a distance and zero relative velocity. If the gravitational field is generated by a point mass Eq.(6.20)], we do expect the A and B trajectories to converge toward the cen- f mass (see Figure 6.5b). But this implies having $D^2\delta x^\alpha/ds^2 \neq 0$—that is, a vanishing Riemann-Christoffel tensor. It follows that the space-time is not flat. e have to conclude that the point mass produces a curvature of the space-time.

In the light of these considerations, we can now better grasp the meaning of the adverb *locally* in the statement of the Equivalent Principle: *locally* means that the unavoidable tidal effects of gravity described by the geodesic deviation can be safely neglected only for small regions of space and short interval of times that is, for $\delta x^\alpha \to 0$. From a geometrical point of view, *locally* means that we have to remain very near to the point of tangency between the pseudo-Riemannian manifold and its tangent Minkowski space-time.

6.9 RIEMANN-CHRISTOFFEL TENSOR IN THE ROTATING FRAME

On the basis of what has been discussed in the last Section, we expect that the Riemann-Christoffel tensor evaluated in the rotating reference frame of Section 3.2 must vanish. Let's see if this is actually the case. Given the metric of Eq.(3.6), the non-vanishing Christoffel symbols are those given by Eq.(3.33) and Eq.(3.34). Thus, given Eq.(5.39), we should consider only the components $R^1{}_{\beta\mu\nu}$ and $R^2{}_{\beta\mu\nu}$, all the others being identically zero as $\Gamma^0{}_{\beta\gamma} = \Gamma^3{}_{\beta\gamma} = 0$. Given its symmetries [*c.f.* Eq.(5.9)], the Riemann-Christoffel tensor has only two independent components

$$\begin{aligned} R^1{}_{010} &= \Gamma^1{}_{00,1} - \Gamma^1{}_{01,0} + \Gamma^1{}_{\sigma 1}\Gamma^\sigma{}_{00} - \Gamma^1{}_{\sigma 0}\Gamma^\sigma{}_{01} \\ R^2{}_{020} &= \Gamma^2{}_{00,2} - \Gamma^2{}_{02,0} + \Gamma^2{}_{\sigma 2}\Gamma^\sigma{}_{00} - \Gamma^2{}_{\sigma 0}\Gamma^\sigma{}_{02} \end{aligned} \tag{6.31}$$

that are indeed both equal to zero, as it is immediate to verify.

To conclude, we are now in the position of answering the question asked at the end of Section 3.8. Eq.(6.31) tells us that the space-time observed from the merry-go-round is *intrinsically* flat, even if the explicit form of the metric is not Minkowskian [*c.f.* Eq.(3.6)]. This result should have been expected: we can always step down from a merry-go-round to find ourselves in the inertial lab. frame—that is, in a homogeneous and isotropic space-time fully described by a Minkowski metric.

6.10 RIEMANN-CHRISTOFFEL TENSOR AND GRAVITY

Let's consider the case of space-time with a point mass M at the origin of the spatial reference frame. In this case, the components of the Riemann-Christoffel tensor are *not* expected to be identically equal to zero. Clearly, we don't need to evaluate all the 20 independent components of the Riemann-Christoffel tensor [*c.f.* Section 5.9]. It would be enough to show that at least one of its components is different from zero. To facilitate the calculations, let's write the metric given in Eq.(6.20) in terms of the *Schwarzschild radius* $\mathscr{R}_S \equiv 2GM/c^2$ and of a new variable $x \equiv \mathscr{R}_S/r$. Assuming to be far enough from the point mass ($x \ll 1$) would help in keeping our treatment to first order and to neglect higher order terms. With these definitions, the metric becomes

$$ds^2 = (1-x)\,dx^{0^2} - (1+x)\frac{\mathscr{R}_S^2}{x^4}dx^2 - \frac{\mathscr{R}_S^2}{x^2}\left[d\theta^2 + \sin^2\theta\, d\phi^2\right] \tag{6.32}$$

Now, let's limit ourselves to the spatial components of the Riemann-Christoffel tensor. It is easy to show that the *spatial* components of the Christoffel symbols are given by the following expression [see Exercise A.28].

$$\Gamma^1_{jk} = \mathrm{diag}\left(-\frac{3x+4}{2x^2+2x}, \frac{x}{1+x}, \frac{x}{x+1}\sin^2(\theta)\right) \tag{6.33}$$

and

$$\begin{aligned} \Gamma^2_{jk} &= \begin{pmatrix} 0 & -1/x & 0 \\ -1/x & 0 & 0 \\ 0 & 0 & -\sin\theta\cos\theta \end{pmatrix} \\ \Gamma^3_{jk} &= \begin{pmatrix} 0 & 0 & -1/x \\ 0 & 0 & \cot\theta \\ -1/x & \cot\theta & 0 \end{pmatrix} \end{aligned} \tag{6.34}$$

where the Latin indexes j and k go, as usual, from 1 to 3. In order to evaluate the Riemann-Christoffel tensor, we also need the first derivatives of the Christoffel symbols. Given the symmetry of the problem, we restrict to the derivatives *w.r.t.* x. Then, we have

$$\Gamma^1_{jk,1} = \mathrm{diag}\left(\frac{3x^2+8x+4}{2x^2(x+1)^2}, \frac{1}{(x+1)^2}, \frac{\sin^2(\theta)}{(x+1)^2}\right) \tag{6.35}$$

and

$$\begin{aligned} \Gamma^2_{jk,1} &= \begin{pmatrix} 0 & 1/x^2 & 0 \\ 1/x^2 & 0 & 0 \\ 0 & 0 & 0 \end{pmatrix} \\ \Gamma^3_{jk,1} &= \begin{pmatrix} 0 & 0 & 1/x^2 \\ 0 & 0 & 0 \\ 1/x^2 & 0 & 0 \end{pmatrix} \end{aligned} \tag{6.36}$$

The *spatial* components of the Riemann-Christoffel tensor, $R^i{}_{jkl}$, are 81. Because of its symmetry [see Section 5.9], only six components are independent. However, given the isotropy of the problem, the number reduces to 3 [51]. To first order in x, we get

$$\begin{aligned} R^1{}_{212} &= \Gamma^1_{22,1} + \Gamma^1_{11}\Gamma^1_{22} - \Gamma^1_{22}\Gamma^2_{21} = -\frac{x}{2} \\ R^1{}_{313} &= \Gamma^1_{33,1} + \Gamma^1_{11}\Gamma^1_{33} - \Gamma^1_{33}\Gamma^3_{31} = -\frac{x}{2}\sin^2\theta \\ R^2{}_{323} &= \Gamma^2_{33,2} + \Gamma^2_{12}\Gamma^1_{33} - \Gamma^2_{33}\Gamma^3_{32} = x\sin^2\theta \end{aligned} \tag{6.37}$$

It follows that the components of the Riemann-Christoffel tensor are not identically equal to zero. As expected, the presence of a point mass induces a curvature in space-time, whose geometry is described by Eq.(6.20).

Part II

From Curvature to Observations

7 Observational Test of the Equivalence Principle

7.1 INTRODUCTION

In the last chapter, we have shown how the Equivalence Principle naturally connects the physics of gravity with the curvature of space-time. However, the conclusions of the last chapter were mostly based on plausibility arguments. The goal of this chapter is to discuss the observational implications of the Equivalence Principle, as well as the experimental tests that proved it over the years.

7.2 INERTIAL VS. GRAVITATIONAL MASSES

Testing the equality of inertial and gravitational masses has a long history. The first breakthrough occurred with the classical experiment by Eötvös with a torsion balance. Eötvös reported his results at the Hungarian Academy of Sciences in January 1889, concluding: "*I herewith assert that, if there is any difference between gravity of bodies of equal mass but of different composition, it is less than one part in twenty millionth in the case of brass, glass, antimonite and corkwood and it is undoubtedly less than one part in one hundred thousandth in the case of air*" [28][1]. Eötvös's conclusion shows the equivalence between inertial and gravitational masses of objects of different chemical composition. This implies that nuclear and electromagnetic interactions contribute equally to the two masses. Thus, we can say that *the motion of a test-particle subjected only to gravitational forces does not depend on its composition*. New results were reported by Eötvös in 1909 at the 16th International Geodesic Conference in London, where Eötvös quoted the result with an increased precision: 1 part in 10^8 [29].

This precision has been improved over the years by several ground-based experiments: in 1935, by Renner, a former students of Eötvös, down to ~ 3 part in 10^9; in 1964, by Roll, Krotkov and Dicke down to one part in 10^{11} [79]. Very recently, the CNES MicroSCOPE satellite was able to improve this precision even further. This space-borne experiment consisted of two differential accelerometers. In the first one, the test masses are made by different materials (one in titanium and the other in a platinum–rhodium alloy). In the second one, the two test masses are of the same material (platinum–rhodium alloy). The second accelerometer provided a measurement reference for the experiment. The test masses remained motionless with respect to the satellite inside the two accelerometers. So, they were subjected to the same control acceleration, as expected from the Equivalence Principle. The MicroSCOPE satellite

[1] See [27] for the English version of the orignal Eötvös's paper.

DOI: 10.1201/9781003141259-7

was able to push the difference between inertial and gravitational mass down to one part in 10^{15} [92].

7.3 GRAVITATIONAL TIME DILATION

The assumed validity of the Equivalence Principle led us to write the first-order metric of a space-time containing a single point particle of mass M. Eq.(6.20) has an important and peculiar implication: in the presence of the point mass, time doesn't flow uniformly over the space-time. Indeed, the relation between proper and coordinate times depends now from where we are [*c.f.* Eq.(6.17)]

$$\boxed{d\tau \equiv \sqrt{1-\frac{\mathscr{R}_S}{r}}dt} \tag{7.1}$$

where $\mathscr{R}_S = 2GM/c^2$ is the Schwarzschild radius. Note the for $r \to \infty$, we have $d\tau = dt$. Then, the coordinate time used in Eq.(6.20) is nothing else than the proper time of an observer at an infinite distance from M. However, for an observer that gets nearer and nearer to M, time flows slower and slower. This is the so-called *gravitational time dilation* phenomenon, and it has several observable consequences. Many experiments were designed to measure this effect, with the specific goal of testing the Equivalence Principle. Let's mention here the space-borne Gravity Probe A (GP-A) experiment. The idea was to compare the proper time of two hydrogen maser clocks, stable to one part in 10^{15}, one on Earth and the other on the probe. GP-A was launched in 1976 and reached an altitude $h = 10,000\ km$. According to Eq.(7.1), the ratio of the proper times measured by the two clocks, one on the probe at its maximum height and that other on Earth, is given by

$$\frac{\Delta\tau|_{probe}}{\Delta\tau|_{Earth}} = \frac{\sqrt{1-\dfrac{2GM_\oplus/c^2}{R_\oplus+h}}}{\sqrt{1-\dfrac{2GM_\oplus/c^2}{R_\oplus}}} \simeq 1+4\cdot 10^{-10} \tag{7.2}$$

where $M_\oplus = 6\times 10^{27} g$ and $R_\oplus \simeq 6373 km$ are the mass and the mean radius of the Earth. As expected, the clock on the probe runs faster than that on the Earth, gaining $\simeq 4\cdot 10^{-10} s$ per Earth second. The results of the GP-A experiment were published in 1980, probing the predictions based on Eq.(7.1) and, then, the Equivalence Principle with an accuracy of 0.007% [93].

7.4 THE GLOBAL POSITIONING SYSTEM – GPS

In daily life, taking into proper account the gravitational time dilation is crucial for obtaining a good and useful functioning of the Global Positioning System-GPS. The GPS is constituted by 24 satellites orbiting around the Earth with a velocity of $v = 3.8\ km/s$, at a height of $20,000 km$ and with a revolution period of $\simeq 12h$.

Each satellite is provided with high-precision atomic clocks that accumulate an error of $0.86ns$ per day. At least four satellites are always visible from any point on the terrestrial globe. Each of them sends a signal with two crucial information: the exact time when the signal is sent; the satellite position at that time. To find a position on the Earth's surface, we then need to resolve a system of four equations

$$c(t - t_i) = \sqrt{(x - x_i)^2 + (y - y_i)^2 + (z - z_i)^2}; \qquad i = 1,4 \tag{7.3}$$

where the index i now identifies the position $\{x_i, y_i, z_i\}$ of the i-th satellite at the time t_i when the signal was sent to Earth, whereas the coordinates $\{x, y, z\}$ are associated to a specific geographical location on the globe.

The positions of the four satellites are known to high accuracy. So, we have to have the same accuracy in evaluating the *lhs* of Eq.(7.3)—that is, the difference $t - t_i$ between the emission and receiving times of the satellite signals. To do this, we need to take into account two effects, each of them affecting the $t - t_i$ estimates far beyond the clock sensitivity. The first one has to do with the orbital velocity of a GPS satellite. On the basis of Special Relativity [*c.f.* Eq.(2.22)], we have to conclude that the clock on the satellite is slower than the one on Earth. Since

$$\sqrt{1 - \left(\frac{\mathrm{v}}{c}\right)^2} = 1 - 8.17 \cdot 10^{-11} \tag{7.4}$$

a clock on the GPS satellite indeed loses $8.17 \cdot 10^{-11}\,s$ per Earth second, yielding a lag of $7.07\,\mu s$ per Earth day. The second effect has to do with the Equivalence Principle. The satellite orbits at a height of $20,000\ km$. Because of the gravitational time dilation [*c.f.* Eq.(7.1)], we have to conclude that [*c.f.* Eq.(7.2), but with $h = 20,000\,\mathrm{km}$]

$$\frac{\Delta\tau|_{Satellite}}{\Delta\tau|_{Earth}} \simeq 1 + 5 \cdot 10^{-10} \tag{7.5}$$

Indeed, a clock on the GPS satellite gains $\simeq 5 \cdot 10^{-10}\,s$ per Earth second, corresponding to $45.65\,\mu s$ per Earth day (larger than both the orbital velocity effect and the intrinsic clock precision). Both these relativistic effects are actually taken into account, among others, for resolving the system given in Eq.(7.3) and for locating a point on the Earth surface with the required accuracy. In conclusion, whenever we use a GPS navigator we are verifying (and, most of all, exploiting) the Equivalence Principle.

7.5 GRAVITATIONAL REDSHIFT

Consider two observers, A and B, at distance r_A and r_B from M. Suppose that A sends a series of N light pulses to B with a frequency ν_A in a given proper time interval $\Delta\tau_A$. B receives these light pulses at a frequency ν_B in a proper time interval $\Delta\tau_B$. Since the number of light pulses is conserved, we can write that

$$\nu_A \Delta\tau_A = \nu_B \Delta\tau_B \tag{7.6}$$

Note that here we use proper time intervals and not coordinate time intervals, because the measurement of the frequency ν_A (ν_B) is done in the same spatial position of A (B) [see Section 3.6]. Then, Eq.(7.6) provides

$$\frac{1}{c}\nu_A\sqrt{g_{00}(A)}\Delta x_A^0 = \frac{1}{c}\nu_B\sqrt{g_{00}(B)}\Delta x_B^0 \tag{7.7}$$

For light-like intervals, $cdt = dr/(1-\mathscr{R}_S/r)$, where $\mathscr{R}_S$ is the Schwarzschild radius already introduced in Section 6.10. Then, the first light pulse, emitted by A at t_A, is received by B at time

$$x_B^0 = x_A^0 + \left[(r_B - r_A) + \mathscr{R}_S \ln\left(\frac{r_B}{r_A}\right)\right] \tag{7.8}$$

where r_A and r_B are the radial coordinates of the two observers, both supposed to be much larger than $\mathscr{R}_S$ in agreement with the assumptions of Eq.(6.10). Since the time delay is constant, it follows that $\Delta x_B^0 = \Delta x_A^0$: the coordinated time intervals elapsed between the emission (by A) and the receiving (by B) of two consecutive light pulses is the same. Then, Eq.(7.8) provides

$$\frac{\nu_B}{\nu_A} = \sqrt{\frac{g_{00}(A)}{g_{00}(B)}} \simeq 1 - \frac{GM}{c^2 r_A} + \frac{GM}{c^2 r_B} \tag{7.9}$$

It follows that the fractional change in frequency experienced by light traveling in a weak gravitational field is given by the difference in the Newtonian potential between the emitter and the observer

$$\boxed{\frac{\Delta\nu}{\nu} \equiv \frac{\nu_B - \nu_A}{\nu_A} \simeq \frac{\Delta U}{c^2}} \tag{7.10}$$

where $\Delta U \equiv U(A) - U(B)$. So, light climbing out of a gravitational well ($r_B \gg r_A$) is redshifted; light falling into a potential well ($r_B \ll r_A$) is blueshifted. This is a nice and simple theoretical prediction based *only*, let's stress it again, on the assumed validity of the Equivalence Principle and, then, on Eq.(6.17).

7.6 THE LONG GRAVITATIONAL REDSHIFT HUNT

Einstein predicted the gravitational redshift of light in 1907 [19]. Soon after, the idea was developed to measure such an effect in white dwarf stars, because of their intense gravitational field. These stars are indeed very dense and compact, with roughly a solar mass in a volume comparable to that of the Earth. The nearest white dwarf is Sirius B, only 8.6 light years away. The initial attempts to measure the gravitational redshift of the spectral lines of Sirius B go back to the 20s of the last century [5], but a reliable measurement of this effect arrived only in 1971, when Greenstein *et al.* [41] found a value perfectly consistent with the Equivalence Principle. These results have been confirmed in 2005 by accurate measurements done with the Hubble Space

Telescope [8]. A nice review of the experimental efforts to measure the gravitational redshift of Sirius B can be found in [43].

It is interesting to note that a clever method for the detection of the gravitational redshift in a lab. was proposed in 1959 by Pound and Rebka [73]. The idea was to use the total usable height ($h = 22,6m$) of the Jefferson Physical Laboratory tower at Harvard University. An emitter and a receiver were positioned at the top ($r_{emitter} = R_\oplus + h$) and at the bottom ($r_{receiver} = R_\oplus$) of the tower. Photons sent downwards experienced a gravitational blueshift. According to Eq.(7.10), the effect amounts to

$$\left.\frac{\Delta\nu}{\nu}\right|_{downwards} = -\frac{GM_\oplus}{c^2}\left(\frac{1}{r_{receiver}} - \frac{1}{r_{emitter}}\right) \simeq \frac{gh}{c^2} \simeq 2.5\cdot 10^{-15} \tag{7.11}$$

By changing the setting, with the emitter at the bottom and the receiver at the top of the tower, we expect a gravitational redshift: $\Delta\nu/\nu|_{upwards} = -2.5\cdot 10^{-15}$. Combining the two different configurations yields a cumulative effect of

$$\left.\frac{\Delta\nu}{\nu}\right|_{downwards} - \left.\frac{\Delta\nu}{\nu}\right|_{upwards} \simeq 5\cdot 10^{-15} \tag{7.12}$$

Detecting such tiny frequency shifts is not easy. The idea was to exploit the Mossbauer effect [60] and to use γ-ray photons of $14.4keV$ to have high enough spectral resolution. Pound and Rebka reported the following result

$$\left.\frac{\Delta\nu}{\nu}\right|^{(exp)}_{downwards} - \left.\frac{\Delta\nu}{\nu}\right|^{(exp)}_{upwards} \simeq (5.13 \pm 0.51)\cdot 10^{-15} \tag{7.13}$$

in agreement with the prediction of the Equivalence Principle at the 10% level [71]. This result was further confirmed with an improved precision (at the 0.76% level) in 1965 by Pound and Snider [72]. It is worth mentioning that the first paper reporting a terrestrial measurement of the gravitational redshift (although at the 50% level) was the one by Cranshaw, Shiffer and Whitehead [15], published one month and a half before the one by Pound and Rebka. An interesting review of the measurements of gravitational redshift between 1959 and 1971 is given in [42].

To conclude, let's stress that all these measurements constitute the experimental confirmation of the correctness of the Equivalence Principle and, then, of Eq.(6.17).

7.7 THE NORDTVEDT RFFECT

Let's conclude this chapter by further extending our discussion on the Equivalence Principle. As seen in Section 7.2, we can conclude that nuclear and electromagnetic interactions contribute equally well to the inertial and gravitational masses. We have just discussed a number of experimental tests that support this formulation. However, we didn't make any reference to the gravitational binding energy W of a body. So, it is worth asking what happens if such an energy doesn't contribute equally to the inertial and to the gravitational masses. Let's then write the fractional difference between the two masses as follows:

$$\frac{M_g - M_i}{M_i} \equiv \varepsilon = \eta \frac{|W|}{M_i c^2} \tag{7.14}$$

This fractional difference, ε, is usually expressed in terms of the so-called *Nordtvedt parameter*, η, and of the gravitational binding energy, $W = -G\int_0^{M_g} m(r)dm(r)/r$. Assuming a uniform density distribution, we recover the standard result $W = -3GM_g^2/(5R)$. It is common to take into account non-uniform density distribution by writing $|W|/(M_i c^2) = \alpha GM/(c^2 R)$, where α is a parameter and to the lowest order we assume $M = M_g = M_i$. In any case, the binding energy is a growing function of the mass. Unfortunately, for laboratory-sized objects, we are dealing with vanishingly small numbers. Just for having an order of magnitude, 100 g of water in a dimension of 5 cm (a typical water goblet) would provide $\varepsilon \simeq 9 \times 10^{-28}$ $(\eta = 1)$. So, to test Eq.(7.14), we have to consider very large masses with a non-negligible binding energy. We have then to move to planetary scale.

Consider the system composed by the Sun and Jupiter, of mass $M_\odot$ and M_J, orbiting around the common center of mass at distances $r_\odot$ and r_J, respectively. If *gravitational* and *inertial* masses are not alike, then

$$\omega^2 r_\odot = \frac{GM_J}{(r_\odot + r_J)^2}(1 + \varepsilon_\odot) \tag{7.15a}$$

$$\omega^2 r_J = \frac{GM_\odot}{(r_\odot + r_J)^2}(1 + \varepsilon_J) \tag{7.15b}$$

where $\varepsilon_\odot$ and ε_J are the Sun and Jupiter epsilon values [*c.f.* Eq.(7.14)]. The previous equation leads to the following relations:

$$\omega^2 = \frac{G}{(r_\odot + r_J)^3})[M_\odot(1 + \varepsilon_J) + M_J(1 + \varepsilon_\odot)] \tag{7.16a}$$

$$\frac{M_\odot r_\odot}{M_J r_J} = \frac{1 + \varepsilon_\odot}{1 + \varepsilon_J} \tag{7.16b}$$

Consider now a test-particle in L, at distances r_1 from the Sun, r_2 from Jupiter and r from the center of mass of the system [see Figure 7.1]. We can ask under which conditions the test-particle can be in equilibrium in L. To answer to this question, let's project the gravitational forces exerted by $M_\odot$ and M_J along two directions: one orthogonal to and the other aligned with the r direction [see Figure 7.1]. We then have

$$\frac{GM_\odot}{r_1^2}\sin\alpha_1 = \frac{GM_J}{r_2^2}\sin\alpha_2 \tag{7.17a}$$

$$\frac{GM_\odot}{r_1^2}\cos\alpha_1 + \frac{GM_J}{r_2^2}\cos\alpha_2 = \omega^2 r \tag{7.17b}$$

Because of the Law of Sines, $r_\odot/r_1 = \sin\alpha_1/\sin\theta$ and $r_J/r_2 = \sin\alpha_2/\sin\theta$. Then, Eq.(7.17a) provides

$$M_\odot r_\odot r_1^{-3} = M_J r_J r_2^{-3} \tag{7.18}$$

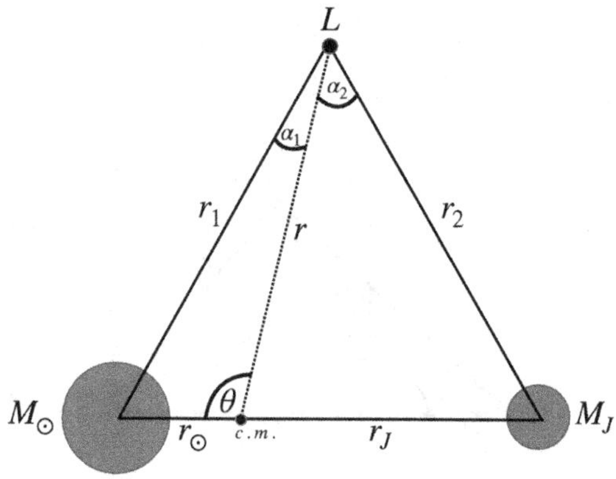

Figure 7.1 The Lagrangian point L is a stable equilibrium point of the Sun–Jupiter system. If inertial and gravitational masses are alike, the Sun, Jupiter and L are at the vertexes of an equilateral triangle. This triangle rigidly rotates around the center of mass, *c.m.*, of the Sun–Jupiter system [see text].

and, because of Eq.(7.16b),

$$(1+\varepsilon_\odot)r_1^{-3} = (1+\varepsilon_J)r_2^{-3} \tag{7.19}$$

It is easy to verify that $r = r_1\cos\alpha_1 + r_\odot\cos\theta = r_2\cos\alpha_2 - r_J\cos\theta$ (*c.f.* Figure 7.1). Then, Eq.(7.17b) becomes

$$r\left(\frac{GM_\odot}{r_1^3}+\frac{GM_J}{r_2^3}\right)+G\left(-\frac{M_\odot r_\odot}{r_1^3}+\frac{M_J r_J}{r_2^3}\right)\cos\theta = \omega^2 r \tag{7.20}$$

The second term on the *lhs* vanishes because of Eq.(7.18), whereas the term on the *rhs* can be rewritten using Eq.(7.16a). Thus, at the end, we can write

$$\frac{M_\odot}{r_1^3}+\frac{M_J}{r_2^3} = \frac{M_\odot(1+\varepsilon_J)+M_J(1+\varepsilon_\odot)}{(r_\odot+r_J)^3} \tag{7.21}$$

The system provided by Eq.(7.19) and Eq.(7.21) admits as solution

$$\begin{aligned} r_1 &= (r_\odot+r_J)(1+\varepsilon_J)^{-1/3} \simeq (r_\odot+r_J)(1-\frac{1}{3}\varepsilon_J) \\ r_2 &= (r_\odot+r_J)(1+\varepsilon_\odot)^{-1/3} \simeq (r_\odot+r_J)(1-\frac{1}{3}\varepsilon_\odot) \end{aligned} \tag{7.22}$$

If $\eta = 0$, then the epsilons are zero [*c.f.* Eq.(7.14)], and we reproduce the classical result of Lagrange, $r_1 = r_2 = r_\odot + r_J$: the test-particle is in stable equilibrium in the Lagrangian point L at the vertex of an equilateral triangle, the other two vertexes being occupied by $M_\odot$ and M_J, respectively. This triangle rigidly rotates around the

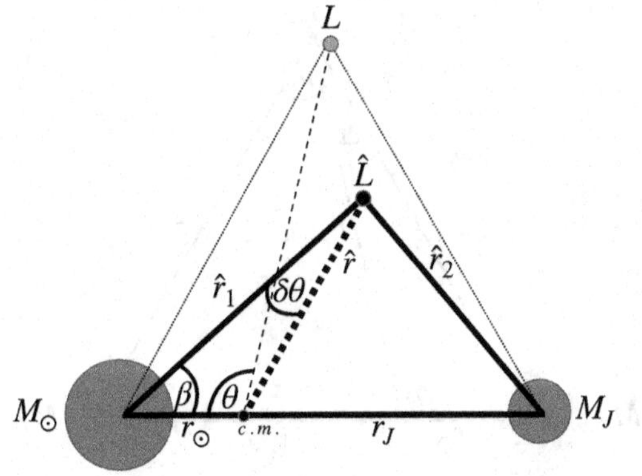

Figure 7.2 The Lagrangian point $\hat{L}$ is a stable equilibrium point of the Sun–Jupiter system. If inertial and gravitational masses are not alike, the Sun, Jupiter and $\hat{L}$ are not at the vertexes of an equilateral triangle [see text].

center of mass of the Sun–Jupiter system. In the case of Figure 7.1, test-particles in L "precede" Jupiter in its orbital motion[2].

If $\eta \neq 0$, the triangle will not be equilateral anymore [*c.f.* Eq.(7.22)] and the position of the stable Lagrangian point changes [see Figure 7.2]. In particular, we have

$$\begin{aligned} \hat{r}_1 \sin\beta &= \hat{r}\sin(\theta+\delta\theta) \\ \hat{r}_1 \cos\beta &= r_\odot - \hat{r}\cos(\theta+\delta\theta) \end{aligned} \tag{7.23}$$

leading to

$$\tan(\theta+\delta\theta) = \frac{\hat{r}_1 \sin\beta}{r_\odot - \hat{r}_1\cos\beta} \tag{7.24}$$

In addition, for the Law of Cosines, we have

$$\hat{r}_2^2 = (r_\odot + r_J)^2 + \hat{r}_1^2 - 2(r_\odot + r_J)\hat{r}_1\cos\beta \tag{7.25}$$

On the basis of Eq.(7.22), we can write $\hat{r}_1 = R(1-\delta r_1/R)$ and $\hat{r}_2 = R(1-\delta r_2/R)$, where $R \equiv r_\odot + r_J$. Then, to first order, Eq.(7.25) provides

$$\cos\beta = \frac{R^2+\hat{r}_1^2-\hat{r}_2^2}{2R\hat{r}_1^2} \simeq \frac{1}{2} - \frac{1}{2}\frac{\delta r_1}{R} + \frac{\delta r_2}{R} \tag{7.26}$$

It follows that, to the same order,

$$\sin\beta = \sqrt{1-\left(1-\frac{1}{2}\frac{\delta r_1}{R}+\frac{\delta r_2}{R}\right)^2} \simeq \frac{\sqrt{3}}{2} + \frac{\delta r_1}{2\sqrt{3}R} - \frac{\delta r_2}{\sqrt{3}R} \tag{7.27}$$

[2] There is another stable Lagrangian point, L', in a symmetric configuration obtained by rotating the equilateral triangle around the line joining the Sun with Jupiter. In this configuration, test-particles in L' will rather "follow" Jupiter in its orbital motion.

Let's then write Eq.(7.24) to first order

$$\tan(\theta+\delta\theta) = -\frac{\sqrt{3}R}{R-2r_\odot} + \frac{4(2R-r_\odot)\delta r_2}{\sqrt{3}(R-2r_\odot)^2} - \frac{4(R+r_\odot)\delta r_1}{\sqrt{3}(R-2r_\odot)^2} \tag{7.28}$$

Now, $\tan(\theta+\delta\theta) = \tan\theta + (1-\tan^2\theta)\delta\theta$ and, to zeroth-order, $\tan\theta = -\sqrt{3}R/(R-2r_\odot)$ [*c.f.* Eq.(7.28)]. So, remembering that $R \equiv r_\odot + r_J$, we can write that

$$\delta\theta = \frac{\delta r_2(r_\odot + 2r_J) - \delta r_1(2r_\odot + r_J)}{\sqrt{3}\left(r_\odot^2 + r_\odot r_J + r_J^2\right)} \tag{7.29}$$

At this point, we can exploit the fact that $r_\odot \ll r_J$, as the center of mass of the Sun–Jupiter system is defined by the Sun. It follows that $R \simeq r_J$. Then, we can write that $\delta r_1/r_J = \varepsilon_J/3$ and $\delta r_2/r_j = \varepsilon_\odot/3$. Under the further assumption of having $\varepsilon_\odot >> \varepsilon_{Jup}$, Eq.(7.29) becomes

$$\delta\theta = \frac{2}{3\sqrt{3}}\varepsilon_\odot \tag{7.30}$$

For the Sun, $\alpha = 15/2$ and $\varepsilon_\odot = 1.59\times10^{-5}\eta$. So, for $\eta = 1$, we expect a value of $\delta\theta \simeq 1''$, a large value for astrometric standards. A systematic analysis of the Trojan asteroid's motion—occurring around the Lagrangian point—provides an estimate of the Nordtvedt parameter: $\eta = 0.0\pm0.5$ [62] and $\eta = -0.56\pm0.48$ [63]. Both these estimates are consistent with zero. For completeness, let's mention other estimates of the Nordtvedt parameter: one is coming from the NASA MESSENGER mission, providing $\eta = (-6.6\pm7.2)\times10^{-5}$ [39]; few others are coming from the Lunar Laser Ranging, providing $\eta = 0.00\pm0.03$ (see, *e.g.*, [99]).

In conclusion, we can safely state that binding energy equally contributes to the inertial and gravitational masses. The term *Strong Equivalence Principle* is often used to underline that it applies to all laws of nature, including all the form of energies.

Let us now write Eq. (7.24) [illegible] or

[illegible] (7.26)

Now [illegible] $\tan\theta$ [illegible] [illegible], we can write [illegible]

[illegible] (7.27)

[illegible] the fact that [illegible] is the center of [illegible] [illegible] the Sun [illegible] [illegible] the [illegible]

[illegible]

[illegible] the Sun [illegible]

[illegible]

This [illegible] we can [illegible] Principle [illegible]

8 Field Equations in the "vacuum"

8.1 INTRODUCTION

In Chapter 6, we derived at first-order, the metric of a space-time containing a single point mass. This was done heuristically, on the basis of the Equivalence Principle alone, rather than using rigorous solutions of well-established field equations. Einstein introduced his General Theory of Gravity in 1915, publishing in that year four papers [20–23]. In the last of them, he proposed the field equations of his theory of gravitation. The purpose of this chapter is to discuss the equations needed to derive the geometry of the space-time produced by a point mass or, as it is often said, in the "vacuum".[1]

8.2 FIELD EQUATIONS IN THE "VACUUM": REQUIREMENTS

The field equations needed to find the metric tensor components have to fulfill a number of requirements. First, accordingly to the Principle of General Covariance, using tensors, for them to be valid in any arbitrarily chosen reference frame [*c.f.* Section 4.7]. Secondly, in the weak field limit, they should be consistent with the Laplace equation, $\nabla^2 U = 0$. Since $g_{00} = 1 + 2U/c^2$ [*c.f.* Eq.(6.18)], this would imply $\sum_{i=1}^{3} g_{00,ii} = 0$. As a consequence, and this is the third and final requirement, they should be second-order differential equations that are linear in the second derivatives of the metric coefficients.

All these criteria seem to suggest to adopt field equations based on the vanishing of the Riemann tensor: $R^{\alpha}{}_{\beta\mu\nu} = 0$. This would immediately satisfy the first and third requirements: we have covariant expressions and second-order differential equations, linear in the second derivatives of the metric tensor components. There are, however, two obvious drawbacks. First, we would have 20 differential equations [the number of the independent components of the Riemann's tensor, *c.f.* Eq.(5.39)], whereas we have only ten unknowns [the independent components of the symmetric metric tensor, $g_{\alpha\beta}$]. Secondly, and most importantly, requiring the Riemann tensor to be zero forces the space-time to be flat, and this is incompatible with the presence of gravitational fields [see Section 6.9 and Eq.(6.30)]. Given the strong link between gravity

[1] In this context, the term "vacuum" refers to the spatial regions outside the matter distribution responsible for the curvature of the space-time. In this chapter, we will use the point mass approximation.

DOI: 10.1201/9781003141259-8

and curvature, we should still work with the Riemann tensor, but using somehow less stringent conditions.

8.3 THE RICCI TENSOR

Instead of imposing $R^{\alpha}{}_{\beta\mu\nu} = 0$, let's require the vanishing not of each of the Riemann tensor components, but rather of a *linear combination* of them. The only way of combining tensor components in a covariant way consists in contracting the tensor indexes [see *e.g.* Eq.(4.18)]. If we do this with the Riemann tensor, we obtain a tensor of rank two, the *Ricci tensor*: $R_{\beta\nu} \equiv R^{\mu}{}_{\beta\mu\nu}$ [74, 75]. On the basis of Eq.(5.39), the Ricci tensor can be written as follows:

$$\boxed{R_{\beta\nu} = \Gamma^{\mu}_{\beta\nu,\mu} - \Gamma^{\mu}_{\beta\mu,\nu} + \Gamma^{\mu}_{\sigma\mu}\Gamma^{\sigma}_{\beta\nu} - \Gamma^{\mu}_{\sigma\nu}\Gamma^{\sigma}_{\beta\mu}} \tag{8.1}$$

The symmetry properties of the Ricci tensor can be highlighted by rewriting Eq.(8.1) in a slightly simpler way. To do so, let's first note that $\partial g/\partial x^{\beta} = \mathrm{adj}(g_{\mu\nu})g_{\mu\nu,\beta}$ where g and $\mathrm{adj}(g_{\mu\nu}) \equiv \partial g/\partial g_{\mu\nu}$ are the determinant and adjugate of the metric tensor. On the other hand, the inverse of the metric tensor is given by $g^{\mu\nu} = \mathrm{adj}(g_{\mu\nu})/g$. Then,

$$g^{\mu\nu} g_{\mu\nu,\beta} = \frac{\partial \ln(-g)}{\partial x^{\beta}} \tag{8.2}$$

By using this result and Eq.(3.24), we get

$$\boxed{\Gamma^{\mu}_{\beta\mu} = \frac{1}{2} g^{\mu\rho} g_{\mu\rho,\beta} = \frac{\partial \ln \sqrt{-g}}{\partial x^{\beta}}} \tag{8.3}$$

Thus, the Ricci tensor can more conveniently be written in the following form.

THE RICCI TENSOR: A GENERAL EXPRESSION

$$\boxed{R_{\beta\nu} = \Gamma^{\mu}_{\beta\nu,\mu} - \frac{\partial^2 \ln \sqrt{-g}}{\partial x^{\nu} \partial x^{\beta}} + \frac{\partial \ln \sqrt{-g}}{\partial x^{\sigma}} \Gamma^{\sigma}_{\beta\nu} - \Gamma^{\mu}_{\sigma\nu}\Gamma^{\sigma}_{\beta\mu}} \tag{8.4}$$

Eq.(8.4) clearly shows that the Ricci tensor is symmetric. So, in a four-dimensional space-time, it has only ten independent components, as the metric tensor does. The writing of Eq.(8.4) can be further simplified when the metric tensor does not depend explicitly on a given coordinate. Consider, for example, the metric of Eq.(6.17): the metric tensor depends neither on time nor on the equatorial angle, ϕ. So, let's assume, for sake of generality, that the metric tensor doesn't depend on the coordinate x^{β} – that is, $g_{\mu\nu,\beta} = 0$. Under these circumstances, the Ricci tensor can be written either in a fully covariant

$$R_{\beta\nu} = \Gamma^{\mu}_{\beta\nu,\mu} + \frac{\partial \ln \sqrt{-g}}{\partial x^{\sigma}} \Gamma^{\sigma}_{\beta\nu} - \Gamma^{\mu}_{\sigma\nu}\Gamma^{\sigma}_{\beta\mu} \tag{8.5}$$

or in a mixed form by raising one of its indexes

$$R^{\alpha}{}_{\beta} = g^{\alpha\nu}R_{\nu\beta} = g^{\alpha\nu}\Gamma^{\mu}_{\beta\nu;\mu} + g^{\alpha\nu}\frac{\partial \ln\sqrt{-g}}{\partial x^{\sigma}}\Gamma^{\sigma}_{\beta\nu} - g^{\alpha\nu}\Gamma^{\mu}_{\sigma\nu}\Gamma^{\sigma}_{\beta\mu} \tag{8.6}$$

The covariant derivative of the metric tensor vanishes [*c.f.* Eq.(5.33)]. Since, $(g^{\alpha\sigma}g_{\sigma\beta})_{;\mu} = \delta^{\alpha}{}_{\beta;\mu} = 0$, it vanishes also the covariant derivative of the inverse of the metric tensor [*c.f.* Eq.(5.30)]. Then, $g^{\alpha\mu}{}_{;\sigma} = g^{\alpha\mu}{}_{,\sigma} + g^{\alpha\nu}\Gamma^{\mu}_{\sigma\nu} + g^{\nu\mu}\Gamma^{\alpha}_{\sigma\nu} = 0$. Multiplying this relation by $\Gamma^{\sigma}_{\beta\mu}$, we get: $-g^{\alpha\nu}\Gamma^{\mu}_{\sigma\nu}\Gamma^{\sigma}_{\beta\mu} = g^{\alpha\mu}{}_{,\sigma}\Gamma^{\sigma}_{\beta\mu} + g^{\nu\mu}\Gamma^{\alpha}_{\sigma\nu}\Gamma^{\sigma}_{\beta\mu}$. Consider now the last term on the *rhs* of this equation. It actually vanishes. In fact,

$$g^{\nu\mu}\Gamma^{\alpha}_{\sigma\nu}\Gamma^{\sigma}_{\beta\mu} = g^{\nu\mu}\Gamma^{\alpha}_{\sigma\nu}\left[\frac{1}{2}g^{\sigma\rho}\left(-g_{\beta\mu,\rho} + g_{\rho\beta,\mu}\right)\right] = 0 \tag{8.7}$$

as both ρ and μ (but also σ and ν) are dummy indexes. Then, $-g^{\alpha\nu}\Gamma^{\mu}_{\sigma\nu}\Gamma^{\sigma}_{\beta\mu} = g^{\alpha\mu}{}_{,\sigma}\Gamma^{\sigma}_{\beta\mu}$. This allows us to write Eq.(8.6) in a very compact way.

THE RICCI TENSOR WHEN $g_{\mu\nu,\beta} = 0$

$$R^{\alpha}{}_{\beta} = \frac{1}{\sqrt{-g}}\frac{\partial}{\partial x^{\mu}}\left[g^{\alpha\nu}\Gamma^{\mu}_{\beta\nu}\sqrt{-g}\right] \tag{8.8}$$

Eq.(8.8) is valid, let's repeat it again, *if and only if* the metric does not explicitly depend on the β coordinate, *i.e.* $g_{\mu\nu,\beta} = 0$. If this is not the case, we have to use the full expression given in Eq.(8.4)

8.4 GRAVITATIONAL FIELD EQUATIONS IN THE "VACUUM"

Let's consider again the simple case of the point mass considered in Section 6.4. The metric of Eq.(6.17) does not depend on time: $g_{\mu\nu,0} = 0$. Then, we can write the Ricci tensor by using Eq.(8.8) with $\beta = 0$

$$R^{\alpha}{}_{0} = \frac{1}{\sqrt{-g}}\frac{\partial}{\partial x^{k}}\left[g^{\alpha\nu}\Gamma^{k}_{0\nu}\sqrt{-g}\right] \tag{8.9}$$

In the non-relativistic limit, the metric tensor can be written to first order by perturbing the Minkowski metric: $g_{\mu\nu} = \eta_{\mu\nu} + h_{\mu\nu}(x^{i})$ [*c.f.* Eq.(6.10)]. In the same limit, $ds \simeq cdt$, $\dot{x}^{0} \simeq 1$ and $\dot{x}^{k} \ll 1$. Now, consider a test-particle subjected only to the action of the gravitational field of the point mass. From what we have discussed in Section 6.4, the test-particle will move along geodesic of the space-time. In the non-relativistic limit, the spatial components of the geodesic equations [*c.f.* Eq.(3.30)] yield

$$\frac{1}{c^2}\frac{d^2x^k}{dt^2} + \Gamma^{k}_{00} = 0 \tag{8.10}$$

clearly showing that, to first order, there are only three non-vanishing Christoffel symbols: $\Gamma^k_{00} \simeq h_{00,k}/2$, with $k = 1,3$ [*c.f.* Eq.(3.24)]. The comparison of Eq.(8.10) with the classical result

$$\frac{d^2x^k}{dt^2} = -\frac{\partial U}{\partial x^k} \tag{8.11}$$

shows that $h_{00} = 2U/c^2$, as found on the basis of the Equivalence Principle [*c.f.* Eq.(6.20]. Since Γ^k_{00} is in this case a first-order quantity, we must consider all the other terms in Eq.(8.9) to the zeroth order: $g^{\alpha\nu} \to \eta^{\alpha\nu}$; $\sqrt{-g} \to 1$. Then, Eq.(8.9) becomes

$$R^0{}_0 = \frac{\partial}{\partial x^k}\left[\Gamma^k_{00}\right] = \frac{1}{2}h_{00,kk} \simeq \frac{1}{c^2}\nabla^2 U = 0 \tag{8.12}$$

where the last equality follows from the Laplace equation of classical mechanics. All this suggests to postulate that good field equations in the "vacuum" could be obtained by imposing the vanishing of the Ricci tensor components.

FIELD EQUATIONS USING THE RICCI TENSOR

$$\boxed{R_{\alpha\beta} = 0} \tag{8.13}$$

Indeed, these field equations fulfill all the requirements we have discussed in Section 8.2: they are written in a tensorial form; in the weak field limit, they admit the Newtonian solution [*c.f.* Eq.(8.12)]; they provide ten, second-order differential equations that are linear in the second derivative of the metric coefficients.

The field equations in the "vacuum" can be modified by introducing a Cosmological Constant, Λ. A positive value of the Cosmological Constant is needed to account for the present acceleration phase of our universe [see Section 15.18]. To consistently take into account Λ, we should consider modified field equations in the "vacuum".

FIELD EQUATIONS USING THE RICCI TENSOR WITH $\Lambda \neq 0$

$$\boxed{R_{\alpha\beta} + \Lambda g_{\alpha\beta} = 0} \tag{8.14}$$

Note that by introducing a non-vanishing Cosmological Constant, we are in principle loosing the weak field limit of Eq.(8.12). However, the values of Λ needed in Cosmology are of the order of $\simeq 10^{-56} cm^{-2}$ [see Section 15.18]. So, the effect of a non-vanishing Cosmological Constant is relevant only on scales $\gtrsim \Lambda^{-1/2}$, much larger than any planetary scale. Nonetheless, for didactical purposes, in this chapter we will use the modified field equations of Eq.(8.14).

8.5 THE EINSTEIN TENSOR

The "vacuum" field equations, $R_{\alpha\beta} = 0$, may be written in an alternative way by means of the *Einstein tensor*:

$$\boxed{G_{\alpha\beta} \equiv R_{\alpha\beta} - \frac{1}{2}g_{\alpha\beta}R} \tag{8.15}$$

where

$$R \equiv g^{\mu\nu}R_{\mu\nu} \tag{8.16}$$

is the *Ricci scalar*, the trace of the Ricci tensor. Such a scalar provides, point by point, the simplest curvature invariant of a pseudo-Riemannian space-time. Let's write Eq.(8.15) in a mixed form: $G^{\alpha}{}_{\beta} = R^{\alpha}{}_{\beta} - \delta^{\alpha}{}_{\beta}R/2$. The index contraction provides $G = -R$ where $G \equiv G^{\mu}{}_{\mu}$ is the trace of the Einstein tensor. Then, by using Eq.(8.15), we can write the Ricci tensor in terms of the Einstein tensor:

$$R_{\alpha\beta} = G_{\alpha\beta} - \frac{1}{2}g_{\alpha\beta}G \tag{8.17}$$

If $R_{\alpha\beta} = 0$, then $R = 0$ and $G_{\alpha\beta} = 0$ [*c.f.* Eq.(8.15)]. Vice versa, if $G_{\alpha\beta} = 0$, then $G = 0$ and $R_{\alpha\beta} = 0$ [*c.f.* Eq.(8.17)]. So, in the "vacuum", both $R_{\alpha\beta} = 0$ and $G_{\alpha\beta} = 0$ are completely equivalent field equations. If we want to take into account a non-vanishing Cosmological Constant, we have to use either Eq.(8.14) or the following expression.

FIELD EQUATIONS USING THE EINSTEIN TENSOR WITH $\Lambda \neq 0$

$$\boxed{G_{\alpha\beta} - \Lambda g_{\alpha\beta} = 0} \tag{8.18}$$

Let's conclude this section, by showing a useful property of the Einstein tensor. Consider the Bianchi identity given in Eq.(5.52). Contracting over the co- and contravariant indexes yield

$$R_{\beta\nu;\lambda} - R_{\beta\lambda;\nu} + R^{\mu}{}_{\beta\nu\lambda;\mu} = 0 \tag{8.19}$$

Raise the β index and contract again to obtain $R_{;\lambda} - R^{\nu}{}_{\lambda;\nu} - R^{\mu}{}_{\lambda;\mu} = 0$ or, equivalently

$$R_{;\lambda} = 2R^{\nu}{}_{\lambda;\nu} \tag{8.20}$$

It follows that the covariant divergence of the Einstein tensor

$$\boxed{G^{\alpha\beta}{}_{;\beta} = R^{\alpha\beta}{}_{;\beta} - \frac{1}{2}g^{\alpha\beta}R_{;\beta} = 0} \tag{8.21}$$

vanishes because of Eq.(8.20) in all the frames. This is an important result that we will use later in the book, in Section 13.7.

8.6 HILBERT'S ACTION

It is useful to derive the field equations also from another perspective. Let's define the action of the gravitational field as an integral over the proper volume of an invariant Lagrangian density, $\mathscr{L}$. The infinitesimal element of the four-dimensional proper volume, $d^4\mathscr{V}_p$, can be written in the *locally* Minkowskian frame $\widehat{\mathscr{K}}$ of coordinates ξ^α. The transformation from $\widehat{\mathscr{K}}$ to a generic frame $\mathscr{X}$ of curvilinear coordinates x^α is given by the Jacobian matrix, $J^\alpha{}_\mu \equiv (\partial\xi^\alpha/\partial x^\mu)$, so that

$$d^4\mathscr{V}_p \equiv d\xi^0 d\xi^1 d\xi^2 d\xi^3 = J\,dx^0 dx^1 dx^2 dx^3 \tag{8.22}$$

where $J \equiv \det(J^\alpha{}_\mu)$. On the other hand, the metric tensor transforms as a tensor of rank two: $g_{\mu\nu} = \eta_{\alpha\beta} J^\alpha{}_\mu J^\alpha{}_\nu$ [*c.f.* Eq.(3.4)]. The determinant of the product of matrixes is equal to the product of the determinants of each of the matrixes. Thus, $g = -J^2$, where $g = \det(g_{\mu\nu})$ and $\det(\eta_{\alpha\beta}) = -1$. It follows that $d^4\mathscr{V}_p = \sqrt{-g}d^4x$. We can then write the action as follows:

$$S_H = \int \mathscr{L}\sqrt{-g}d^4x \tag{8.23}$$

where the subscript H stands for Hilbert [45]. The Lagrangian density should be a function of the metric coefficients, so that the variation of the action can provide second-order differential equations. This would actually be the case if $\mathscr{L}$ contained only the first derivatives of the metric tensor. Unfortunately, these derivatives are associated with the Christoffel symbols that are not invariant quantities. Thus, the most obvious invariant to use is the Ricci scalar [*c.f.* Eq.(8.16)]. However, R contains already second derivatives of metric tensor. It follows that the variational principle associated with the action S_H provides second-order differential equations *if and only if* the higher order derivatives of metric tensor cancel out. Let's see if this is really the case. If $\mathscr{L} = -c^4/(16\pi G)R$, then,

$$\boxed{S_H = -\frac{c^4}{16\pi G}\int g^{\alpha\beta}R_{\alpha\beta}\sqrt{-g}d^4x = 0} \tag{8.24}$$

The field equations should then be derived by a variation principle

$$\delta S_H = -\frac{c^4}{16\pi G}\int d^4x\,\delta\left(\sqrt{-g}g^{\alpha\beta}R_{\alpha\beta}\right) = 0 \tag{8.25}$$

The variation of the round parenthesis provides three terms

$$\delta(\sqrt{-g}\,g^{\alpha\beta}R_{\alpha\beta}) = \delta(\sqrt{-g})\,g^{\alpha\beta}R_{\alpha\beta} + \sqrt{-g}\delta g^{\alpha\beta}R_{\alpha\beta} + \sqrt{-g}\,g^{\alpha\beta}\delta R_{\alpha\beta} \tag{8.26}$$

The variation of the metric determinant results to be $\delta g = g\,g^{\alpha\beta}\delta g_{\alpha\beta}$ [*c.f.* Eq.(8.2)]. Then,

$$\delta\sqrt{-g} = -\frac{1}{2}\frac{\delta g}{\sqrt{-g}} = \frac{1}{2}\sqrt{-g}\,g^{\alpha\beta}\delta g_{\alpha\beta} \tag{8.27}$$

On the other hand, $g^{\alpha\beta}g_{\alpha\beta} = \delta^{\alpha}{}_{\alpha}$ and $\delta g^{\alpha\beta}g_{\alpha\beta} + g^{\alpha\beta}\delta g_{\alpha\beta} = 0$. Then

$$\delta\sqrt{-g} = -\frac{1}{2}\sqrt{-g}\, g_{\alpha\beta}\,\delta g^{\alpha\beta} \tag{8.28}$$

Eq.(8.26) can then be rewritten as follows:

$$\delta(\sqrt{-g}g^{\alpha\beta}R_{\alpha\beta}) = \sqrt{-g}\left\{-\frac{1}{2}g_{\alpha\beta}\delta g^{\alpha\beta}R + \delta g^{\alpha\beta}R_{\alpha\beta} + g^{\alpha\beta}\delta R_{\alpha\beta}\right\} \tag{8.29}$$

Substituting this expression in Eq.(8.25) provides

$$\delta S_H = -\frac{c^4}{16\pi G}\left\{\int \delta g^{\alpha\beta}G_{\alpha\beta}\sqrt{-g}\, d^4x + \int g^{\alpha\beta}\delta R_{\alpha\beta}\sqrt{-g}\, d^4x\right\} = 0 \tag{8.30}$$

where we used Eq.(8.15). The second term on the *rhs* of the previous equation could in principle produce third-order derivatives of the metric tensor. However, this term can actually be written as the divergence of a vector. To see this point, remember that in the local inertial reference frame, the Christoffel symbols can *locally* be set to zero. Then, in $\widehat{\mathscr{K}}$, the variation of the Ricci tensor [*c.f.* Eq.(8.1)] provides

$$\delta\widehat{R}_{\alpha\beta} = \frac{\partial}{\partial x^{\mu}}\delta\widehat{\Gamma}^{\mu}{}_{\alpha\beta} - \frac{\partial}{\partial x^{\beta}}\delta\widehat{\Gamma}^{\mu}{}_{\alpha\mu} \tag{8.31}$$

Being the metric *locally* Minkowskian, in $\widehat{\mathscr{K}}$ we can write, to first order,

$$\widehat{g}^{\alpha\beta}\delta\widehat{R}_{\alpha\beta} = \frac{\partial}{\partial x^{\mu}}\left[\eta^{\alpha\beta}\delta\widehat{\Gamma}^{\mu}{}_{\alpha\beta} - \eta^{\alpha\mu}\delta\widehat{\Gamma}^{\tau}{}_{\alpha\tau}\right] \equiv \frac{\partial\delta\widehat{V}^{\mu}}{\partial x^{\mu}} \tag{8.32}$$

When we move to a generic frame, we have to use the covariant derivative: $g^{\alpha\beta}\delta R_{\alpha\beta} = \delta V^{\mu}{}_{;\mu}$. Because of Eq.(8.3), the covariant four-divergence of δV^{μ} [*c.f.* Eq.(5.20)] can be written in a very compact way

$$\delta V^{\mu}{}_{;\mu} = \frac{1}{\sqrt{-g}}\frac{\partial}{\partial x^{\mu}}\left(\sqrt{-g}\,\delta V^{\mu}\right) \tag{8.33}$$

So, we can rewrite Eq.(8.30) as follows:

$$\delta S_H = -\frac{c^4}{16\pi G}\left\{\int \delta g^{\alpha\beta}G_{\alpha\beta}\sqrt{-g}d^4x - \int \frac{\partial}{\partial x^{\mu}}\left(\sqrt{-g}\,\delta V^{\mu}\right)d^4x\right\} = 0 \tag{8.34}$$

The second integral can be transformed into an integral over the hypersurface containing the integration volume $\mathscr{V}$. But, on the frontier of the integration region, the boundary conditions constrain δV^{μ} to be zero. This implies that the second integral in Eq.(8.34) vanishes and that the variational principle of Eq.(8.25) yields [*c.f.* Eq.(8.30)]

$$\boxed{G_{\alpha\beta} = 0} \tag{8.35}$$

consistently with Eq.(8.18) derived for a vanishing Cosmological Constant.

8.7 THE ACTION FOR A COSMOLOGICAL CONSTANT

To take into account the presence of a non-vanishing Cosmological Constant, we have to add a new contribution, S_ϕ say, to the action of the gravitational field. In line with Eq.(8.23), let's write the action of a scalar field as follows:

$$S_\phi = \int \mathscr{L}_\phi \sqrt{-g} d^4x \tag{8.36}$$

The new invariant Lagrangian density, $\mathscr{L}_\phi$, can be written in terms of an another invariant, the Cosmological Constant: $\mathscr{L}_\phi = -c^4\Lambda/(8\pi G)$. Then,

$$S_\phi = -\frac{c^4\Lambda}{8\pi G}\int \sqrt{-g} d^4x \tag{8.37}$$

The variation of S_ϕ *w.r.t.* the metric coefficients yields

$$\delta S_\phi = \frac{c^4\Lambda}{16\pi G}\int \delta g^{\mu\nu} g_{\mu\nu}\sqrt{-g} d^4x = 0 \tag{8.38}$$

where we have used Eq.(8.28). It follows that the variation of the total action $S_{tot} = S_H + S_\phi$ provides [*c.f.* Eq.(8.34) and Eq.(8.38)]:

$$\delta S_{tot} = \delta S_{EH} + \delta S_\phi = -\frac{c^4}{16\pi G}\int \delta g^{\alpha\beta}\left(G_{\alpha\beta} - \Lambda g_{\alpha\beta}\right)\sqrt{-g} d^4x = 0 \tag{8.39}$$

that is,

$$\boxed{G_{\alpha\beta} - \Lambda g_{\alpha\beta} = 0} \tag{8.40}$$

in line with Eq.(8.18).

It is interesting to note that the action S_ϕ can be associated with a scalar field ϕ that evolves under the potential $V(\phi)$. The invariant Lagrangian density of a minimally coupled scalar field is given by the following expression

$$\mathscr{L}_\phi \equiv \left[\frac{1}{2} g^{\rho\sigma}\frac{\partial\phi}{\partial x^\rho}\frac{\partial\phi}{\partial x^\sigma} - V(\phi)\right] \tag{8.41}$$

Note that $\mathscr{L}_\phi$ depends only on the metric and not on the metric derivatives. We can then use ordinary derivatives instead of the functional derivatives. It follows that the variation of S_ϕ [*c.f.* Eq.(8.36)] provides

$$\delta S_\phi = \int \delta g^{\mu\nu}\frac{\partial(\sqrt{-g}\mathscr{L}_\phi)}{\partial g^{\mu\nu}} d^4x \tag{8.42}$$

On the other hand, from Eq.(8.41), we get

$$\begin{aligned}\frac{\partial(\sqrt{-g}\mathscr{L}_\phi)}{\partial g^{\mu\nu}} &= -\frac{1}{2}\sqrt{-g} g_{\mu\nu}\left(\frac{1}{2} g^{\rho\sigma}\frac{\partial\phi}{\partial x^\rho}\frac{\partial\phi}{\partial x^\sigma} - V(\phi)\right) + \sqrt{-g}\frac{1}{2}\frac{\partial g^{\rho\sigma}}{\partial g^{\mu\nu}}\frac{\partial\phi}{\partial x^\rho}\frac{\partial\phi}{\partial x^\sigma}\\ &= \frac{1}{2}\sqrt{-g}\left[\frac{\partial\phi}{\partial x^\mu}\frac{\partial\phi}{\partial x^\nu} - \frac{1}{2} g_{\mu\nu} g^{\rho\sigma}\frac{\partial\phi}{\partial x^\rho}\frac{\partial\phi}{\partial x^\sigma} + g_{\mu\nu}V(\phi)\right]\end{aligned} \tag{8.43}$$

where we have used again Eq.(8.28). The evolution of the scalar field ϕ is described by the Klein-Gordon equation

$$\boxed{\Box\phi + V'(\phi) = 0} \tag{8.44}$$

where the covariant expression of the D'Alambertian operator is given by the following expression [*c.f.* Eq.(8.3)]

$$\Box\phi \equiv \left(g^{\mu\nu}\frac{\partial\phi}{\partial x^\nu}\right)_{;\mu} = \frac{1}{\sqrt{-g}}\frac{\partial}{\partial x^\mu}\left(\sqrt{-g}g^{\mu\nu}\frac{\partial\phi}{\partial x^\nu}\right) \tag{8.45}$$

Now, let's consider the very simple case of a scalar field, trapped at the minimum ϕ_0 of the potential: $V'(\phi)|_{\phi_0} = 0$. In this condition, the equation of motion [*c.f.* Eq.(8.44)] is clearly satisfied and Eq.(8.43) simplifies in

$$\frac{\partial(\sqrt{-g}\mathscr{L}_\phi)}{\partial g^{\mu\nu}} = \frac{1}{2}\sqrt{-g}g_{\mu\nu}V(\phi_0) \tag{8.46}$$

It follows that the variation principle of Eq.(8.42) now provides

$$\delta S_\phi = \int \delta g^{\mu\nu} g_{\mu\nu}\left(\frac{1}{2}V(\phi_0)\right)\sqrt{-g}d^4x \tag{8.47}$$

Comparing Eq.(8.47) with Eq.(8.38) yields

$$g\,\boxed{V(\phi_0) = \frac{\Lambda c^4}{8\pi G}} \tag{8.48}$$

This suggests to interpret the Cosmological Constant as an average *vacuum energy* density. This energy is not associated with any existing physical system, but it is rather a form of energy that uniformly permeates all the space. Understanding the origin of the Cosmological Constant and its role within the theories of elementary particles is one of the most pressing challenges of modern theoretical physics. This is particularly so, because the "vacuum" energy density needed to explain state-of-the-art cosmological observations is very low, $\lesssim 6\cdot 10^{-9} erg/cm^3$ [97].

8.8 THE GEOMETRY OF SPACE-TIME IN THE "VACUUM"

Finding analytical solutions of Einstein's equation is in general a very difficult task. In fact, even in the "vacuum", the field equations are differential equations linear in the second derivatives, but *non-linear* in the first derivatives of the metric tensor. Because of this, they are in general not easy to solve. To simplify the job, it would be advisable to follow four different logical steps. We have to:

1. Simplify as much as possible the expression of the metric, by exploiting all the symmetries of the problem and by choosing an appropriate reference frame;

2. Write the Lagrangian of the problem [*c.f.* Eq.(3.25)] and use Euler-Lagrangian equations to find the non-vanishing Christoffel symbols [see *e.g.* Section 3.5];
3. Use Eq.(8.4) or Eq.(8.8) to write the Ricci tensor;
4. Use Eq.(8.13) or Eq.(8.14) to find the geometry of the space-geometry in the "vacuum"—that is, the ten independent components of the metric tensor.

• STEP 1: A METRIC SIMPLIFIED EXPRESSION

The metric of the space-time can be written in very general terms as follows:

$$ds^2 = g_{00}(x_\star^\tau)dx_\star^{0\,2} + 2g_{0i}(x_\star^\tau)dx_\star^0 dx_\star^i + g_{ik}(x_\star^\tau)dx_\star^i dx_\star^k \tag{8.49}$$

where the metric tensor depends in principle both on the time and space coordinates, $x_\star^\tau$. However, it is always possible to redefine the time coordinate by writing

$$dx^0 = dx_\star^0 + \frac{g_{0i}(x_\star^\tau)}{g_{00}(x_\star^\tau)}dx_\star^i \tag{8.50}$$

In terms of this new time coordinate, x^0, the metric can be written in a very convenient way

$$ds^2 = g_{00}(x^0, x_\star^m)dx^{0\,2} - d\ell^2 \tag{8.51}$$

where

$$d\ell^2 = \left[-g_{ik}(x^0, x_\star^m) + \frac{g_{0i}(x^0, x_\star^m)g_{0k}(x^0, x_\star^m)}{g_{00}(x^0, x_\star^m)}\right]dx_\star^i dx_\star^k \tag{8.52}$$

is the spatial metric defined in Eq.(3.43). Let's use polar coordinates. If we assume an isotropic mass distribution, we do expect the spatial line element $d\ell$ to be invariant under inversion of the angular coordinates: $d\theta_\star \to -d\theta_\star$ and/or $d\phi_\star \to -d\phi_\star$. Thus,

$$d\ell^2 = \mathscr{A}_\star(t, r_\star)dr_\star^2 + \mathscr{B}_\star(t, r_\star)d\theta_\star^2 + \mathscr{C}_\star(t, r_\star)\sin^2\theta_\star d\phi_\star^2 \tag{8.53}$$

Without loss of generality, we can also consider $\mathscr{B}_\star = \mathscr{C}_\star$. In fact, due to the isotropy, a displacement of an angle $d\theta_\star$ along a meridian (*e.g.* at fixed $r_\star$ and $\phi_\star$) must be indistinguishable by an equal displacement $d\phi_\star$ along the equator (*e.g.* at fixed $r_\star$ and $\theta_\star = \pi/2$). In fact, in terms of proper lengths, $d\ell = \sqrt{\mathscr{B}_\star}d\theta_\star\big|_{\phi_\star = const} = \sqrt{\mathscr{C}_\star}d\phi_\star\big|_{\theta_\star = \pi/2}$. In the light of these considerations, we can write

$$d\ell^2 = \mathscr{A}_\star(t, r_\star)dr_\star^2 + \mathscr{B}_\star(t, r_\star)\left[d\theta_\star^2 + \sin^2\theta_\star d\phi_\star^2\right] \tag{8.54}$$

We can now take advantage of the freedom we have in choosing the coordinate system. In particular, we can choose a new radial coordinate by imposing that the angular part of the space metric is the same as in an Euclidean space: $r = \sqrt{\mathscr{B}_\star(t, r_\star)}$. It follows that

$$\begin{aligned} dr &= \mathscr{B}_\star^{1/2}\frac{d\ln\mathscr{B}_\star^{1/2}}{dr_\star}dr_\star \\ \mathscr{A}_\star(t, r_\star)dr_\star^2 &= \mathscr{A}_\star(t, r_\star)\mathscr{B}_\star^{-1}\left(\frac{d\ln\mathscr{B}_\star^{1/2}}{dr_\star}\right)^{-2}dr^2 = \mathscr{A}(t, r)dr^2 \end{aligned} \tag{8.55}$$

Thus, we can finally write the spatial metric in the following form

$$d\ell^2 = \mathscr{A}(t,r)dr^2 + r^2\left[d\theta^2 + \sin^2\theta d\phi^2\right] \tag{8.56}$$

where we dropped the subscript $\star$ from all the coordinates. Thus, at the end, the metric of the space-time can be written in a diagonal form with only two unknowns

$$ds^2 = \mathrm{e}^{\nu(x^0,r)}c^2dt^2 - \mathrm{e}^{\lambda(x^0,r)}dr^2 - r^2\left[d\theta^2 + \sin^2\theta d\varphi^2\right] \tag{8.57}$$

where $\mathrm{e}^{\nu(x^0,r)} \equiv g_{00}(r,t)$ and $\mathrm{e}^{\lambda(x^0,r)} \equiv \mathscr{A}(t,r)$.

• STEP 2: THE CHRISTOFFEL SYMBOLS

To evaluate the non-vanishing Christoffel symbols, we follow the same method outlined in Section 3.4. So, let's write the Lagrangian of the problem [*c.f.* Eq.(3.25)]

$$L = \mathrm{e}^{\nu(r,t)}\left(\dot{x}^0\right)^2 - \mathrm{e}^{\lambda(r,t)}\dot{r}^2 - r^2\dot{\theta}^2 - r^2\sin^2\theta\dot{\varphi}^2 \tag{8.58}$$

The Euler-Lagrangian equations [*c.f.* Eq.(3.27)] provide the four components of the geodesic equations

$$\begin{aligned}
\ddot{x}^0 + \frac{1}{2}\frac{\lambda_t}{c}e^{\lambda-\nu}\dot{r}^2 + \nu_r\dot{r}\dot{x}^0 + \frac{1}{2}\frac{\nu_t}{c}(\dot{x}^0)^2 &= 0\\
\ddot{r} + \frac{1}{2}\lambda_r\dot{r}^2 + \frac{\lambda_t}{c}\dot{r}\dot{x}^0 - e^{-\lambda}r\dot{\theta}^2 - e^{-\lambda}r\sin^2\theta\dot{\phi}^2 + \frac{1}{2}e^{\nu-\lambda}\nu_r(\dot{x}^0)^2 &= 0\\
\ddot{\theta} - \sin\theta\cos\theta\dot{\phi}^2 + \frac{2}{r}\dot{\theta}\dot{r} &= 0\\
\ddot{\phi} + 2\cot\theta\dot{\theta}\dot{\phi} + \frac{2}{r}\dot{r}\dot{\phi} &= 0
\end{aligned} \tag{8.59}$$

Here the subscripts r and t indicate partial derivatives *w.r.t.* to the radial and time coordinates, whereas the dot sign indicates a derivative *w.r.t.* to the line element ds. From these, we can immediately recognize the non-vanishing Christoffel symbols:

$$\begin{aligned}
&\Gamma^0_{10} = \frac{\nu_r}{2};\quad \Gamma^0_{00} = \frac{\nu_t}{2c} &&\Gamma^0_{11} = \frac{\lambda_t}{2c}e^{\lambda-\nu};\quad \Gamma^1_{00} = \frac{\nu_r}{2}e^{\nu-\lambda};\\
&\Gamma^1_{01} = \frac{\lambda_t}{2c};\quad \Gamma^1_{11} = \frac{\lambda_r}{2}; &&\Gamma^1_{22} = -re^{-\lambda};\quad \Gamma^1_{33} = -re^{-\lambda}\sin^2\theta;\\
&\Gamma^2_{12} = \frac{1}{r};\quad \Gamma^2_{33} = -\sin\theta\cos\theta;\quad \Gamma^3_{13} = \frac{1}{r}; &&\Gamma^3_{23} = \cot\theta.
\end{aligned} \tag{8.60}$$

• STEP 3: THE RICCI TENSOR COMPONENTS

The metric given in Eq.(8.57) depends on both the time and the radial coordinates. So, we have to use Eq.(8.4) to evaluate the non-vanishing Ricci tensor components:

$$R_{00} = -\frac{1}{4}\left[\frac{\lambda_t^2}{c^2} - \frac{\nu_t\lambda_t}{c^2} + 2\frac{\lambda_{tt}}{c^2}\right] + \frac{1}{2}e^{\nu-\lambda}\left[\frac{1}{2}\nu_r^2 - \frac{1}{2}\lambda_r\nu_r + 2\frac{\nu_r}{r} + \nu_{rr}\right] \quad (8.61a)$$

$$R_{01} = \frac{\lambda_t}{cr} \quad (8.61b)$$

$$R_{11} = \frac{1}{4}e^{\lambda-\nu}\left[\frac{\lambda_t^2}{c^2} - \frac{\nu_t\lambda_t}{c^2} + 2\frac{\lambda_{tt}}{c^2}\right] - \frac{1}{2}\left[\frac{\nu_r^2}{2} - 2\frac{\lambda_r}{r} - \frac{\lambda_r\nu_r}{2} + \nu_{rr}\right] \quad (8.61c)$$

$$R_{33} = \sin^2\theta R_{22} = \sin^2\theta\left\{1 - e^{-\lambda}\left[1 + \frac{1}{2}r(\nu_r - \lambda_r)\right]\right\} \quad (8.61d)$$

• STEP 4: FIELD EQUATIONS

Consider the field equations given in Eq.(8.14). Since the metric is diagonal, the Cosmological Constant must be added only to the diagonal terms of the Ricci tensor. Then the time-space component of the field equations provides

$$R_{01} = \frac{\lambda_t}{cr} = 0 \quad (8.62)$$

showing that the unknown function λ cannot be a function of time: $\lambda = \lambda(r)$. This simplifies the writing of Eq.(8.61a) and Eq.(8.61c). The corresponding field equations become

$$R_{00} + \Lambda g_{00} = 0 \quad \Rightarrow \quad \frac{1}{2}e^{\nu-\lambda}\left(-\frac{\lambda_r\nu_r}{2} + \frac{\nu_r^2}{2} + \frac{2\nu_r}{r} + \nu_{rr}\right) + \Lambda e^{\nu} = 0 \quad (8.63a)$$

$$R_{11} + \Lambda g_{11} = 0 \quad \Rightarrow \quad \frac{1}{2}\left(-\frac{\lambda_r\nu_r}{2} - \frac{2\lambda_r}{r} + \frac{\nu_r^2}{2} + \nu_{rr}\right) + \Lambda e^{\lambda} = 0 \quad (8.63b)$$

By subtracting Eq.(8.63b) from Eq.(8.63a), one gets

$$\nu_r(r,t) + \lambda_r(r) = 0 \quad (8.64)$$

with solution

$$e^{\nu(r,t)} = f(t)e^{-\lambda(r)} \quad (8.65)$$

Here $f(t)$ is an arbitrary function of time that can actually be eliminated by defining a new time coordinate, $x^{0'}$, such that $dx^{0'} = f(t)dx^0$. With this new time coordinate, the metric of Eq.(8.57) can be written in a *stationary* form that does not explicitly depend on time:

$$ds^2 = \mathrm{e}^{-\lambda(r)}dx'^{02} - \mathrm{e}^{\lambda(r)}dr^2 - r^2\left[d\theta^2 + \sin^2\theta d\varphi^2\right] \quad (8.66)$$

Now, let's substitute Eq.(8.64) in Eq.(8.63b) to eliminate ν_r, and to get a single differential equation for $\lambda(r)$

$$-\lambda_{rr}e^{-\lambda} + \lambda_r^2 e^{-\lambda} - 2\frac{\lambda_r}{r}e^{-\lambda} + 2\Lambda = 0 \qquad (8.67)$$

This equation can be conveniently rewritten in the following form

$$\frac{1}{r}\frac{d^2}{dr^2}(re^{-\lambda}) + 2\Lambda = 0 \qquad (8.68)$$

providing the solution

$$e^{-\lambda} = \mathscr{C}_1 + \frac{\mathscr{C}_2}{r} - \frac{1}{3}\Lambda r^2 \qquad (8.69)$$

Here $\mathscr{C}_1$ and $\mathscr{C}_2$ are two integration constants to be determined by our boundary conditions. Now, for sufficiently small values of r, we do expect $\Lambda r^2/3 \ll \mathscr{C}_2/r$. Then, to recover the Newtonian limit [*c.f.* Eq.(6.18)], we need $\mathscr{C}_1 = 1$ and $\mathscr{C}_2 = -\mathscr{R}_S \equiv -2GM/c^2$. Thus, in the end, we can state the following.

THE POINT-MASS SOLUTION IN THE "VACUUM"

The geometry of space-time outside an isotropic mass distribution is described by the following stationary metric

$$ds^2 = \left[1 - \frac{\mathscr{R}_S}{r} - \frac{1}{3}\Lambda r^2\right] dx^{0^2} - \left[1 - \frac{\mathscr{R}_S}{r} - \frac{1}{3}\Lambda r^2\right]^{-1} dr^2 - r^2(d\theta^2 + \sin^2\theta d\varphi^2) \qquad (8.70)$$

Note that in Eq.(8.70) we omit the $'$ sign in the dx^0 differential. Note also that if $\Lambda = 0$, the weak field limit of Eq.(8.70) leads to Eq.(6.20), derived on the basis of the Equivalence Principle alone, as in this case $\mathscr{R}_S/r \ll 1$.

9 Test-Particles in the Schwarzschild Space-Time

9.1 INTRODUCTION

In the previous chapters, we derived a simple and analytical solution of the field equations in the "vacuum", under two assumptions: i) the curvature of the space-time is generated by a single point mass and ii) the Cosmological Constant is not zero. The first to derive the metric of the space-time around a point mass was Schwarzschild in 1916 [81].[1] Schwarzschild didn't take into account the Cosmological Constant, introduced by Einstein only one year later, in 1917, in his famous cosmological paper [1]. The Schwarzschild solution is indeed an exact, analytical solution of the field equations of General Relativity in the "vacuum", and it represents one of the main achievements of General Relativity in the framework of celestial mechanics. The goal of this and of the forthcoming chapters is to discuss both the impact and the consequences of this solution.

9.2 THE SCHWARZSCHILD SOLUTION FOR A POINT MASS

The Sun is the dominant body of the solar system, with a radius $R_\odot \simeq 7 \cdot 10^5 km$ and a mass $M_\odot \simeq 2 \cdot 10^{33} g$. In geometric units ($G = 1$ and $c = 1$), the mass has the dimension of a length[2]. For the Sun, $m_\odot \equiv GM_\odot/c^2 \simeq 1.5\, km$, implying a Schwarzschild radius $\mathscr{R}_{S\odot} \equiv 2m_\odot \simeq 3\, km$. Let's consider as a typical size of the Solar System the distance of Pluto from the Sun: $r_♇ \simeq 6 \times 10^9\, km$. So, on a planetary scale, we have $\mathscr{R}_{S\odot}/r_♇ \simeq 5 \times 10^{-10}$. On the other hand, observations of Supernovae Ia [66, 76] and of the Cosmic Microwave Background [70] highlight that, at the present, the Universe is going through an accelerated expansion phase, described by a non-vanishing Cosmological Constant. However, the observed amount of acceleration requires $\Lambda \simeq 1.2 \times 10^{-46} km^{-2}$ [see Section 15.18]. So, if we restrict to planetary scales, $\Lambda r_♇^2/3 \simeq 1.4 \cdot 10^{-27} \ll \mathscr{R}_{S\odot}/r_♇$ and the Λ term can be safely neglected. This leads to the *Schwarzschild solution* of the field equations in the "vacuum" [*c.f.* in Eq.(8.70)]

THE METRIC OF THE SCHWARZSCHILD SPACE-TIME

$$ds^2 = (1 - \mathscr{R}_S/r)\, dx^{0^2} - (1 - \mathscr{R}_S/r)^{-1} dr^2 - r^2(d\theta^2 + \sin^2\theta d\phi^2) \qquad (9.1)$$

[1] An English translation of the original Schwarzschild paper can be found in [82].

[2] The *geometric mass* will be hereafter indicated by m, not to be confused with the mass of a test-particle.

DOI: 10.1201/9781003141259-9

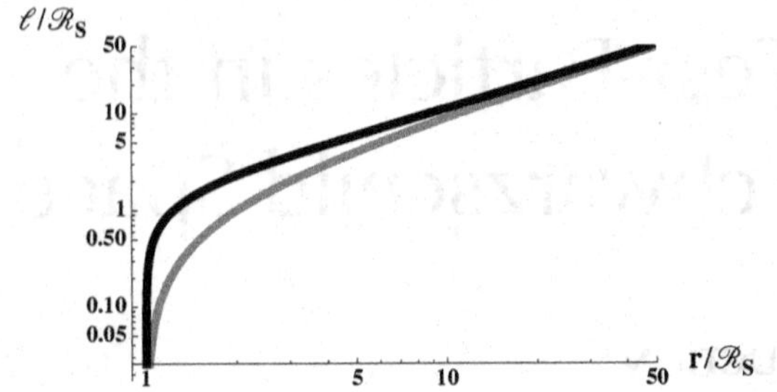

Figure 9.1 Proper (continuous black line) and coordinate (continuous gray line) radial distance from the Schwarzschild radius of the central point mass

Note that this metric becomes singular at $r = \mathscr{R}_S$. So the region where $r \lesssim \mathscr{R}_S$ has to be handle with care. This is what we will do in Chapter 12. Until then, we will restrict our discussion to the case $r > \mathscr{R}_S$.

We arrived to write the metric given in Eq.(8.50) by making a judicious choice of the time and radial coordinates. The coordinate time is the proper time of an observer at infinity [*c.f.* Eq.(7.1)]. The radial coordinate is chosen to have the proper length of great circles of coordinate radius R equal to $2\pi R$, as expected in the Euclidean geometry.[3] However, this choice has a clear impact on the proper radial distance from $\mathscr{R}_S$: $\ell_{\parallel}(r) = \int_{\mathscr{R}_S}^{r} dr'/\sqrt{1-\mathscr{R}_S/r'}$ [*c.f.* Eq.(3.48) and Eq.(3.49)]. Then,

$$d^{(p)}(x) = \left[\sqrt{(x-1)x} + \frac{\log(x)}{2} + \log\left(\sqrt{\frac{x-1}{x}} + 1\right)\right] \tag{9.2}$$

where $d^{(p)}(x) \equiv \ell_{\parallel}/\mathscr{R}_S$ and $x = r/\mathscr{R}_S$ are the proper distance and the radial coordinate, both in units of $\mathscr{R}_S$. Note that $d^{(p)}(x) \to 0$ when $x \to 1$, that is when $r = \mathscr{R}_S$. Just for comparison, the coordinate distance from the Schwarzschild radius in units of $\mathscr{R}_S$ is given by $d^{(c)} = x - 1$. As shown in Figure 9.1, $d^{(p)}(x) \gtrsim d^{(c)}(x)$ for $1 \lesssim x \lesssim 10$. We can say that in a Schwarzschild space-time proper lengths are "stretched" along the radial direction. However, at large distances, proper and coordinate radial distances must become more and more alike, as expected in a flat space-time.

9.3 THE "EMBEDDING" PROCEDURE

The 2D surface ($t = const$ and $\theta = \pi/2$) of the Schwarzschild space-time can be written as follows [*c.f.* Eq.(9.1)]:

$$d\sigma^2 = \left(1 - \frac{\mathscr{R}_S}{r}\right)^{-1} dr^2 + r^2 d\phi^2 \tag{9.3}$$

---

[3] Likewise, the surface of a sphere of coordinate radius R has an area equal to $4\pi R^2$, as in the Euclidean case.

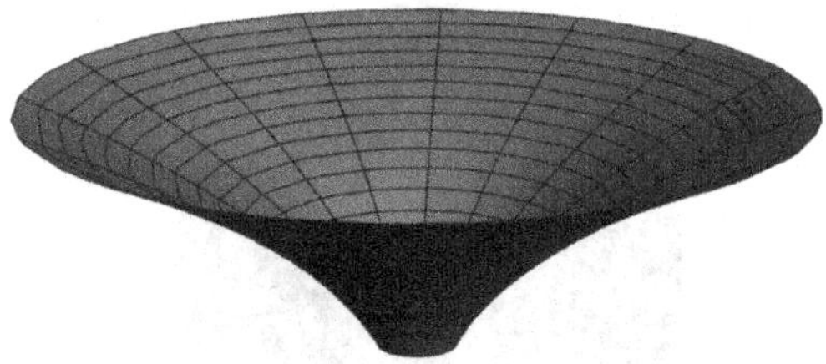

Figure 9.2 The 2D surface $z(r, \phi)$ embedded in a 3D Euclidean space with cylindrical coordinates [*c.f.* Eq.(9.7)].

Let's visualize this surface by "embedding" it into a 3D Euclidean space, whose metric

$$d\ell^2 = d\bar{r}^2 + \bar{r}^2 d\bar{\phi}^2 + d\bar{z}^2 \tag{9.4}$$

can be written in terms of cylindrical coordinates: $\bar{r}$, $\bar{\phi}$ and $\bar{z}$. After comparing the angular part of Eq.(9.3) and Eq.(9.4), we can conclude that $\bar{r} = r$ and $\bar{\phi} = \phi$. Then, Eq.(9.4) becomes

$$d\ell^2 = \left[1 + \left(\frac{\partial \bar{z}}{\partial r}\right)^2\right] dr^2 + r^2 d\phi^2 \tag{9.5}$$

Equating the radial part of Eq.(9.3) and Eq.(9.5) yields

$$\left(\frac{\partial \bar{z}}{\partial r}\right) = \sqrt{\left(1 - \frac{\mathscr{R}_S}{r}\right)^{-1} - 1} \tag{9.6}$$

Integrating Eq.(9.6) leads to write

$$\boxed{\bar{z}(r) = 2\sqrt{\mathscr{R}_S(r - \mathscr{R}_S)}} \tag{9.7}$$

The 2D surface described by Eq.(9.7) is shown in Figure 9.2. Thus, the embedding procedure allows us to show the geometry of the 2D equatorial ($\theta = \pi/2$) slice of the 3D Schwarzschild hypersurface at an arbitrary coordinate time t. Note that $\bar{z}(\mathscr{R}_S) = 0$ and that we restric our analysis to values $r > \mathscr{R}_S$.

9.4 THE JEBSEN-BIRKHOFF THEOREM

The Schwarzschild solution [81, 82] was derived for a single point mass—that is, under the assumption of a static and isotropic gravitational field. However, in the derivation of Eq.(8.70), we didn't make any assumption neither on the actual size of the central object (we used the point mass approximation) nor on its status of motion (*e.g.* hydrostatic equilibrium). We only assumed spherically symmetry and, as discussed above, we did a specific choice for the time and radial coordinates. From these assumptions followed that the metrics given in Eq.(8.70) and Eq.(9.1) do not depend on time. We have then to conclude that the Schwarzschild solution

Figure 9.3 A spherical shell of matter of mass M, empty inside. Outside the shell, the metric is Schwarzschild ; inside the shell, the metric is Minkowski.

should hold even if the central mass distribution is in a dynamical state. However, for this to be true, three conditions must be fulfilled: i) the central body is spherically symmetric; ii) it can collapse or expand, but always in an isotropic way; and iii) we are probing the geometry of the space-time in the "vacuum"—that is, outside the isotropic mass distribution.

The first to demonstrate that this is indeed the case was Jebsen in 1921 [49].[4] In 1923, Birkhoff, independently, arrived to the same result.

THE JEBSEN-BIRKHOFF THEOREM

The space-time geometry outside a spherical body does not know neither its actual dimension nor its internal dynamical status.

There is a corollary to the Birkhoff's theorem. Consider a spherical cavity surrounded by a shell of matter [see Figure 9.3]. Outside the shell, the geometry is still given by Eq.(9.1), as if the mass was concentrated at the origin of our reference frame. Inside the cavity, no matter is present. Then, the Schwarzschild metric should hold, but with $\mathscr{R}_S = 0$: in this case the integration constant $\mathscr{C}_2$ [*c.f.* Eq.(8.69)] must vanish in order to avoid divergences at the origin. We can, therefore, conclude that the metric within a spherical cavity surrounded by a spherical distribution of matter is strictly Minkowskian [see Figure 9.3].

9.5 FIRST INTEGRALS IN THE SCHWARZSCHILD SPACE-TIME

According to the Equivalence Principle, a test-particle subjected only to a gravitational field moves along geodesics of the space-time. So, let's start from the

[4] An English translation of the original Jebsen's paper can be find in [48].

Lagrangian of our problem, given by [*c.f.* Eq.(3.25) and Eq.(9.1)].

$$L = \left(1 - \frac{\mathscr{R}_S}{r}\right)\dot{x}^{0\,2} - \left(1 - \frac{\mathscr{R}_S}{r}\right)^{-1}\dot{r}^2 - r^2\left[\dot{\theta}^2 + \sin^2\theta\dot{\phi}^2\right] \tag{9.8}$$

Now, we can use the Euler-Lagrangian equations [*c.f.* Eq.(3.27)] to find

$$\alpha = 0) \qquad \frac{d}{ds}\left[\left(1 - \frac{2m}{r}\right)\dot{x}^0\right] = 0 \tag{9.9a}$$

$$\alpha = 2) \qquad \frac{d}{ds}\left(r^2\dot{\theta}\right) - r^2\sin\theta\cos\theta\dot{\phi}^2 = 0 \tag{9.9b}$$

$$\alpha = 3) \qquad \frac{d}{ds}\left(r^2\sin^2\theta\dot{\phi}\right) = 0 \tag{9.9c}$$

These equations show the emergence of some conserved quantities, very useful for the discussion of the following Sections. Indeed, because of the choice of the radial coordinate, the angular part of the Lagrangian in Eq.(9.8) is equal (beside an irrelevant factor -1/2) to the angular part of Eq.(1.42). Then, it shouldn't be a surprise that, also in the relativistic case, the motion of the test-particle: i) occurs on the equatorial plan plane, *e.g.* $\theta = \pi/2$; ii) exhibits a conserved quantity

$$r^2\dot{\phi} = h \tag{9.10}$$

Note that in the weak field limit ($ds \approx cdt$)

$$h \simeq r^2\frac{1}{c}\frac{d\phi}{dt} \simeq \frac{H}{c} \tag{9.11}$$

where H is the classical angular momentum per unit mass of the test-particle [*c.f.* Eq.(1.45)]. Last but not least, Eq.(9.9a) provides another first integral,

$$\left(1 - \frac{\mathscr{R}_S}{r}\right)\dot{x}^0 = \mathscr{C} \tag{9.12}$$

where $\mathscr{C}$ is a constant that will be evaluated in the next section.

9.6 ENERGY CONSERVATION IN GR

From what has been just discussed, we can draw a very general conclusion: whenever the metric, and then the Lagrangian, does not depends on one coordinate, x^α say, the corresponding covariant component of the particle four-velocity is conserved. This is an immediate consequence of the structure of the Euler-Lagrangian equations, that in this case provide $d\left(g_{\alpha\beta}\dot{x}^\beta\right)/ds = 0$, and then $\dot{x}_\alpha = const$. If the metric does not depends on time, then $\dot{x}_0 = const$. In the specific case of the Schwarzschild metric, Eq.(9.12) yields

$$\dot{x}_0 = \left(1 - \frac{\mathscr{R}_S}{r}\right)\frac{dx^0}{ds} = \mathscr{C} \tag{9.13}$$

whereas the metric writes

$$ds^2 = \left(1 - \frac{\mathscr{R}_S}{r}\right) dx^{02} \left(1 - \beta^2\right) \tag{9.14}$$

where $\beta = (1 - \mathscr{R}_S/r)^{-1/2} d\ell/dx^0$ is the proper velocity of a test particle in units of the speed of light. Substituting Eq.(9.14) in Eq.(9.13) yields

$$\dot{x}_0 = \frac{\sqrt{1 - \mathscr{R}_S/r}}{\sqrt{1-\beta^2}} \equiv \frac{E}{m_p c^2} \tag{9.15}$$

where we set $\mathscr{C} = E/(m_p c^2)$, m_p and E being the rest mass and the total energy of test-particle. Does it make sense? This is indeed the case of a flat space-time, where the previous equation reduces to the well-known expression of Special Relativity, $E = m_p c^2/\sqrt{1-\beta^2}$ [*c.f.* Eq.(2.42)]. In the weak field approximation, $\mathscr{R}_S/r \ll 1$ and $\beta^2 \ll 1$, Eq.(9.15)] reduce to the classical expression of the energy, $E = m_p c^2 + m_p \mathrm{v}^2/2 - GMm_p/R$, modified to take into account the rest energy of the test-particle.

9.7 THE RADIAL INFALL

Eq.(9.15) has an interesting application. Consider the radial infall of a test-particle in the Schwarzschild space-time. Let's assume that at some initial time, $t_{in} = 0$ say, the particle is at a coordinate distance $r_{in} \gg \mathscr{R}_S$ with velocity $\mathrm{v}(t_{in}) \simeq 0$. According to Eq.(9.15), we have

$$\frac{\sqrt{1 - \mathscr{R}_S/r}}{\sqrt{1 - \mathrm{v}^2(r)/c^2}} \simeq 1 \tag{9.16}$$

which leads to

$$\mathrm{v}(r) \simeq -c\sqrt{\frac{\mathscr{R}_S}{r}} \tag{9.17}$$

Here we have chosen the minus sign because the test-particle is moving inwards, while the radial axis is oriented outwards. Remember that in Eq.(9.16), $\mathrm{v}(r)$ is a proper velocity. Then,

$$\mathrm{v}(r) \equiv \frac{d\ell}{d\tau} = \frac{1}{1 - \mathscr{R}_S/r} \frac{dr}{dt} \tag{9.18}$$

Let's integrate the equation obtained by combining Eq.(9.17) and Eq.(9.18). We get

$$\frac{1}{\mathscr{R}_S} \int_{t_{in}}^{t} c dt = \int_{r}^{r_{in}} \frac{dr/\mathscr{R}_S}{1 - \mathscr{R}_S/r} \sqrt{\frac{r}{\mathscr{R}_S}} = \int_{x}^{x_{in}} \frac{x^{3/2} dx}{x-1} \tag{9.19}$$

where $x = r/\mathscr{R}_S$. Eq.(9.19) provides

$$\frac{ct(x)}{\mathscr{R}_S} = \frac{2}{3}\left(x_{in}^{3/2} - x^{3/2}\right) + 2\left(\sqrt{x_{in}} - \sqrt{x}\right) + \log\left(\frac{\sqrt{x}+1}{\sqrt{x}-1}\frac{\sqrt{x_{in}}-1}{\sqrt{x_{in}}+1}\right) \tag{9.20}$$

If we assume $x_{in} \gg 1$ and $x_{in} \gg x$, then

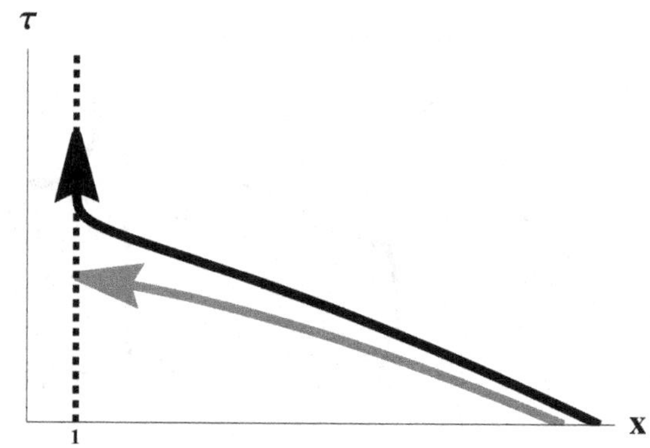

Figure 9.4 A test-particle is radially infalling toward the central mass. For an observer at infinity, the proper time needed for the particle to reach the Schwarzschild radius is infinite (black line). On the contrary, for an observer in free fall with the test-particle, the proper time needed to reach $\mathscr{R}_S$ is finite (gray line). Here $x = r/\mathscr{R}_S$, see text.

$$\frac{ct(x)}{\mathscr{R}_S} = \frac{2}{3}x_{in}^{3/2} + 2\sqrt{x_{in}} + \log\left(\frac{\sqrt{x}+1}{\sqrt{x}-1}\right) \tag{9.21}$$

Remember that in the Schwarzschild solution, the coordinate time, t, is the proper time of an observer at the infinity [*c.f.* Eq.(9.4)]. Thus, for such an observer, the test-particle never reaches $x = 1$, as the particle would need an infinite coordinate time to reach the Schwarzschild radius [see Figure 9.4].

The situation is completely different for the observer in free-fall with the test-particle: for her/him the proper time necessary to cross $r = \mathscr{R}_S$ is indeed *finite*. In fact, we can write (always in the limit $x_{in} >> 1$)

$$d\tau = \frac{ds}{c} = \frac{ds}{dx^0} \times \frac{dt}{dr} \times dr = -\frac{dr}{c}\sqrt{\frac{r}{\mathscr{R}_S}} \tag{9.22}$$

where we exploited Eq.(9.14), Eq.(9.16), and Eq.(9.17). Then,

$$\frac{c\tau}{\mathscr{R}_S} = \int_x^{x_{in}} dx\sqrt{x} = \frac{2}{3}\left(x_{in}^{3/2} - x^{3/2}\right) \tag{9.23}$$

where $x = r/\mathscr{R}_S$. So, for the observer in free-fall, the proper time τ_{ff} needed to reach $r = \mathscr{R}_S$ is given by $\tau_{ff} = 2r_{in}^{3/2}/(3c\mathscr{R}_S^{1/2})$ [see Figure 9.4].

9.8 ORBITS IN A SCHWARZSCHILD GEOMETRY

The conserved quantities of Eq.(9.10) and Eq.(9.15) provide very powerful tools for a qualitative study of the motion of a test-particle. In the Schwarzschild space-time, the test-particle radial and transverse proper velocities are given by $\mathrm{v}_{\parallel} \equiv d\ell_{\parallel}/d\tau$ and

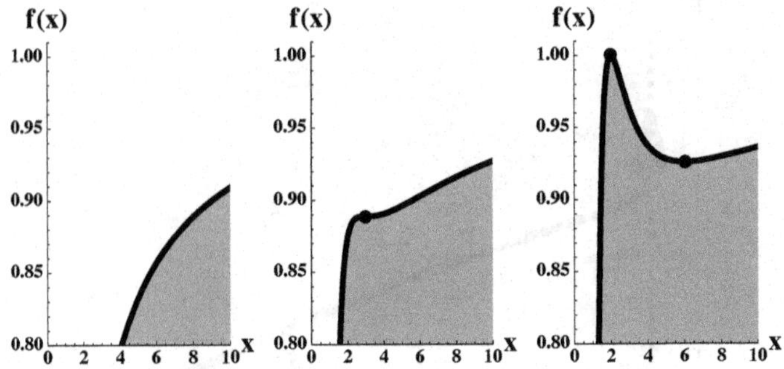

Figure 9.5 *Left panel:* for $\alpha < \sqrt{3}$, $f(x)$ is monotonically increasing, from 0 at $x = 1$ up to 1 for $x \to \infty$. *Central panel:* for $\alpha = \sqrt{3}$, $f(x)$ shows an inflection point at $x = 3$. *Right panel:* for $\alpha > \sqrt{3}$, $f(x)$ has a maximum and a minimum. For $\alpha = 2$, $x_{max} = 2$ and $x_{min} = 6$.

$v_\perp \equiv r d\phi/d\tau$, as the motion occurs in the equatorial plane ($\theta = \pi/2$). It follows that

$$\beta^2 = \frac{1}{c^2}\left[\left(\frac{d\ell}{d\tau}\right)^2 + r^2\left(\frac{d\phi}{d\tau}\right)^2\right] \tag{9.24}$$

So, let's rewrite Eq.(9.15) as follows:

$$\left(\frac{m_p c^2}{E}\right)^2\left(1 - \frac{\mathscr{R}_S}{r}\right) = 1 - \frac{1}{c^2}\left[\left(\frac{d\ell}{d\tau}\right)^2 - r^2\left(\frac{d\phi}{d\tau}\right)^2\right] \tag{9.25}$$

If the metric tensor is diagonal, as in the Schwarzschild solution, $ds = c d\tau\sqrt{1-\beta^2}$ [*c.f.* Eq.(9.14)]. So, Eq.(9.10) can be written in the following way

$$h = r^2\frac{d\phi}{c d\tau}\frac{1}{\sqrt{1-\beta^2}} = r^2\frac{d\phi}{c d\tau}\frac{E}{m_p c^2}\frac{1}{\sqrt{1-\mathscr{R}_S/r}} \tag{9.26}$$

After substituting this equation in Eq.(9.25), one gets

$$\frac{1}{c^2}\left(\frac{d\ell}{d\tau}\right)^2 = 1 - \left(\frac{m_p c^2}{E}\right)^2\left(1 - \frac{\mathscr{R}_S}{r}\right)\left[1 + \frac{h^2}{r^2}\right] \tag{9.27}$$

The requirement $(d\ell/d\tau)^2 \geq 0$ implies

$$\left(\frac{E}{m_p c^2}\right)^2 \geq f(x) \equiv 1 - \frac{1}{x} + \frac{\alpha^2}{x^2} - \frac{\alpha^2}{x^3} \tag{9.28}$$

where $x = r/\mathscr{R}_S$ and $\alpha = h/\mathscr{R}_S$. In the weak field limit, $E \simeq m_p c^2$, we can Taylor expand Eq.(9.28) to find

$$\frac{E}{m_p c^2} \geq 1 - \frac{1}{2x} + \frac{\alpha^2}{2x^2} - \frac{\alpha^2}{2x^3} \tag{9.29}$$

On the other hand, we can rewrite Eq.(1.47) by adding a term describing the rest energy of the test-particle, and by using the same notation of Eq.(9.28). If we do it, we get

$$\frac{E^{(N)}}{m_p c^2} \geq 1 - \frac{1}{2x} + \frac{\alpha^2}{2x^2} \tag{9.30}$$

Here the superscript (N) stands for Newtonian. Thus, it is the last term in Eq.(9.28) that provides the contribution due to General Relativity. Since this term is proportional to x^{-3} and negative, we do expect quite different and new constraints on the motion of a test-particle. In order to attack the problem, let's first study the behavior of $f(x)$. This function has a maximum at x_{max} and a minimum at x_{min} given by the following expression

$$\begin{pmatrix} x_{max} \\ x_{min} \end{pmatrix} = \alpha^2 \left[1 \begin{pmatrix} - \\ + \end{pmatrix} \sqrt{1 - \frac{3}{\alpha^2}} \right] \tag{9.31}$$

- $\alpha < \sqrt{3}$

Eq.(9.31) doesn't have real solutions. The function $f(x)$ has neither maxima nor minima. It starts being zero at $x = 1$ and tends to unity as $x \to \infty$ [see Figure 9.4a]. Unlike the classical case, there is not any "effective potential" even if $\alpha \neq 0$. In this case, a test-particle coming from infinity is always gravitationally captured by the central body.

- $\alpha = \sqrt{3}$

Eq.(9.31) has two real, but coincident solutions: $x_{max} = x_{min} = 3$, where the function f has an inflection point [see Figure 9.4b]. Since $f(3) = \sqrt{8/9}$, the minimum energy that a test-particle at $x = 3$ can have is given by $E = mc^2\sqrt{8/9}$ [*c.f.* Eq.(9.28)]. Note that in this case $h = \sqrt{3}\mathscr{R}_S$.

- $\alpha > \sqrt{3}$

Eq.(9.31) has two distinct and real solutions. The function $f(x)$ has a maximum at x_{max} and a minimum at x_{min} [see Figure 9.4c]. There are two equilibrium positions, one unstable in x_{max} and the other stable in x_{min}. Note that $dx_{min}/d\alpha > 0$: increasing α moves x_{min} to arbitrarily larger values. On the contrary, $dx_{max}/d\alpha < 0$: increasing α moves x_{max} to lower values. Note, however, that $\lim_{\alpha\to\infty} x_{max} = 3/2$ (see Figure 9.6).

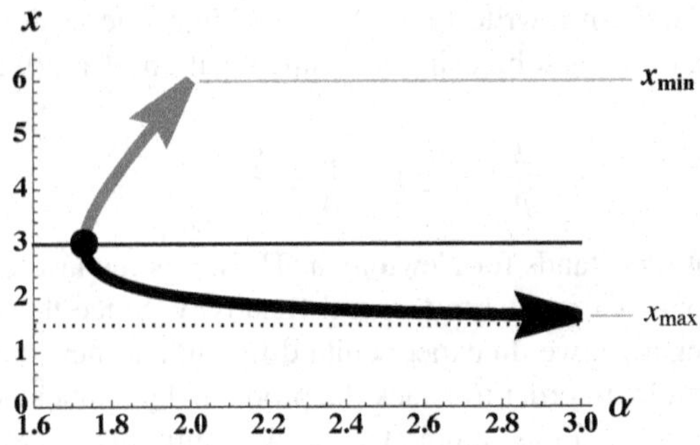

Figure 9.6 The values of x_{max} (black line) and x_{min} (gray line) for $\alpha \geq \sqrt{3}$. See text.

9.9 STABLE CIRCULAR ORBITS: $\alpha > \sqrt{3}$

In light of the considerations of the previous section, it is clear that stable orbits can occur around x_{min} only if $\alpha > \sqrt{3}$. A more detailed discussion of the planetary motion will be given in the next chapter [*c.f.* Section 10.2]. Here we want to concentrate on circular orbits that can occur with a radius $r = x_{min}\mathscr{R}_S$. The smaller, among the possible, circular orbits are obtained for $\alpha = \sqrt{3}$, which implies, as seen in the previous section, $r = 3\mathscr{R}_S$, $E = m_p c^2\sqrt{8/9}$ and $h = \sqrt{3}\mathscr{R}_S$. In this case, Eq.(9.26) can be written as follows:

$$\sqrt{3}\mathscr{R}_S = \frac{9\mathscr{R}_S^2}{c}\frac{d\phi}{d\tau}\sqrt{\frac{8}{9}}\frac{1}{\sqrt{1-(1/3)}} \tag{9.32}$$

The proper angular velocity and orbital period are then given by the following expressions

$$\frac{d\phi}{d\tau} = \frac{1}{6}\frac{c}{\mathscr{R}_S} \tag{9.33a}$$

$$T^{(p)} = 12\pi\frac{\mathscr{R}_S}{c} \tag{9.33b}$$

Note that for an observer at infinity, the proper period of a circular orbit with $x_{min} = 3$ is given by

$$T^{(\infty)} = \frac{2\pi}{(d\phi/dt)} = \frac{1}{\sqrt{g_{00}}}\frac{2\pi}{(d\phi/d\tau)} = \sqrt{\frac{3}{2}}T^{(p)} \tag{9.34}$$

Note also that the proper orbital velocity for a circular orbit of radius $r = 3\mathscr{R}_S$ is highly relativistic

$$\mathrm{v}_\perp = \frac{\ell}{T^{(p)}} = \frac{2\pi(3\mathscr{R}_S)}{12\pi\mathscr{R}_S/c} = \frac{c}{2} \tag{9.35}$$

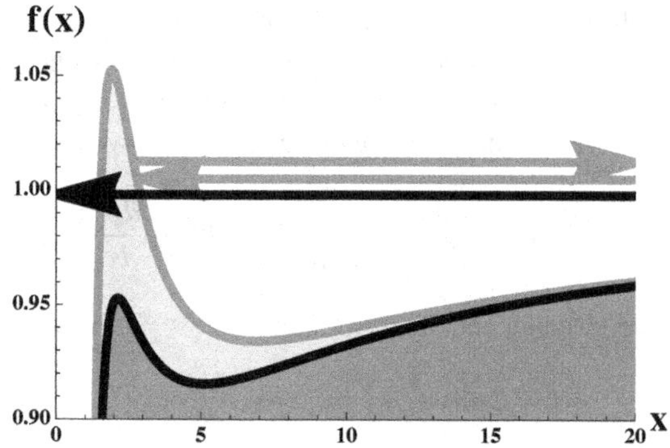

Figure 9.7 The function $f(x)$ is plotted for $\alpha < 2$ (black line) and $\alpha > 2$ (gray line). The forbidden regions are indicated by the shaded areas, black for $\alpha < 2$ and gray for $\alpha > 2$. A quasi-parabolic test-particle will return to infinity if $\alpha > 2$ (gray arrows). In the opposite case ($\alpha < 2$), the particle will be captured by the central mass (black arrow).

9.10 THE CASE OF A NON-RADIAL INFALL: $\alpha > \sqrt{3}$

Consider a test-particle moving toward the point mass with an impact parameter b, an energy E, and a constant $h = b\mathrm{v}_\infty/c$, where v_∞ is the velocity of the test-particle at infinity. For a given value of $\alpha > \sqrt{3}$, the function $f(x)$ has a maximum in x_{max}. If the energy of the test-particle is larger than $E(x_{max}) = m_p c^2 \sqrt{f(x_{max})}$, then the particle will be captured, in spite of having an h value different from zero. In the opposite case, $E < E(x_{max})$, the particle will return back to infinity. This is what expected in the Newtonian case for a particle with a non-vanishing angular momentum.

In the parabolic case, $E \simeq m_p c^2$, as $\mathrm{v}_\infty \simeq 0$ and $\lim_{r\to\infty} g_{00}(r) \simeq 1$ [*c.f.* Eq.(9.15)]. Consider now, for sake of simplicity, the case of a test-particle with $\alpha = 2$. The maximum of the function $f(x)$ occurs in this case at $x_{max} = 2$, where $f(x_{max}) = 1$ and, then, $E(x_{max}) = m_p c^2$ [*c.f.* Eq.(9.28)]. It follows that a particle in a parabolic motion toward M will return to infinity if $\alpha > 2$. In this case, in fact, $f(x_{max}) > 1$ and $E(x_{max}) > m_p c^2$. On the contrary, the same parabolic particle will be captured by the central mass if $\alpha < 2$, as in this case $f(x_{max}) < 1$ and $E(x_{max}) < m_p c^2$ [see Figure 9.7].

To conclude, let's define a cross section $\sigma \equiv \pi b^2$ for the capture of particles by the central point mass. From the definition of α, it follows that for $\alpha \leq 2$ we have $b\mathrm{v}_\infty/c \leq 2\mathscr{R}_S$. Then,

$$\sigma = 4\pi\mathscr{R}_S^2\beta_\infty^{-2} \tag{9.36}$$

where $\beta_\infty = \mathrm{v}_\infty/c$.

9.11 PHOTONS IN THE SCHWARZSCHILD SPACE-TIME

We can extend the considerations of the previous section to study photons traveling in a Schwarzschild space-time. Of course we don't have a Newtonian analogous.

However, we can still use the same formalism of the previous sections, after substituting the energy of the test-particle, $m_p c^2/\sqrt{1-\beta^2}$, with the energy of the photon, $h\nu$. Consider the angular momentum of a test particle: $L = m_p ch$, which reduces to $L = m_p H$ in the weak field limit. Remembering Eq.(9.26), we can write

$$L = m_p r^2 \frac{d\phi}{d\tau} \frac{1}{\sqrt{1-\beta^2}} \Rightarrow 2\pi\hbar\nu \frac{r^2}{c^2} \frac{d\phi}{d\tau} \tag{9.37}$$

where $\hbar$ is the reduced Planck constant. So,

$$L = 2\pi\hbar\nu \frac{r}{c^2} \mathrm{v}_\perp = 2\pi\hbar\nu_\infty \frac{r_\infty}{c^2} \mathrm{v}_{\perp,\infty} \tag{9.38}$$

Here $\mathrm{v}_\perp = r d\phi/d\tau$, whereas ν_∞, r_∞ and $\mathrm{v}_{\perp,\infty}$ refer to a photon at infinity, with $r \gg \mathscr{R}_S$. Given these premises, following the same line of reasoning of Section 9.10, we can write Eq.(9.15) as follows:

$$E = \sqrt{1 - \frac{\mathscr{R}_S}{r}} 2\pi\hbar\nu = 2\pi\hbar\nu_\infty \tag{9.39}$$

After dividing Eq.(9.38) by Eq.(9.39), we get

$$r\mathrm{v}_\perp = \frac{Lc^2}{E} \sqrt{1 - \frac{\mathscr{R}_S}{r}} = bc\sqrt{1 - \frac{\mathscr{R}_S}{r}} \tag{9.40}$$

where $b = r_\infty \sin\chi$ is the impact parameter, $\pi - \chi$ is the angle between the photon and the radial directions at infinity and $\mathrm{v}_{\perp,\infty} = c\sin\chi$. On the other hand, $\mathrm{v}_\parallel^2 + \mathrm{v}_\perp^2 = c^2$. Then,

$$\mathrm{v}_\parallel^2 = c^2 \left[1 - \frac{b^2}{r^2}\left(1 - \frac{\mathscr{R}_S}{r}\right)\right] \tag{9.41}$$

The condition $\mathrm{v}_\parallel^2 \geq 0$ yields a constraint on the impact parameter

$$b \leq \frac{r}{\sqrt{1 - \mathscr{R}_S/r}} \tag{9.42}$$

The *rhs* of the previous equation plays the role of an "effective potential" [see Figure 9.8]. It has its minimum value at $r_{min} = 3\mathscr{R}_S/2$, corresponding to $b_{min} = 3\sqrt{3}\mathscr{R}_S/2$. It is interesting to note that the photon radial velocity vanishes for $r = r_{min}$ and $b = b_{min}$ [*c.f.* Eq.(9.41)]. Thus, a photon can be in a circular, although unstable, orbit of radius r_{min}, a result similar to the one obtained in Section 9.9 for a massive test-particle. Note that b_{min} constitutes an important threshold, for the incoming photon to be captured and for outgoing photon to escape to infinity.

Let's discuss the case of outgoing photons, emitted in the region $\mathscr{R}_S < r < 3\mathscr{R}_S/2$. Such photons can escape to infinity only if $b \leq b_{min}$ [see Figure 9.8]. To be more quantitative, let's consider Eq.(9.40) and Eq.(9.41) to find

$$\left(\frac{\mathrm{v}_\parallel}{\mathrm{v}_\perp}\right)^2 = \frac{1 - b^2(1 - \mathscr{R}_S/r)/r^2}{b^2(1 - \mathscr{R}_S/r)/r^2} \tag{9.43}$$

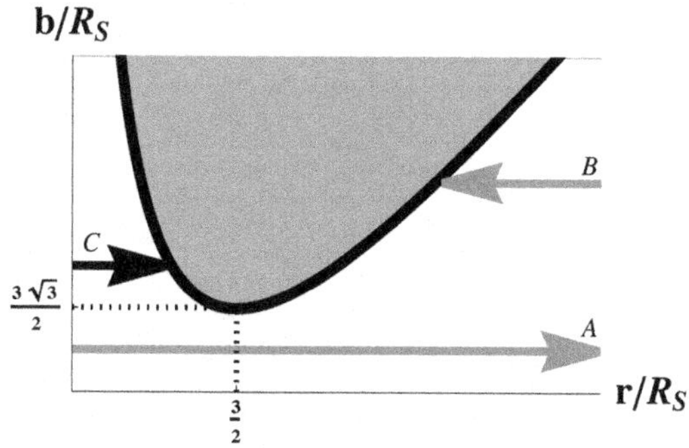

Figure 9.8 The *rhs* of Eq.(9.42) is plotted vs. r. The minimum value, $b_{min} = 3\sqrt{3}\mathscr{R}_S/2$,. corresponds to $r_{min} = 3\mathscr{R}_S/2$. The shaded area is the forbidden region of the parameter space, given the constraint of Eq.(9.42)

We can use this equation to find the impact parameter and, then, to impose the condition $b < b_{min}$. Thus,

$$b^2 = \frac{r^2}{(1-\mathscr{R}_S/r)\left[1+(\mathrm{v}_{\parallel}/\mathrm{v}_{\perp})^2\right]} \leq b^2_{min} = \frac{27}{4}\mathscr{R}_S^2 \tag{9.44}$$

It is straightforward to verify that Eq.(9.44) yields

$$\left(\frac{\mathrm{v}_{\parallel}}{\mathrm{v}_{\perp}}\right)^2 \geq \left[\frac{4}{27}\frac{r^2}{\mathscr{R}_S^2} - \left(1-\frac{\mathscr{R}_S}{r}\right)\right]\frac{1}{1-\mathscr{R}_S/r} \tag{9.45}$$

Since $\tan\chi = \mathrm{v}_{\perp}/\mathrm{v}_{\parallel}$, we can state that the photon will go to infinity only if [*c.f.* Eq.(9.45)]

$$\chi \leq \chi_\star \equiv \tan^{-1}\left\{\left(1-\frac{\mathscr{R}_S}{r}\right)^{1/2}\left[\frac{4}{27}\left(\frac{r}{\mathscr{R}_S}\right)^2 - 1 + \frac{\mathscr{R}_S}{r}\right]^{-1/2}\right\} \tag{9.46}$$

Note that the value of χ at r_{min} is $\pi/2$. This is consistent with the fact that at r_{min} the radial velocity vanishes when $b = b_{min}$ [*c.f.* Eq.(9.41)].

10 The Classical Tests of General Relativity

10.1 INTRODUCTION

The importance of the Schwarzschild solution stands on two pillars. From the theoretical side, it has been the first analytical solution of the field equations for the point mass approximation. From the experimental side, it was crucial to verify the predictions of General Relativity against the extremely precise, astronomical observations of our Solar system. The goal of this chapter is to present and discuss three classical tests of General Relativity. The other ones, based on the gravitational time dilation and the gravitational redshift, can indeed be considered tests of the Equivalence Principle and, as such, they were presented and discussed in Chapter 6.

10.2 PLANETARY MOTION

As discussed in the last chapter, the motion of a test-particle in a Schwarzschild space-time can be qualitatively understood in terms of three constraints: i) the planar nature of the orbit; ii) the energy conservation; and iii) the conservation of the relativistic angular momentum per unit mass. Stable orbits can occur around the minima of the *effective potential* [see Figure 9.5c]. They can be either circular [see Section 9.9] or "elliptical". In order to better understand why the term "elliptical" is between quote marks, we have to fully resolve the problem of the orbital motion of a test-particle in a Schwarzschild geometry. We could do it by using the results of Section 9.5 and by resolving the Euler-Lagrangian equations also for the radial component of the geodesic equation. However, for the sake of simplicity, let's exploit the Lagrangian of the test-particle, which is by definition normalized to unity [*c.f.* Eq.(3.25) and Eq.(9.1)]

$$\left(1-\frac{\mathscr{R}_S}{r}\right)\dot{x}^{0\,2}-\left(1-\frac{\mathscr{R}_S}{r}\right)^{-1}\dot{r}^2-r^2\left[\dot{\theta}^2+\sin^2\theta\,\dot{\varphi}^2\right]=1 \tag{10.1}$$

This is somehow equivalent to what was done in Chapter 1, where we used the energy conservation to find the solution for the radial coordinate [*c.f.* Section 1.11]. As done in the classical case, let's define a new variable $u[\varphi(t)] = r^{-1}(\varphi,t)$ and its partial derivative $u' \equiv \partial u/\partial\phi$. Then,

$$\dot{r} \equiv \frac{dr}{ds} = -\frac{u'}{u^2}\dot{\varphi} = -u'h \tag{10.2}$$

DOI: 10.1201/9781003141259-10

This is very similar to Eq.(1.11), a part from the obvious substitutions; $ds \to cdt$ and $h \to H/c$. Thus, Eq.(10.1) becomes

$$1 - \mathscr{R}_S u = \left(\frac{E}{m_p c^2}\right)^2 - (u')^2 h^2 - u^2 h^2 (1 - \mathscr{R}_S u) \tag{10.3}$$

where we used Eq.(9.12). Unfortunately, this is a non-linear differential equation in our unknown $u(\phi)$, that is not easy to handle analytically. Then, instead of explicitly resolving this equation, let's derive it *w.r.t.* φ. Each term will be proportional to u', that we can eliminate as we are not interested to circular solutions in the context of planetary orbits. So, at the end, Eq.(10.3) yields the following second-order differential equation.

$$u'' + u = \frac{GM}{c^2 h^2} + 3\frac{GM}{c^2} u^2 \tag{10.4}$$

Note that the first term on the *rhs* contains the classical solution: in fact, in the weak-field limit, $ch \simeq H$ [*c.f.* Eq.(9.11)]. So, the novelty introduced by General Relativity is all included in the second term on the *rhs* of Eq.(10.4). In the case under study, this term is small *w.r.t.* the classical one. In fact, in the case of our planetary system,

$$\frac{3GMu^2/c^2}{GM/c^2h^2} = 3h^2u^2 \simeq \frac{3r^4}{c^2}\left(\frac{d\varphi}{dt}\right)^2 \frac{1}{r^2} = 3\left(\frac{\mathrm{v}_{planet}}{c}\right)^2 \ll 1 \tag{10.5}$$

where $\mathrm{v}_{planet} \ll c$ is the orbital velocity of a planet around the Sun. Therefore, it makes sense to resolve Eq.(10.4) following a perturbative approach. So, after defining

$$A \equiv \frac{GM}{c^2 h^2} \qquad \varepsilon \equiv 3\frac{GM}{c^2} A \tag{10.6}$$

we can rewrite Eq.(10.4) as follows:

$$u'' + u = A + \frac{\varepsilon}{A} u^2 \tag{10.7}$$

We want to find perturbative solutions to first order in ε:

$$u = u_0 + \varepsilon \mathrm{v} \tag{10.8}$$

So let's use the last equation in Eq.(10.7) to write $(u_0 + \varepsilon\mathrm{v})'' + (u_0 + \varepsilon\mathrm{v}) = A + \varepsilon u_0^2/A$. Note that, to keep the first order in ε, we have to consider the zeroth order solution for u in the last term on the *rhs* of Eq.(10.7). Now, for $\varepsilon \to 0$, we recover the classical result [*c.f.* Eq.(1.51)]:

$$\boxed{u_0 = A + B\cos\varphi} \tag{10.9}$$

After eliminating the zeroth order solution, we are left with the following differential equation:

$$\mathrm{v}'' + \mathrm{v} = \left(A + \frac{B^2}{2A}\right) + \frac{B^2}{2A}\cos 2\varphi + 2B\cos\varphi \tag{10.10}$$

A particular solution of this equation is given by [see Exercise A.29]

$$\mathrm{v} = A + \frac{B^2}{2A} - \frac{B^2}{6A}\cos 2\varphi + B\varphi \sin\varphi; \tag{10.11}$$

Then, after substituting Eq.(10.9) and Eq.(10.11) in Eq.(10.8), we find

$$u = \left[A + \varepsilon\left(A + \frac{B^2}{2A}\right)\right] - \frac{B^2\varepsilon}{6A}\cos 2\varphi + B\cos\varphi + \varepsilon B\varphi\sin\varphi \tag{10.12}$$

As in the classical case [*c.f.* Eq.(1.51)], there is a constant term, now given by the Newtonian solution plus a small relativistic correction, proportional to ε: $K = A + \varepsilon(A + B^2/2A)$. The second term, also provided by General Relativity, is periodic in φ: $P(\varphi) = -\left(B^2/6A\right)\cos 2\varphi$. The third term is the classical one, while the fourth provides another relativistic correction. These two terms can be combined together to provide $B\cos[(1-\varepsilon)\varphi]$ –that is, a *non*-periodic function of φ. At last, we can write

$$\boxed{u(\varphi) = K + \varepsilon P(\varphi) + B\cos\left[(1-\varepsilon)\varphi\right]} \tag{10.13}$$

The point of least distance from the Sun, the perihelion, is obtained by minimizing r—that is, by maximizing u. This requires $\cos\left[(1-\varepsilon)\varphi\right] = 1$, $(1-\varepsilon)\varphi = 2\pi n$ or, to first order, $\varphi \simeq 2\pi n(1+\varepsilon)$. Now, choose $\varphi = 0$ to identify the perihelion at some given initial time. After the completion of one orbit, $\varphi = 2\pi$: note that we are not yet back to the perihelion. The perihelion is indeed reached when $\varphi = 2\pi + \Delta\varphi$. In other words, at each round of the planet around the Sun, the angular position of the perihelion moves by an angle

$$\boxed{\Delta\varphi = 2\pi\varepsilon} \tag{10.14}$$

This is the so-called *perihelion shift*. Note that because of this effect, the orbit of the planet is not a closed ellipse, as in the classical case, but rather an open, "elliptical" trajectory that never closes on itself. This is what produce the so-called Rosette orbit [see Figure 10.1].

10.3 THE PERIHELION SHIFT OF MERCURY

In order to maximize the relativistic effects, it would be advisable to study the planet with the larger orbital velocity [*c.f.* Eq.(10.5)]. Because of the Second Kepler's law, this requirement is obtained by studying the orbit of the inner planet, Mercury. It must be stressed that from an observational point of view, studying Mercury is not an easy task. The best time to observe it is either at the sunset or at the sunrise. In spite of these difficulties, the orbit of Mercury is known to great accuracy: the position of maximum approach to the Sun moves by 574″ every century [14]. A perihelion shift is naturally expected, even in a Newtonian framework, once one takes into account the presence, and the relative gravitational effects, of all the other planets. Celestial mechanics, fully based on Newtonian theory of gravity, was able to enucleate these

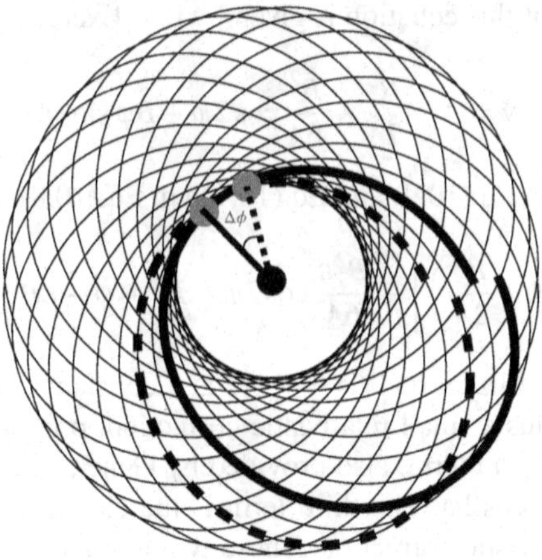

Figure 10.1 The positions of least distances from the Sun after completing one orbit (gray dots). The perihelion moves, orbit after orbit, by $\Delta\phi = 2\pi\varepsilon$, see text.

contributions: 278″/century due to Venus; 153″/century due to Jupiter; 90″/century due to the Earth; 10″/century due to all the other planets. Summing up, the total shift that celestial mechanics could explain considering the effects of all the planets is of 531″/century, still far from what observed. The difference is measured with extremely high precision: $(42.980 \pm 0.001)''$/century (see *e.g.* [67]). Let's calculate what Eq.(10.14) provides in the case of Mercury

$$\Delta\varphi = 2\pi \frac{3GM}{c^2} \frac{GM}{c^2 h^2} = 2\pi \frac{3}{4} \left(\frac{\mathscr{R}_S}{r_M}\right)^2 \left(\frac{c}{\mathrm{v}_M}\right)^2 \tag{10.15}$$

where $r_M = 5.8 \times 10^7$ km, $\mathrm{v}_M = 47.36$ km/s, and $T_M = 88$ days are the radius of the orbit, the orbital velocity, and the revolution period of Mercury. Using Eq.(10.15) with the orbital parameters of Mercury and with $\mathscr{R}_{S\odot} \simeq 3$ km for the Sun, one finds that General Relativity predicts a shift of the perihelion of Mercury of 43″ per century, in spectacular agreement with the observations [14]. It is interesting to note that the first to report on the slow precession of Mercury's orbit around the Sun was Urbain Le Verrier, known for having predicted the existence of Neptune. Le Verrier clearly presented, back in 1859, the difficulty to explain the perihelion shift of Mercury in terms of the perturbations induced by the known planets [55].

10.4 LIGHT RAY'S DEFLECTION

In Section 10.2, we derived the equation of motion of a massive test-particle in a Schwarzschild space-time. We didn't consider circular orbits and we landed to Eq.(10.4) to describe the planetary motions. We may now ask what happens if we consider photons instead of test-particles. In Section 9.7, we have already discussed, although qualitatively, how photons can move in a Schwarzschild space-time. Here we want to discuss a more specific configuration that was studied and used to provide another classical test of General Relativity, the *deflection of light*. So, let's go back to Eq.(10.4) to note that the classical term, the first one on the *rhs*, can be rewritten in the following form

$$\frac{1}{2}\frac{\mathscr{R}_S}{h^2} = \frac{1}{2}\frac{\mathscr{R}_S}{r^4}\left(\frac{ds}{d\varphi}\right)^2 \tag{10.16}$$

Now, if we follow a light ray along its trajectory, $d\varphi \neq 0$. However, since photons move along null geodesics, $ds = 0$. So, it is reasonable to assume that the equation describing a light ray propagation in a Schwarzschild space-time is given by:

$$u'' + u = \frac{3}{2}\mathscr{R}_S u^2 \tag{10.17}$$

as the first term on the *rhs* of Eq.(10.4) vanishes. A more formal derivation of Eq.(10.17) is given in Box 10.1 Note that Eq.(10.17), as already discussed in Section 9.7, shows the possibility for a photon to be in circular orbit at a distance $r = 3\mathscr{R}_S/2$ from the point mass. Note also that it is the *rhs* term in Eq.(10.17) that contains the relativistic correction. Then, neglecting this term, we should recover the classical result. Indeed, in this limit, Eq.(10.17) reduces to $u_0'' + u_0 = 0$, with solution

$$u_0 = \frac{1}{b}\cos\varphi \tag{10.18}$$

Here b is the classical impact parameter, and we set $\varphi = 0$ at the point of maximum approach to the central body (see Figure 10.2). Eq.(10.18) states that, in the classical limit, light rays propagate along straight lines. We can rephrase this statement saying that free-streaming photons propagate along geodesics of the Minkowski space-time that are indeed straight lines. At great distances from the central body, $\lim_{r\to+\infty} u_0 = 0$. This implies that incoming and outgoing light rays have $\varphi_{-\infty} = -\pi/2$ and $\varphi_{+\infty} = +\pi/2$, respectively (see Figure 10.2).

Before resolving the relativistic case, note that on planetary scales the term on the *rhs* of Eq.(10.17) is expected to be small *w.r.t.* u. In fact, the ratio of these two terms provides

$$\frac{3}{2}\frac{\mathscr{R}_S}{r} \ll 1 \tag{10.19}$$

For the Sun, $\mathscr{R}_{S\odot} \simeq 3$ km. Even if the light travels passing on the verge of Sun's photosphere, $\mathscr{R}_{S\odot}/R_\odot \simeq 10^{-6}$, as the Sun's radius is $R_\odot \simeq 700{,}000$ km. This allows us to look for a solution of Eq.(10.17) in a perturbative way. Let's write again $u = u_0 + \varepsilon v$, where

$$\varepsilon \equiv \frac{3}{2}\frac{\mathscr{R}_S}{b} \ll 1 \tag{10.20}$$

BOX 10.1 A MORE FORMAL DERIVATION OF EQ.(10.17)

Consider a test-particle in a Schwarzschild space-time and use Eq.(9.15) to write

$$ds = \left(1 - \frac{\mathscr{R}_S}{r}\right)\left(\frac{m_p c^2}{E}\right) dx^0 \qquad \text{(B10.1.a)}$$

Its angular momentum is $L = m_p c r^2 d\phi/ds$ [*c.f.* Eq.(9.37)]. Use Eq.(B10.1.a) to get

$$d\phi = \frac{Lc}{E}\left(1 - \frac{\mathscr{R}_S}{r}\right) dx^0 \qquad \text{(B10.1.b)}$$

Rewrite the Lagrangian given in Eq.(10.1) in the following form

$$\left(\frac{dr}{ds}\right)^2 = \left(\frac{E}{m_p c^2}\right)^2 - \left(1 - \frac{\mathscr{R}_S}{r}\right)\left[1 + \left(\frac{L}{m_p c}\right)^2 \frac{1}{r^2}\right] \qquad \text{(B10.1.c)}$$

where we used Eq.(9.12). Take the square root with a minus sign, appropriate for an incoming test-particle and use Eq.(B10.1.a) to get

$$dr = -\left(1 - \frac{\mathscr{R}_S}{r}\right)\sqrt{1 - \left(1 - \frac{\mathscr{R}_S}{r}\right)\left[1 + \left(\frac{L}{m_p c}\right)^2 \frac{1}{r^2}\right]\left(\frac{m_p c^2}{E}\right)^2} dx^0 \qquad \text{(B10.1.d)}$$

If $m_p \to 0$, Eq.(B10.1.d) simplifies in

$$dr = -\left(1 - \frac{\mathscr{R}_S}{r}\right)\sqrt{1 - \left(1 - \frac{\mathscr{R}_S}{r}\right)\left(\frac{Lc}{E}\right)^2} dx^0 \qquad \text{(B10.1.e)}$$

We can now eliminate dx^0 by using Eq.(B10.1.b)

$$-\frac{1}{r^2}\frac{dr}{d\phi} = \frac{E}{Lc}\sqrt{1 - \left(1 - \frac{\mathscr{R}_S}{r}\right)\left(\frac{Lc}{E}\right)^2 \frac{1}{r^2}} \qquad \text{(B10.1.f)}$$

Remembering that $u(\phi) = 1/r$ and $u' = -(dr/d\phi)/r^2$, the previous equation can be rewritten as follows:

$$u'^2 = \left(\frac{E}{Lc}\right)^2 - u^2 + \mathscr{R}_S u^3 \qquad \text{(B10.1.g)}$$

So, at last, deriving once *w.r.t.* to ϕ, we get Eq.(10.17), QED.

Then, Eq.(10.17) provides $u'' + u \simeq \varepsilon b u_0^2$. After subtracting the zeroth order solution provided by Eq.(10.18), we are left with the following differential equation

$$\mathrm{v}'' + \mathrm{v} = \frac{1}{b}\cos^2\varphi \qquad (10.21)$$

Let's look for solutions like $\mathrm{v} = \alpha + \beta\cos^2\varphi$. By direct substitution, we get:

$$\alpha + 2\beta - 3\beta\cos^2\varphi = \frac{1}{b}\cos^2\varphi \qquad (10.22)$$

which is satisfied if $\alpha + 2\beta = 0$ and $-3\beta = 1/b$. This implies $\alpha = 2/(3b)$ and $\beta = -1/(3b)$. So, the solution of Eq.(10.4) can be written as follows:

$$u = \frac{1}{b}\cos\varphi + \varepsilon\left[\frac{2}{3b} - \frac{1}{3b}\cos^2\varphi\right] \qquad (10.23)$$

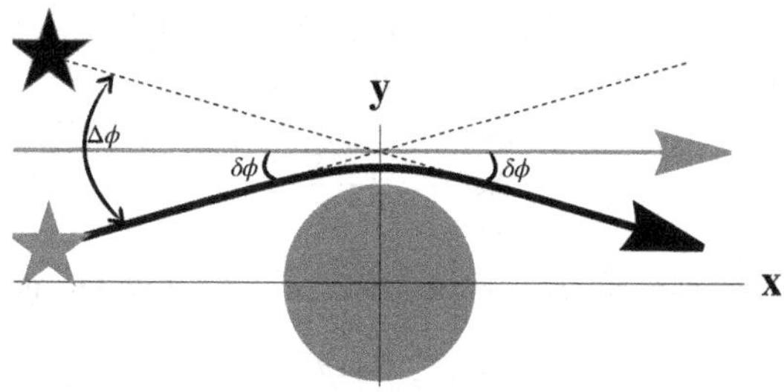

Figure 10.2 Light moves along geodesic of the Schwarzschild space-time, which is curved. Light is then deflected by an angle $\delta\phi = \mathscr{R}_S/r_{min}$. Because of this effect, the apparent position of a source in the sky changes by an angle $\Delta\phi = 2\delta\phi$, see text.

The closest approach to the Sun, r_{min}, is no longer given by the classical impact parameter b, since we have to take into account the relativistic correction, even if it is small. Indeed, when $\varphi = 0$, $r_{min} = b(1-\varepsilon/3)$ [see Figure 10.2]. Now, by definition, $\lim_{r\to\infty} u = 0$. In this limit, Eq.(10.23) provides a second-order algebraic equation in $\varphi_{\mp\infty}$, the polar angle identifying the incoming and outcoming photon directions:

$$\cos^2\varphi_{\mp\infty} - \frac{3}{\varepsilon}\cos\varphi_{\mp\infty} - 2 = 0 \tag{10.24}$$

with solutions

$$\cos\varphi_{\mp\infty} = \frac{3}{2\varepsilon}\left[1 \pm \sqrt{1+\frac{8}{9}\varepsilon^2}\right] \tag{10.25}$$

Only the solution with the minus sign makes sense. So, after linearizing in ε, we find

$$\cos\varphi_{\mp\infty} \simeq -\frac{2}{3}\varepsilon = -\frac{\mathscr{R}_S}{b} \simeq -\frac{\mathscr{R}_S}{r_{min}} \tag{10.26}$$

with two solutions for $\varphi_{\mp\infty}$. In fact, at very large distance from the central mass (the Sun in our case), the initial and final directions of the incoming and outgoing light ray are given by

$$\varphi_{-\infty} = -\frac{\pi}{2} - \frac{\mathscr{R}_S}{r_{min}} \tag{10.27a}$$

$$\varphi_{+\infty} = +\frac{\pi}{2} + \frac{\mathscr{R}_S}{r_{min}} \tag{10.27b}$$

The first conclusion is that light rays do not propagate along straight lines. This result is expected on the basis of the Equivalence Principle: free test-particles and free-streaming photons move along geodesic of a curved, pseudo-Riemannian space-time, that are not straight lines. This is indeed the case of the Schwarzschild space-time. The second point is that if we are infinitely far from the central body and we see the light coming from direction $\varphi_{+\infty}$, we associate this virtual angular position to the source. But the source has position $\varphi_{-\infty}$. So the total deflection angle, $\Delta\varphi = \varphi_{+\infty} - \varphi_{-\infty}$, is therefore given by (see Figure 10.2)

$$\Delta\varphi \equiv = 2\frac{\mathscr{R}_S}{r_{min}} \tag{10.28}$$

Inserting the appropriate values for the Sun, $\mathscr{R}_{S\odot} = 2.96$ km and $r_{min} \simeq R_\odot = 696{,}000$ km, yields

$$\Delta\varphi = 1.75'' \tag{10.29}$$

again perfectly consistent with the observations. These started in 1919, just after the first World War, with the work by Eddington and collaborators [95], providing the correct result but with an uncertainty of about 30%. This was taken as an experimental test of Einstein new theory of gravity. More recently, VLBI light deflection measurements provide consistency with the General Relativity prediction with a precision of five parts in 1000 [78].

10.5 GRAVITATIONAL LENSING

The deflection of light is not just one of the classic tests of General Relativity. In fact, it has become a very useful tool to study very distant objects. Before discussing this point, let's introduce the concept of angular diameter distance. Consider a ruler in the equatorial plane ($\theta = \pi/2$), disposed perpendicularly to the line of sight. If ϕ is the angle subtended by the ruler, then the angular diameter distance is defined to be: $D = \ell_\perp/\phi$, where $\ell_\perp$ is the transverse proper length of the ruler. In the Schwarzschild space-time $\ell_\perp = r\phi$. So, the angular diameter distance coincide with the coordinate distance r. This is in general not true for space-times characterized by different metrics, as in the case of a Friedman metric [see Section 15.16]. So, for sake of generality, let's then indicate D as the angular diameter distances.

Consider now the configuration of Figure 10.3. There is a background source S, a foreground spherical mass distribution M, and an observer O. The source, the mass M, and the observer are all on the same equatorial plane $\theta = \pi/2$. The observer is at the origin of the chosen polar coordinate system. The foreground mass M is at an angular diameter distance D_M on the polar axis $\phi = 0$. The background source is at an angular diameter distance D_{SM} and D_S from the mass M and from the observer, respectively. In the absence of any deflection, we should observe the source where it actually is, in the angular direction $\phi = \sigma$. However, we know that in the Schwarzschild space-time light is deflected. The foreground object M acts then as a gravitational lens: let's call it *the lens*. As derived in the previous section, the lens induces a deflection angle $\Delta\phi = 2\mathscr{R}_S/b$, where b is the impact parameter [see Figure 10.3]. Let's assume, as it is often the case, that the background source is very far away from the lens, and that the latter is very far away from the observer. Then, both D_{SM} and D_M are much larger than the impact parameter b. In this situation, the light emitted by the source spends most of its time to go from the source to the lens, and from the lens to the observer, but very little time to be deflected. Let's

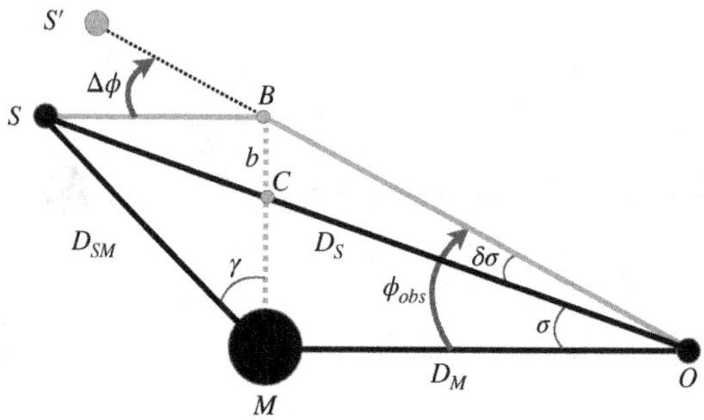

Figure 10.3 A background source S, a point mass M, and the observer O are in the equatorial plane of the chosen polar coordinate system (*i.e.* $\theta = \pi/2$). The point mass is on the polar axis, whereas σ and ϕ_{obs} are the angular positions in the sky of the source S and of its lensed image. The deflection of light is shown by the gray line $\overline{SB}$ and $\overline{BO}$. The quantity $b = \overline{MB}$ defines the impact parameter, see text.

then consider that the deflection occurs instantaneously at the time of maximum approach to the lens, that is, at the point B of Figure 10.3. This is the so-called *thin lens approximation*.

Because of the deflection, the observer will see the source in the position S', an angle ϕ_{obs} away from the lens. Clearly, ϕ_{obs} and the deflection angle $\Delta\phi$ are related. In our thin lens approximation, $\overline{SB} \simeq D_{SM}$ and $\overline{BO} \simeq D_M$, as $\gamma \to \pi/2$ and $\sigma \to 0$ [see Figure 10.3]. Then, $D_S \delta\sigma \simeq D_{SM} \Delta\phi$. Being $\phi_{obs} = \sigma + \delta\sigma$, we have

$$\sigma = \phi_{obs} - \Delta\phi \frac{D_{SM}}{D_S} \tag{10.30}$$

The deflection angle can be written as $\Delta\phi = 2\mathscr{R}_S/(D_M \phi_{obs})$, as, in the thin lens approximation, $b \simeq D_M \phi_{obs}$. Then, the *scalar lens equation* can be written as follows:

$$\boxed{\sigma = \phi_{obs} - \frac{2\mathscr{R}_S}{\phi_{obs}} \frac{D_{SM}}{D_S D_M}} \tag{10.31}$$

There is a particular interesting configuration, the one with $\sigma = 0$. In this case, S, M and O are completely aligned, and Eq.(10.31) reduces to an algebraic equation for ϕ_{obs}. We get two solutions, $\phi_{obs} = \pm\phi_E$, where the angular *Einstein radius* is defined as follows:

$$\phi_E = 2\sqrt{m \frac{D_{SM}}{D_S D_M}} \tag{10.32}$$

These solutions are shown in Figure 10.4a: the *same* source is seen at two different symmetric positions in the sky, at an angle $+\phi_E$ and $-\phi_E$ from the lens [see Figure 10.4a]. Given the axial symmetry of the problem, we can rotate the plane of Figure 10.4a around the S-M-O polar axis by an angle α, repeating in the new plane exactly the same considerations: the same source is seen again at two different symmetric

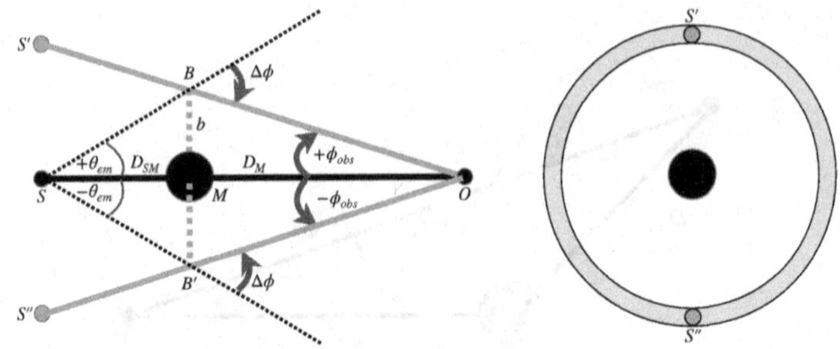

Figure 10.4 *Left panel:* A background source S, a point mass M, and the observer O are aligned along the polar axis of the chosen polar coordinate system. The light emitted in the direction $\pm\theta_{em}$ is observed coming from directions $\pm\phi_{obs} = \pm\phi_E$. *Right panel:* Given the symmetry around the polar axis, a rotation of the equatorial plane by an angle $0 \leq \alpha \leq \pi$ will move the lensed images S' and S'' around a ring, the Einstein ring [see text].

position in the sky, an angle α away from the previous ones. It is easy to convince ourselves that a continuous rotation around the polar angle implies that the apparent positions of the *same* source move along a circle, forming the so-called *Einstein Ring* [see Figure 10.4b]. The occurrence of such a ring requires form one hand a very good alignment of S, M and O, and, on the other hand, a specific arrangement in their relative distances (D_{SM}, D_S and D_M) as shown in Eq.10.30. Discovered for the first time in 1988 [44], many other Einstein rings have been discovered since then. Figure 10.5 shows the lensed image of the *SDP*81 galaxy, at redshift $z = 3.042$, obtained with the Atacama Large Millimeter Array-ALMA [6].

Let's conclude this section by stressing two characteristics of the gravitational lensing phenomenon. First, the deflection angle derived in Eq.(10.28) doesn't know anything about the frequency of the light that has been deflected. It follows that the gravitational lensing is frequency independent, *i.e.* it is *achromatic*. Secondly, as it is apparent from Figure 10.4b, the angular size of the lensed image is larger than that of the source itself. This affects the luminosity of the lensed image. In fact, the energy flux received from a source of uniform brightness[1] is given by $F_\nu = I_\nu \Omega_S$, where Ω_S is the solid angle subtended by the source. If the background source is lensed, the energy flux will be given by $F_\nu^{(L)} = I_\nu \Omega_S^{(L)}$, where $\Omega_S^{(L)}$ is the angular size of the lensed image. If $\Omega_S^{(L)} = \mu \Omega_S$, the energy flux received from the lensed image will be greater than that received by the source itself: $F_\nu^{(L)} = \mu F_\nu$. The prefactor μ is the so-called *magnification factor*. All this line of reasoning is based on the fact that the brightness of the background source and that of the lensed image are the same. This is indeed the case if photons are able to free-stream from the source up to us without suffering any scattering. If so, the Liouville equation holds: $dI_\nu/d\rho = 0$,

[1] The brightness (or specific intensity) of a source is the energy received per unit area, unit time, unit solid angle, and unit frequency: $I_\nu = dE/(dAdtd\Omega d\nu)$.

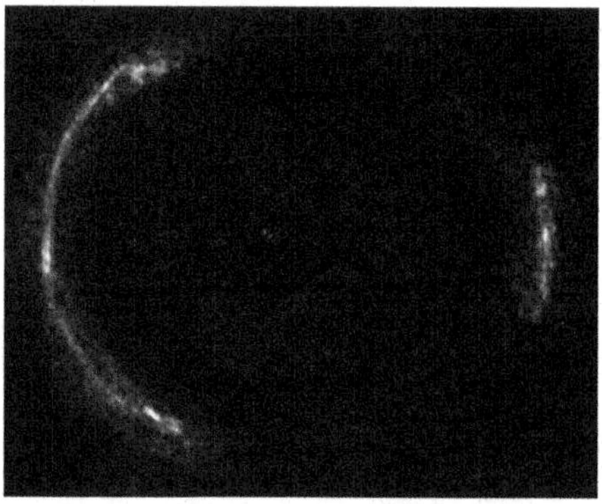

Figure 10.5 The lensed image of the galaxy SDP81 at redshift $z = 3$ obtained with the ALMA's Long Baseline Campaign appears as an almost perfect Einstein ring. Credit: ALMA (NRAO/ESO/NAOJ)/ Y.Tamura (The University of Tokyo). See `https://www.eso.org/public/images/eso1522c/`.

where $d\rho$ is a differential element of length along the null-geodesic followed by the photons.

10.6 THE SCHWARZSCHILD METRIC IN TOTALLY ISOTROPIC FORM

It is useful to write the Schwarzschild solution in the so-called *totally isotropic form*

$$ds^2 = A(\rho)dx^{0^2} + B(\rho)\left[d\rho^2 + \rho^2\left(d\theta^2 + \sin^2\theta d\varphi^2\right)\right] \tag{10.33}$$

with the same *scale factor* for the proper radial and transverse lengths. By comparing Eq.(10.33) with Eq.(9.1), one finds

$$B(\rho)d\rho^2 = \frac{dr^2}{1-2m/r} \tag{10.34a}$$

$$B(\rho)\rho^2 = r^2 \tag{10.34b}$$

where $m = \mathscr{R}_S/2$ is the geometrical mass of the central body. After dividing Eq.(10.34a) by Eq.(10.34b), we get

$$\frac{d\rho}{\rho} = \pm\frac{dr}{\sqrt{r^2-2mr}} \tag{10.35}$$

which yields

$$\ln\rho + \mathscr{C} = \pm\ln\left[\left(r-m\right) + \sqrt{r^2-2mr}\right] \tag{10.36}$$

The constant $\mathscr{C}$ can be determined by imposing appropriate boundary conditions. In particular, we want the metric of Eq.(10.33) to be Minkowskian at large distances from the central body. This naturally implies $\lim_{r\to\infty} B(\rho) = 1$ and that at large distances $\rho = r$. Because of this condition, we have to choose the plus sign in Eq.(10.36) and identify the constant $\mathscr{C}$ with ln 2. Then,

$$2\rho = (r-m) + \sqrt{r^2 - 2mr} \tag{10.37}$$

Note that

$$m^2 = \left[(r-m) + \sqrt{r^2-2mr}\right] \times \left[(r-m) - \sqrt{r^2-2mr}\right] \tag{10.38}$$

Let's divide Eq.(10.38) by Eq.(10.37) to get

$$\frac{m^2}{2\rho} = \left[(r-m) - \sqrt{r^2-2mr}\right] \tag{10.39}$$

Using Eq.(10.37) and Eq.(10.39), we get

$$2\rho + \frac{m^2}{2\rho} = 2(r-m) \tag{10.40}$$

an equation to find the wanted function $r = r(\rho)$:

$$r = \rho\left(1 + \frac{m}{2\rho}\right)^2 \tag{10.41}$$

Thus, the g_{00} term becomes

$$A(\rho) = \frac{(1-m/2\rho)^2}{(1+m/2\rho)^2} \tag{10.42}$$

whereas the function $B(\rho)$ can be derived from Eq.(10.34b)

$$B(\rho) = \left(1 + \frac{m}{2\rho}\right)^4 \tag{10.43}$$

So, at the end, Eq.(10.33) can be explicitly written in the following form

$$\boxed{ds^2 = \left(\frac{1-m/2\rho}{1+m/2\rho}\right)^2 dx^{0 2} - \left(1+\frac{m}{2\rho}\right)^4 \left[d\rho^2 + \rho^2 d\Omega^2\right]} \tag{10.44}$$

10.7 LIGHT TRAVEL TIME IN A SCHWARZSCHILD GEOMETRY

Given the Schwarzschild metric in its totally isotropic form [*c.f.* Eq.(10.44)], we can write the light-like geodesic equation. In the weak filed limit ($2m/\rho \ll 1$), we find

$$\frac{d\rho}{dt} \simeq c\left(1 - \frac{2m}{\rho}\right) \tag{10.45}$$

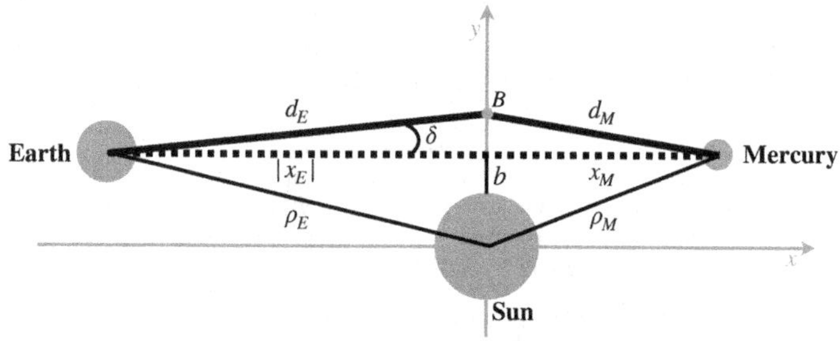

Figure 10.6 Relative position of Earth and Mercury in an almost superior conjunction with the Sun. Radio pulses move back and forth from Earth to Mercury, experiencing the curvature of the space-time induced by the Sun. As a result, the radio pulses' travel time will be longer than the one we would have observed in a flat, Minkowski space-time, see text.

Thus, the coordinate speed of light depends on the curvature of the space-time. Let's stress that Eq.(10.45) provides the speed of light as measured by an observer at infinity that uses his/her proper (Minkowskian) intervals of space and time.[2] All this to say that the light travel time in a (curved) Schwarzschild space-time is expected to be different *w.r.t.* the case of a (flat) Minkowski space-time. It would then be interesting to evaluate the magnitude of the effect and to provide an experimental evidence for it. This effect was discussed for the first time by Shapiro in 1964 in a paper entitled *The fourth test of General Relativity* [85].

The idea is the following. Imagine to send radio-pulses to Mercury and to measure the time interval necessary to receive their echos—that is, their reflected signals. We can then compare this measured time interval with the ratio between the distance traveled by the signal and the velocity of light.

For sake of clarity, let's refer to Figure 10.6. The radio pulses take 22′ to go back and forth from Earth and Mercury. Since this time is much less than their orbital periods, we can safely assume that both the Earth and Mercury are at rest during the radio pulses travel. Given the distances involved, we can again consider that the deflection occurs instantaneously at the time of maximum approach to the lens, that is, at the point B of Figure 10.6. Because of the light rays' deflection, a radio pulses from Earth to Mercury cover a distance $d_E + d_M$. However, $x_E = d_E \cos\delta \simeq d_E(1-\delta^2/2)$, where $\delta \simeq 2m_\odot/b$ is the deflection angle and b the impact parameter. So, to first order, $d_E \simeq x_E$ and $d_M \simeq x_M$. Thus, to first order—let's stress it again—we can approximate the actual light path with the straight, dashed line parallel to the x-axis connecting the Earth with Mercury [see Figure 10.6].

---

[2] Note that the observer at infinity can evaluate the speed of light only by collecting data from *locally* inertial observers. For these observers, the speed of light is by construction always equal to c.

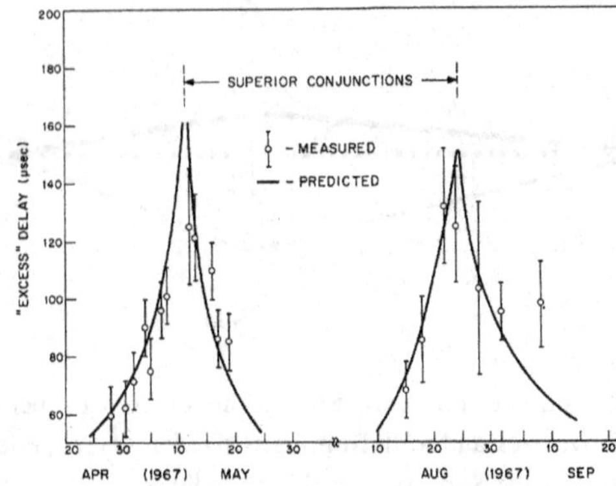

Figure 10.7 This is Figure 3 of the original paper by Shapiro et al. [86], showing a comparison of measured and predicted effects of general relativity on Earth-Mercury time delays.

Consider now the Schwarzschild metric written in its totally isotropic form [*c.f.* Eq.(10.44)], and let's use Cartesian coordinates. To first order in m/ρ, we get

$$ds^2 = \left(1 - \frac{2m}{\rho}\right) dx^{02} - \left(1 + \frac{2m}{\rho}\right) \left[dx^2 + dy^2 + dz^2\right] \qquad (10.46)$$

Consistently with the setting of Figure 10.6, the radio pulses travel in the $x-y$ plane, and their positions along the dashed line are given by $\rho = \sqrt{x^2+b^2}$. The time elapsed for the radio pulses to go back and forth will be twice the time elapsed to go from Earth to Mercury. So, after imposing $ds = 0$, Eq.(10.46) yields

$$\Delta t = \frac{2}{c} \int_{-|x_E|}^{x_M} dx \left[1 + \frac{2m}{\sqrt{x^2+b^2}}\right] \qquad (10.47)$$

that is

$$\Delta t = \frac{2}{c}\left(x_M + |x_E|\right) + \frac{4m}{c} \ln\left(\frac{\rho_M + x_M}{\rho_E - |x_E|}\right) \qquad (10.48)$$

Here $\rho_E = \sqrt{x_E + b^2}$ and $\rho_M = \sqrt{x_M^2 + b^2}$ are the Earth and Mercury radial distances from the Sun, at the origin of our coordinate system. Eq.(10.48) tells us that in flat space-time (*e.g.* $m = 0$), the time elapsed for the radio pulse to go back and forth is given by the distance the radio pulses travel (twice the distance of Mercury from the Earth) divided by the velocity of light. In the presence of the central body, the Sun in our case, the situation changes. In fact, the second term on the *rhs* of Eq.(10.48) describes the *delay* expected in the presence of a non-vanishing curvature of space-time. To maximize this delay, let's minimize the difference $\rho_E - |x_E|$. This can be done when Mercury is almost in superior conjunction with the Sun and the radio pulse approache the Sun photosphere (see Figure 10.6). To zeroth order $\rho_E \simeq |x_E|$ and $\rho_M \simeq |x_M|$. However, to properly evaluate the difference $\rho_E - |x_E|$, we have to go to the next order. To do so let's approximate the obvious relation $\rho_E^2 = |x_E|^2 + R_\odot^2$ with the following $\rho_E \simeq |x_E| + R_\odot^2/(2\rho_E)$, as $R_\odot/|x_E| \ll 1$ and, to zeroth order $\rho_E = |x_E|$. It follows that we can write the *relativistic time delay* as follows:

$$\delta t = \frac{4m}{c} \ln\left(\frac{4\rho_M \rho_E}{R_\odot^2}\right) \tag{10.49}$$

Inserting the numerical values[3] we find $\delta t \approx 220\,\mu s$. The advances in radar astronomy have made possible to realize this experiment already at the end of the 60s of last century. Time-delay measurements of radio pulses traveling between Earth and Mercury or Venus were made when either planet was on the other side of the Sun from the Earth, in the so-called superior-conjunction alignment. We show in Figure 10.7 the data reported by Shapiro and coworkers [86]. The agreement with the General Relativity predictions was reported to be of the order of $\pm 20\%$. The Shapiro time-delay effect was later measured exploiting the passage of the Cassini probe close to the Sun. This measurement confirms the predictions of General Relativity with a precision of few part in 10^5 [11]. So, in conclusion, we can truly count the Shapiro time delay as the fourth classic test of General Relativity.[4]

[3] The values of interest here are $m_\odot = 1.475$ km; $\rho_\oplus = 149 \times 10^6$ km; $\rho_M = 57.909 \times 10^6$ km; $R_\odot = 696,340$ km.

[4] The other three classical tests are the gravitational redshift, the precession of the perihelion and the bending of the light. In this book, however, we prefer to present and discuss the gravitational redshift as a test of the Equivalence Principle, rather than of the field equations of General Relativity.

11 Gravitational Waves in the "Vacuum"

11.1 INTRODUCTION

General Relativity describes gravitational phenomena in geometrical terms, using curved, pseudo-Riemannian space-times. Still, consistently with Special Relativity, gravitational interactions must propagate no faster than the speed of light. This leads to consider the idea of gravitational waves, first predicted and discussed by Einstein in 1916 and 1918 [24, 25], and then detected, indirectly by Hulse and Taylor in 1975 [46] and directly by the LIGO and Virgo collaborations in 2015 [3]. The goal of this chapter is to show that the linearized field equations of General Relativity indeed admit wave solutions and to discuss their main properties in the vacuum.

11.2 LINEARIZED GRAVITY

In Section 6.4, we used the Equivalence Principle to justify the use of the formalism developed in Chapter 3 in the wider context of gravitational phenomena. We argued that quite far away from a given mass distribution, the geometry of space-time should be described by a perturbed Minkowski metric tensor:

$$g_{\mu\nu}(x^\tau) = \eta_{\mu\nu} + h_{\mu\nu}(x^\tau) \tag{11.1}$$

where $|h_{\mu\nu}| \ll |\eta_{\mu\nu}|$. We then heuristically derive the metric form at large distances from a point mass [*c.f.* Eq.(6.20)]. Here, we want to reverse the argument, and to use the field equations in the "vacuum" [*c.f.* Eq.(8.13) or Eq.(8.35)] to find exact solutions for the components of the perturbed metric tensor, $h_{\mu\nu}$.

It is worth noting again that the Einstein field, equations are differential equations, linear in the second derivative of the metric tensor, but non-linear in its first derivatives. This renders the solution of the field equations extremely challenging. However, in the *linearized gravity* regime described by Eq.(11.1), we can make a number of simplifications quite helpful in making the field equations more manageable. For example, we can write the Christoffel symbols [*c.f.* Eq.(3.24)] to first order in the metric perturbation tensor.

$$\boxed{\Gamma^\lambda_{\mu\nu} \simeq \frac{1}{2}\eta^{\lambda\rho}\left[-h_{\mu\nu,\rho} + h_{\rho\mu,\nu} + h_{\nu\rho,\mu}\right]} \tag{11.2}$$

It follows that—again, to first order in the metric perturbation tensor—we can write the Riemann tensor neglecting the second determinant in Eq.(5.39): the products of the Christoffel symbols are clearly of second order in $h_{\mu\nu}$. In this approximation, the

DOI: 10.1201/9781003141259-11

Riemann and Ricci tensors can then be written as follows [see Exercises A.30 and A.31]:

$$R^{\alpha}{}_{\mu\lambda\nu} \simeq \frac{1}{2}\left(-\eta^{\alpha\rho}h_{\mu\nu,\rho\lambda} + h^{\alpha}{}_{\nu,\mu\lambda} + h^{\alpha}{}_{\mu,\lambda\nu} - h^{\alpha}{}_{\lambda,\mu\nu}\right) \tag{11.3a}$$

$$R_{\mu\nu} \equiv R^{\lambda}{}_{\mu\lambda\nu} \simeq \frac{1}{2}\left[-\Box h_{\mu\nu} + \left(h^{\lambda}{}_{\nu,\mu\lambda} + h^{\lambda}{}_{\mu,\lambda\nu} - h_{,\mu\nu}\right)\right] \tag{11.3b}$$

where we use the d'Alambert operator in flat space $\Box := \partial^2/(c^2\partial t^2) - \nabla^2$, so that $\Box h_{\mu\nu} \equiv \eta^{\lambda\rho}h_{\mu\nu,\rho\lambda}$. From Eq.(11.3a) we can evaluate the curvature, or Ricci, scalar [see Exercise A.32]

$$R \equiv R^{\nu}{}_{\nu} \simeq -\Box h + h^{\lambda\mu}{}_{,\lambda\mu} \tag{11.4}$$

and write the Einstein tensor [*c.f.* Eq.(8.15)] to first order in $h_{\mu\nu}$ [see Exercise A.33]

$$G_{\mu\nu} = \frac{1}{2}\left(-\Box h_{\mu\nu} + h^{\lambda}{}_{\nu,\mu\lambda} + h^{\lambda}{}_{\mu,\lambda\nu} - h_{,\mu\nu} + \eta_{\mu\nu}\Box h - \eta_{\mu\nu}h^{\lambda\tau}{}_{,\lambda\tau}\right) \tag{11.5}$$

In the following, it will be useful to work with the so-called *trace-reversed* perturbation tensor

$$\bar{h}_{\mu\nu} = h_{\mu\nu} - \frac{1}{2}\eta_{\mu\nu}h \tag{11.6}$$

its name coming from the fact that $\bar{h} = -h$, as it is easy to verify. In terms of $\bar{h}_{\mu\nu}$, the Einstein tensor becomes [see Exercise A.34]

$$\boxed{G_{\mu\nu} \simeq \frac{1}{2}\left(-\Box \bar{h}_{\mu\nu} + \bar{h}^{\lambda}{}_{\nu,\mu\lambda} + \bar{h}^{\lambda}{}_{\mu,\lambda\nu} - \eta_{\mu\nu}\bar{h}^{\lambda\tau}{}_{,\lambda\tau}\right)} \tag{11.7}$$

11.3 GAUGE TRANSFORMATIONS

Eq.(11.7) can be further simplified by choosing a proper *gauge*. Let's consider the following coordinate transformations

$$x'^{\mu} = x^{\mu} + d^{\mu}(x^{\tau}); \qquad x^{\mu} = x'^{\mu} - d^{\mu}(x'^{\tau}) \tag{11.8}$$

where d^{μ} is an infinitesimal displacement four-vector. According to Eq.(4.15), we have

$$g'_{\mu\nu}(x'^{\tau}) \equiv g_{\mu\nu}(x^{\tau} + d^{\tau}) = g_{\alpha\beta}(x^{\tau})\frac{\partial x^{\alpha}}{\partial x'^{\mu}}\frac{\partial x^{\beta}}{\partial x'^{\nu}} \tag{11.9}$$

Now, let's translate the coordinate system, $x^{\tau} \to x^{\tau} - d^{\tau}$, so that Eq.(11.9) becomes

$$g'_{\mu\nu}(x^{\tau}) = g_{\alpha\beta}(x^{\tau} - d^{\tau})\frac{\partial x^{\alpha}}{\partial x'^{\mu}}\frac{\partial x^{\beta}}{\partial x'^{\nu}} \tag{11.10}$$

Note that $g_{\alpha\beta}(x^{\tau} - d^{\tau}) \simeq g_{\alpha\beta}(x^{\tau})$, as the next term in the Taylor expansion, $g_{\alpha\beta,\sigma}(x^{\tau})d^{\sigma}$, is a second order quantity. Moreover, $\partial x^{\alpha}/\partial x'^{\mu} = \delta^{\alpha}_{\mu} - d^{\alpha}{}_{,\mu}$ [*c.f.* Eq.(11.8)]. Thus, we get

$$g'_{\mu\nu}(x^{\rho}) = g_{\mu\nu}(x^{\tau}) - d_{\nu,\mu} - d_{\mu,\nu} \tag{11.11}$$

This expression, of first order in the displacement four-vector, allows us to compare the metric tensor in the *same* physical point of the space-time. The analogous transformation for the metric perturbation tensor is obtained by using Eq.(11.1) in Eq.(11.11). To first order in d^μ, we get

$$h'_{\mu\nu} \simeq h_{\mu\nu} - d_{\mu,\nu} - d_{\nu,\mu} \tag{11.12}$$

or, in mixed form,

$$\boxed{h'^{\mu}{}_{\nu} \simeq h^{\mu}{}_{\nu} - d^{\mu}{}_{,\nu} - d_{\nu}{}^{,\mu}} \tag{11.13}$$

The contraction of the indexes in Eq.(11.13) provides the relation between the traces of the metric perturbation tensors in the primed and unprimed coordinate systems

$$h' \simeq h - 2d^{\sigma}{}_{,\sigma} \tag{11.14}$$

Then, the transformation of the *trace-reversed* perturbation tensor becomes [see Exercise A.35],

$$\overline{h}'_{\mu\nu} = \overline{h}_{\mu\nu} - d_{\nu,\mu} - d_{\mu,\nu} + \eta_{\mu\nu} d^{\sigma}{}_{,\sigma} \tag{11.15}$$

or, in mixed form,

$$\boxed{\overline{h}'^{\mu}{}_{\nu} = \overline{h}^{\mu}{}_{\nu} - d_{\nu}{}^{,\mu} - d^{\mu}{}_{,\nu} + \delta^{\mu}{}_{\nu} d^{\sigma}{}_{,\sigma}} \tag{11.16}$$

We can now use Eq.(11.15) in Eq.(11.7) to write the Einstein tensor in terms of the trace-reversed perturbation tensor in the primed reference frame. After some straightforward calculations [see Exercise A.36], we get

$$\boxed{G_{\mu\nu} \simeq \frac{1}{2}\left\{-\Box\overline{h}'_{\mu\nu} + \overline{h}'^{\lambda}{}_{\nu,\mu\lambda} + \overline{h}'^{\lambda}{}_{\mu,\lambda\nu} - \eta_{\mu\nu}\overline{h}'^{\lambda\tau}{}_{,\lambda\tau}\right\}} \tag{11.17}$$

11.4 THE LORENTZ GAUGE

Until now, we have not specified in any way the choice of the gauge. There is indeed a class of coordinate systems that are of particular interest for the problem under study. They obey to the *Lorentz gauge condition*

$$\boxed{\overline{h}'^{\mu}{}_{\nu,\mu} = 0} \tag{11.18}$$

So, the question to be asked is the following: given a generic reference frame where $\overline{h}^{\mu}{}_{\nu,\mu} \neq 0$, how can we constrain the transformation of Eq.(11.8) to find a frame where Eq.(11.18) is satisfied? To answer to this question, let's first write the transformation law for the four divergence of the metric perturbation tensor. From Eq.(11.16) we have

$$\overline{h}'^{\mu}{}_{\nu,\mu} = \overline{h}^{\mu}{}_{\nu,\mu} - \Box d_{\nu} \tag{11.19}$$

where we have used the identity $d_\nu{}^{,\mu}{}_\mu = \eta^{\mu\sigma} d_{\nu,\sigma\mu} = \Box d_\nu$. So, we can always satisfy the Lorentz gauge condition in the primed reference frame, provided that the chosen coordinate transformation satisfies the following constraint

$$\Box d_\nu = \bar{h}^\mu{}_{\nu,\mu} \tag{11.20}$$

The advantage of using the Lorentz gauge stands on the fact that all the last three terms in Eq.(11.17) vanishes, as it is immediate to verify. Thus, in the Lorentz gauge, the Einstein tensor reduces to $G_{\mu\nu} = -\Box\bar{h}'_{\mu\nu}/2$, and the field equations in the "vacuum" can be conveniently be written in the form of a wave equation

$$\boxed{\Box\bar{h}'_{\mu\nu} = 0} \tag{11.21}$$

11.5 GRAVITATIONAL WAVES

On the basis of Eq.(11.21), we do expect "ripples" in the otherwise flat geometry of the space-time. This "ripples", described by the metric perturbation tensor, can be seen as the superposition of plane waves traveling at the speed of light. These are indeed *gravitational waves*, first discussed by Einstein in 1916 [24]. Let's consider, for sake of simplicity, a monochromatic case

$$\bar{h}'_{\mu\nu} = \mathscr{A}_{\mu\nu} e^{ik_\mu x^\mu} \tag{11.22}$$

where k^μ is the wavenumber four-vector, $\mathscr{A}_{\mu\nu}$ is the (constant) wave amplitude, whereas the wavefront is identified by the hyper-surface $k_\mu x^\mu = const$. By definition, $\Box\bar{h}'_{\mu\nu} = \eta^{\alpha\beta}\bar{h}'_{\mu\nu,\alpha\beta}$. It follows that Eq.(11.22) in Eq.(11.21) provides

$$\eta^{\alpha\beta} k_\alpha k_\beta \bar{h}'_{\mu\nu} = 0 \tag{11.23}$$

that is,

$$k^\alpha k_\alpha = 0 \tag{11.24}$$

consistently with the fact that k_μ is a light-like vector. Moreover, the Lorentz gauge condition given in Eq.(11.18) provides

$$\eta^{\mu\alpha} \frac{\partial \bar{h}'_{\alpha\nu}}{\partial x^\mu} = \eta^{\mu\alpha} \mathscr{A}_{\alpha\nu} i k_\mu e^{ik_\sigma x^\sigma} = 0 \tag{11.25}$$

that is,

$$\mathscr{A}_{\alpha\nu} k^\alpha = 0 \tag{11.26}$$

So, the metric perturbation tensor is a double-transverse tensor, as the amplitude of the wave is orthogonal to the wavenumber four-vector [see Box 11.1].

BOX 11.1 THE HELMHOLTZ'S THEOREM

Any continuous vector field, with continuous first partial derivatives, can be decomposed in two components: the longitudinal/irrotational component, given by the gradient of a scalar quantity; the transverse and divergence-less component, that cannot be obtained from a scalar.

A demonstration of this theorem can be found in many textbooks (see *e.g.* [7]). Here we want just to clarify the meaning of the terms *longitudinal* and *transverse*. Let's exploit the case of a monochromatic plane wave of unitary amplitude, described by the scalar quantity $Q(x^\tau) = \exp(ik_\mu x^\mu)$. By taking the gradient of this function, we can form a *longitudinal* vector, by definition parallel to the wavenumber four-vector: $V^{(\ell)}_\alpha \equiv (\partial Q(x^\tau)/\partial x^\alpha) = ik_\alpha Q$. Such a vector is by construction irrotational, as it is easy to verify. A *transverse* vector can be written by imposing its orthogonality to k^α: ${V_{(t)}}^\alpha \equiv n^\alpha Q$, with the constraint $n^\alpha k_\alpha = 0$. Then, by construction, such a vector is divergenceless. Similar considerations apply to tensors. We can form a doubly longitudinal ($\ell\ell$) tensor by taking the gradient of a longitudinal vector: ${T^{(\ell\ell)}}_{\alpha\beta} = (\partial {V^{(\ell)}}_\beta/\partial x^\alpha) = k_\alpha k_\beta Q$. We can also form a singly-longitudinal (ℓt) tensor by taking the gradient of a transverse vector ${S^{(\ell t)\beta}}_\alpha = (\partial V^\beta_{(t)}/\partial x^\alpha) \equiv k_\alpha n^\beta Q$. Finally, we can form a doubly-transverse tensor, ${F^{(tt)\alpha}}_\beta = {f^\alpha}_\beta Q$, with the constraint ${f^\alpha}_\beta k_\alpha = 0$.

For a monochromatic plane wave propagating along the z-axis, Eq.(11.21) can be written as follows:

$$\left(\frac{1}{c^2}\frac{\partial^2}{\partial t^2} - \frac{\partial^2}{\partial z^2}\right){\overline{h}'^\mu}_\nu = 0 \tag{11.27}$$

where we use spatial cartesian coordinates.[1] Being solution of a wave equation, the metric perturbation tensor must be a function of $\zeta = t - z/c$. In this way, ${\overline{h}'^\mu}_\nu(\zeta)$ represents a progressive gravitational wave moving in the positive direction of the z-axis. The Lorentz condition given in Eq.(11.18) implies

$${\overline{h}'^\mu}_{\nu,\mu} = \frac{1}{c}\frac{\partial \overline{h}'^0_\nu}{\partial t} + \frac{\partial \overline{h}'^3_\nu}{\partial z} = \frac{1}{c}\frac{\partial \overline{h}'^0_\nu}{\partial \zeta}\frac{\partial \zeta}{\partial t} + \frac{\partial \overline{h}'^3_\nu}{\partial \zeta}\frac{\partial \zeta}{\partial z} = \frac{1}{c}\frac{\partial}{\partial \zeta}\left[\overline{h}'^0_\nu - \overline{h}'^3_\nu\right] = 0 \tag{11.28}$$

This equation can be integrated, yielding

$$\overline{h}'^0_\nu - \overline{h}'^3_\nu = \mathscr{C}_\nu \tag{11.29}$$

where $\mathscr{C}_\nu$ ($\nu = 0, 3$) are four integration constants. Note that if we are interested only to the time behavior of the components of the metric perturbation tensor—that is, at a fixed spatial position $z_\star$—these constants can be set equal to zero, as the time average of both h^0_ν and h^3_ν at $z_\star$ vanishes. On the other hand, if we are interested

[1] We use the standard notation: $x^1 \equiv x$, $x^2 \equiv y$ and $x^3 \equiv z$.

only to the spatial behavior of the components of the metric perturbation tensor—that is, at a given time $t_\star$—then these constants can equally well be set equal to zero: if at a given time and at a given position $\vec{x}$ the metric perturbation tensor is different from zero, then $\lim_{|\vec{y}-\vec{x}|\to\infty} \overline{h}'_{\mu\nu}(\vec{y},t) = 0$, because the space-time is asymptotically flat. Thus, Eq.(11.29) provides the following four conditions

$$\overline{h}_0^{\prime 0} = \overline{h}_0^{\prime 3}; \qquad \overline{h}_1^{\prime 0} = \overline{h}_1^{\prime 3}; \qquad h_2^0 = \overline{h}_2^{\prime 3}; \qquad \overline{h}_3^{\prime 0} = \overline{h}_3^{\prime 3}; \tag{11.30}$$

11.6 TT GAUGE

We can further simplify the situation by changing again reference frame, with a new infinitesimal transformation: $x''^\tau = x'^\tau + d'^\tau$, where d'^τ are the contravariant components of a new (infinitesimal) displacement vector. Since in the "primed" reference frame we already enforced the Lorentz gauge condition, the analogous of Eq.(11.19) reads now $\overline{h}''^\mu{}_{\nu,\mu} = -\Box d'_\nu$. So, the Lorentz gauge condition holds also in the "double primed" reference frame, provided that $\Box d'^\tau = 0$. Thus, from one hand, we still have $\Box \overline{h}''_{\mu\nu} = 0$, and, from the other hand, that Eq.(11.30) is valid also in the "double primed" reference frame:

$$\overline{h}_0^{\prime\prime 0} = \overline{h}_0^{\prime\prime 3} \qquad \overline{h}_1^{\prime\prime 0} = \overline{h}_1^{\prime\prime 3} \qquad \overline{h}_2^{\prime\prime 0} = \overline{h}_2^{\prime\prime 3} \qquad \overline{h}_3^{\prime\prime 0} = \overline{h}_3^{\prime\prime 3} \tag{11.31}$$

Now let's use the analogous of Eq.(11.16)

$$\overline{h}''^\tau{}_\nu = \overline{h}'^\tau{}_\nu - d'_\nu{}^{,\tau} - d'^\tau{}_{,\nu} + \delta^\tau{}_\nu d'^\sigma{}_{,\sigma} \tag{11.32}$$

and choose the four components of the displacement vector d'^τ to set equal to zero four components of the metric perturbation tensor. In particular, let's do it for the time-space components

$$\overline{h}_1^{\prime\prime 0} = \overline{h}_2^{\prime\prime 0} = \overline{h}_3^{\prime\prime 0} = 0 \tag{11.33}$$

and for the combination

$$\overline{h}_1^{\prime\prime 1} + \overline{h}_2^{\prime\prime 2} = 0 \tag{11.34}$$

Then, because of Eq.(11.31), $\overline{h}_0^{\prime\prime 0} = \overline{h}_1^{\prime\prime 3} = \overline{h}_2^{\prime\prime 3} = \overline{h}_3^{\prime\prime 3} = 0$. In conclusion, the *only non-vanishing components* of the metric perturbations tensor are $\overline{h}_2^{\prime\prime 1}$ and $\overline{h}_1^{\prime\prime 1} = -\overline{h}_2^{\prime\prime 2}$.

The gauge that satisfies all these conditions is the so-called *Transverse-Traceless* or, simply, *TT-gauge*: it is transverse because the only non-vanishing components of the metric perturbation tensor are in the plane orthogonal to the direction of propagation of the wave; it is traceless because of the chosen gauge. Under these conditions, the perturbed metric tensor is equal to the *trace-reversed* metric perturbation tensor, and we can forget about the overline. Also, if we agree to work in a *TT-gauge*, we can also forget about the "double prime" superscript and explicitly quote the gauge we are working with: $h^{(TT)}{}_{\mu\nu}$. Note that the Lorentz gauge condition now provides $h^{(TT)i}{}_{j,i} = 0$, showing that the spatial metric perturbation tensor is double- transverse.

At last, we can write

$$h^{(TT)}{}_{\mu\nu} = \begin{pmatrix} 0 & 0 & 0 & 0 \\ 0 & h_+ & h_\times & 0 \\ 0 & h_\times & -h_+ & 0 \\ 0 & 0 & 0 & 0 \end{pmatrix} \tag{11.35}$$

where the two independent components, h_+ and $h_\times$, correspond to the two possible polarization states of the gravitational wave.

11.7 GAUGE INVARIANT APPROACH

Before discussing the observational implications of Eq.(11.35), let's clarify few issues that can arise from the previous discussion. First, it could seem that the metric perturbation tensor defined in Eq.(11.1) *always* obeys to a wave equation. We know that this is not the case: in fact, the same perturbative approach of Eq.(6.10) led us to write the Schwarzschild metric in a linearized regime [*c.f.* Eq.(6.20)]. Secondly, it could seem that the writing given in Eq.(11.35) depends on a very specific choice of the *TT-gauge*, and this would contradict the Principle of General Covariance of General Relativity.

To clarify all these issues, it is convenient to attack the problem in a gauge-independent way. To do so, let's first *SVT* decompose the metric perturbation tensor, $h_{\mu\nu}$, to separate the contributions coming from scalars (*S*), vectors (*V*), and tensors (*T*). Since $ds^2 = (\eta_{\mu\nu} + h_{\mu\nu})dx^\mu dx^\nu$, it is clear that h_{00} has to be a scalar, h_{0i} a vector, and h_{ij} a tensor. This is the only way for having ds^2 invariant. Then,

$$h_{00} = 2\varphi \tag{11.36a}$$

$$h_{0i} = V_i^{(t)} - A_{,i} \tag{11.36b}$$

$$h_{ij} = \frac{1}{3}\delta_{ij}H + \left(B_{,ij} - \frac{1}{3}\delta_{ij}\nabla^2 B\right) + \frac{1}{2}\left(C_{j,i}^{(t)} + C_{i,j}^{(t)}\right) + h_{ij}^{(TT)} \tag{11.36c}$$

Note that the combination $(C^{(t)}{}_{j,i} + C^{(t)}{}_{i,j})/2$ is a necessary one to render h_{ij} symmetric. The previous relations clearly show that scalars, vectors, and tensors contribute differently to the ten independent degrees of freedom-*dof*'s of $h_{\mu\nu}$.

- The scalar h_{00} contributes with one scalar *dof*, φ;
- The vector h_{0i} has been decomposed according to the Helmholtz' theorem [see Box 11.1]: it contributes with one scalar *dof*, A, and with two vector *dof*'s, provided by the transverse and divergenceless vector $V_i^{(t)}$;
- The tensor h_{ij} contributes with: two scalar *dof*'s, H and B; two vector *dof*'s, provided by the transverse and divergenceless vector $C_i^{(t)}$; two tensor *dof*'s, provided by the doubly-transverse, divergenceless, and traceless tensor $h_{ij}^{(TT)}$.

The components of the metric perturbation tensor change under a generic coordinate transformation according to Eq.(11.12)

$$\hat{h}_{00} = h_{00} - 2d_{0,0} \tag{11.37a}$$

$$\hat{h}_{0i} = h_{0i} - d_{0,i} - d_{i,0} \tag{11.37b}$$

$$\hat{h}_{ij} = h_{ij} - d_{i,j} - d_{j,i} \tag{11.37c}$$

Let's apply again the Helmholtz's theorem to express the spatial, covariant components of the displacement four-vector in terms of a transverse vector, $\varepsilon_k^{(t)}$, and of the gradient of a scalar, ξ,

$$d_k = \varepsilon_k^{(t)} - \xi_{,k} \tag{11.38}$$

Then, after substituting Eq.(11.36a) and Eq.(11.36b) in Eq.(11.37a) and Eq.(11.37b), we get

$$\hat{\phi} = \phi - d^0{}_{,0} \tag{11.39a}$$

$$\hat{V}_i^{(t)} - \hat{A}_{,i} = V_i^{(t)} - A_{,i} - d^0{}_{,i} - \varepsilon_{i,0}^{(t)} + \xi_{,i0} \tag{11.39b}$$

as $d_0 = \eta_{0\beta} d^\beta = d^0$. Now, after equating homologous quantities (scalars with scalars, and transverse vectors with transverse vectors), Eq.(11.39b) yields

$$\hat{V}_i^{(t)} = V^{(t)}{}_i - \varepsilon_{i,0}^{(t)} \tag{11.40a}$$

$$\hat{A} = A + d^0 - \xi_{,0} \tag{11.40b}$$

Likewise, from Eq.(11.36c) and Eq.(11.37c), we get

$$\begin{aligned} &\frac{1}{3}\hat{H}\delta_{ij} + \left(\hat{B}_{,ij} - \frac{1}{3}\delta_{ij}\nabla^2\hat{B}\right) + \frac{1}{2}\left(\hat{C}_{j,i}^{(t)} + \hat{C}_{i,j}^{(t)}\right) + \hat{h}_{ij}^{(TT)} = \\ &\frac{1}{3}H\delta_{ij} + \left(B_{,ij} - \frac{1}{3}\delta_{ij}\nabla^2 B\right) + \frac{1}{2}\left(C_{j,i}^{(t)} + C_{i,j}^{(t)}\right) + h_{ij}^{(TT)} - \varepsilon_{i,j}^{(t)} + 2\xi_{,ij} - \varepsilon_{j,i}^{(t)} \end{aligned} \tag{11.41}$$

By equating homologous quantities (scalars with scalars, transverse vectors with transverse vectors, and doubly transverse tensors with doubly transverse tensors), we now obtain

$$\hat{H} - \nabla^2\hat{B} = H - \nabla^2 B \tag{11.42a}$$

$$\hat{B}_{,ij} = B_{,ij} + 2\xi_{,ij} \quad \Rightarrow \quad \hat{B} = B + 2\xi \tag{11.42b}$$

$$\hat{C}_{j,i}^{(t)} = C_{j,i}^{(t)} - 2\varepsilon_{j,i}^{(t)} \quad \Rightarrow \quad \hat{C}_j^{(t)} = C_j^{(t)} - 2\varepsilon_j^{(t)} \tag{11.42c}$$

and

$$\boxed{\hat{h}_{ij}^{(TT)} = h_{ij}^{(TT)}} \tag{11.43}$$

The last result is very important, as it shows that the doubly transverse, traceless tensor $h_{ij}^{(TT)}$ is indeed gauge-invariant. Eq.(11.42a) clearly shows that there is another gauge-invariant quantity, given by the combination of two scalars

$$\boxed{\Psi = H - \nabla^2 B} \tag{11.44}$$

We can now use Eq.(11.40b) and Eq.(11.42b) to eliminate ξ and to express d^0 directly in terms of A and $B_{,0}$

$$d^0 = \left(\hat{A} + \frac{\hat{B}_{,0}}{2}\right) - \left(A + \frac{B_{,0}}{2}\right) \tag{11.45}$$

Eq.(11.39a) can then be re-written as follows

$$\hat{\phi} = \phi - \frac{\partial}{\partial \hat{x}^0}\left(\hat{A} + \frac{\hat{B}_{,0}}{2}\right) + \frac{\partial}{\partial x^0}\left(A + \frac{B_{,0}}{2}\right) \tag{11.46}$$

showing that the quantity

$$\boxed{\Phi = \phi + \frac{\partial}{\partial x^0}\left(A + \frac{B_{,0}}{2}\right)} \tag{11.47}$$

is indeed a gauge-invariant quantity.

Let's conclude this section by writing also the vector *dof*'s in a gauge-invariant form. After using Eq.(11.40a) and Eq.(11.42c) to eliminate $\varepsilon_{j,0}^{(t)}$, we can rewrite Eq.(11.40a)

$$\hat{V}_i^{(t)} = V_i^{(t)} + \frac{1}{2}\left(C_{j,0}^{(t)} - \hat{C}_{j,0}^{(t)}\right) \tag{11.48}$$

showing the existence of another gauge-invariant quantity

$$\boxed{\Theta_i = V_i^{(t)} + \frac{1}{2} C_{j,0}^{(t)}} \tag{11.49}$$

So, at the end, we are left with four gauge-invariant physical quantities (Φ, Ψ, Θ_i, and $h_{ij}^{(TT)}$) corresponding to six *dof*'s. The remaining four *dof*'s are gauge modes, associated with the choice of the displacement vector: two scalar *dof*'s, d^0 and ξ, and two vector *dof*'s, provided by the divergenceless vector $\varepsilon_i^{(t)}$.

11.8 FIELD EQUATIONS

The advantage of the *SVT* decomposition is that scalar, vector, and tensor *dof*'s obey to independent, uncoupled field equations. Then, it makes sense to discuss them separately.

• *Scalar dof's*

If we restrict to scalar *dof*s , we have to deal with four scalar quantities: ϕ, A, H, and B. Let's chose the *longitudinal gauge*, with $A = B = 0$. In this case both $H = \Psi$ and $\phi = \Phi$ are gauge invariant quantities [*c.f.* Eq.(11.44) and Eq.(11.47)]. The components of the metric perturbation tensor can then be written as follows

$$h_{00} = 2\Phi; \qquad h_{ij} = \frac{1}{3}\Psi\delta_{ij} \tag{11.50}$$

If we neglect the contribution of the Cosmological Constant, the field equations in the vacuum are given by $G^{\alpha}{}_{\beta} = 0$ [*c.f.* Eq.(8.18)]. It can be shown [see Exercise A.37] that

$$G^0{}_0 = \boxed{\nabla^2\Psi = 0} \tag{11.51}$$

For $i \neq j$, it can also be shown that

$$G^i{}_j = \frac{\partial^2}{\partial x^i \partial x^j}\left(\Phi - \frac{1}{6}\Psi\right) = 0 \tag{11.52}$$

consistent with having $\Phi = \Psi/6$. Then, because of Eq.(11.51)

$$\boxed{\nabla^2\Phi = 0} \tag{11.53}$$

In addition, we get

$$G^0{}_i = \frac{\partial^2}{\partial x^i \partial x^0}\Psi = 0 \tag{11.54}$$

and

$$\mathrm{Tr}[G^i{}_j] = -\frac{\partial^2}{\partial x^{0^2}}\Psi + 2\nabla^2\Phi - \frac{1}{3}\nabla^2\Psi = 0 \tag{11.55}$$

This indicates, together with Eq.(11.51) and Eq.(11.53), that the gauge-invariant quantity Ψ is time independent.

• *Vector dof's*

In this case, we have to deal with two transverse vectors, $V_i^{(t)}$ and $C_i^{(t)}$. The components of the metric perturbation tensor are given by [*c.f.* Eq.(11.36)]

$$h_{0i} = V_i^{(t)}; \qquad h_{ij} = \frac{1}{2}\left(C_{i,j}^{(t)} + C_{j,i}^{(t)}\right) \tag{11.56}$$

The only non-vanishing Einstein tensor components are found to be

$$G_i^0 = \boxed{\nabla^2\Theta_i = 0} \tag{11.57}$$

• *Tensor dof's*

Let's now consider only the tensor *dof*'s. These are given by the components of the double transverse, traceless tensor $h_{ij}^{(TT)}$. If we keep the same notation of Eq.(11.35), the resulting components of the field equations are found to be

$$G_1^1 = G_2^2 = -\boxed{\Box h_+ = 0}; \qquad G_2^1 = G_1^2 = -\boxed{\Box h_\times = 0} \tag{11.58}$$

From the above discussion, we can conclude that only the doubly transverse tensor, $h_{ij}^{(TT)}$, describes gravitational waves, as *only* tensor *dof*s obey to a wave equation. It must also be stressed that the results given here generalize those given in Section 11.6. In fact, the transverse and traceless tensor of Eq.(11.35), initially derived in the TT- gauge, it is indeed a gauge-invariant quantity, as shown in Eq.(11.43).

Note that only gauge-invariant quantities related to scalar and vector *dof*s obey to Laplace-like equations. Thus, Eq.(11.53) reduces, in the weak field limit, to the classical Laplace equation in the "vacuum": $\nabla^2 U = 0$ [*c.f.* Eq.(6.20)]. It follows that only scalar *dof*s contribute to the overall curvature of the space-time which has been tested observationally in Chapter 10.

11.9 EFFECTS OF A GRAVITATIONAL WAVE

To study the effects of gravitational waves, let's rewrite the metric of the space-time under the assumption that only tensor *dof*s are present [*c.f.* Eq.(11.1) and Eq.(11.35)]

$$ds^2 = c^2 dt^2 - \left[\delta_{ij} - h_{ij}^{(TT)}\right] \tag{11.59}$$

Consider now a test particle, A say, at rest *w.r.t.* the chosen reference frame. The geodesic equations [*c.f.* Eq.(5.26)] provide in this case

$$\frac{d^2 x^\alpha}{dt^2} = 0 \tag{11.60}$$

as $\Gamma_{00}^\alpha = 0$ [*c.f.* Eq.(11.35) and Eq.(11.59)]. This result is perfectly understandable if $h_{ij}^{(TT)} = 0$. However, Eq.(11.60) is definitely valid before, during and after the transit of a gravitational wave. So, we might wrongly conclude that a test particle is unaffected by the wave. This is not the case. In fact, the writing of Eq.(11.59) implicitly implies using a synchronous gauge and comoving coordinates [see Box 11.2]. In this gauge, the test particle keeps its Lagrangian, comoving label even if $h_{ij}^{(TT)} \neq 0$. In other words, the test-particle is simply "free-falling" into the gravitational field produced by the gravitational wave.

BOX 11.2 THE SYNCHRONOUS GAUGE

In the *synchronous* gauge, fundamental observers share the same (universal) time marker. This can happen if two conditions are fulfilled: $g_{00} = 1$ (proper time flows at the same rate for all the observers)[*c.f.* Eq.(2.20)]; $g_{0k} = 0$ (clocks can be fully synchronized even along closed paths)[*c.f.* Eq.(3.43)]. It follows that at a given time, the spatial positions of fundamental observers define a space-like hypersurface. In a sense, we are "slicing" the abstract space-time in spatial hypersurfaces of constant time, recovering the intuitive point of view that events occur somewhere in space at a given time.

In this gauge, spatial coordinates are *comoving* (Lagrangian) coordinates: fundamental observers are at rest, with four velocities $u^\mu \equiv \{1,0,0,0\}$. Note that this statement in strictly related to the $g_{00} = 1$ and $g_{0k} = 0$ requirements. In fact, the geodesic equation reduces to $du^\mu/ds = 0$ [*c.f.* Eq.(11.60)], as in this gauge $\Gamma^\mu_{00} = 0$.

The situation clearly changes if we consider another particle, B say, at a comoving distance $\mathscr{D}^\alpha$ from A. Both particles move in space-time along their geodesics. However, in the presence of a gravitational wave, we don't expect these geodesics to be parallel. So, it make sense to evaluate their *geodesic deviation* as we did in Section 6.9. Let's first rewrite Eq.(6.30)

$$\frac{D^2\mathscr{D}^\alpha}{Ds^2} = R^\alpha{}_{\mu\nu\lambda}u^\mu u^\nu \mathscr{D}^\lambda \tag{11.61}$$

There are several simplifications that we can do here. First, in the synchronous gauge $u^\mu = \{1,0,0,0\}$. Secondly, the coordinate distance $\mathscr{D}^\lambda$ between the A and B particles can be written in a perturbative way

$$\mathscr{D}^\alpha = L^\alpha + \Delta L^\alpha \tag{11.62}$$

where the zeroth order term, L^α, represents the (constant) comoving coordinate distance of B from A in the absence of the gravitational wave. Also, the second absolute derivative can be reduced in this context to an ordinary second derivative. In fact,

$$\frac{D\mathscr{D}^\lambda}{Ds} \simeq \frac{D\Delta L^\alpha}{Ds} \equiv \frac{d\Delta L^\alpha}{ds} + \Gamma^\alpha_{\mu\nu}\Delta L^\mu u^\nu \simeq \frac{d\Delta L^\alpha}{ds} \tag{11.63}$$

as $\Gamma^\alpha_{\mu\nu}\Delta L^\mu$ is a second-order quantity. Finally, the needed components of the Riemann tensor can be derived from Eq.(11.3). We get

$$R^\alpha{}_{00\lambda} \simeq \frac{1}{2}\eta^{\alpha\rho}\left[-h^{(TT)}_{0\lambda,\rho 0} + h^{(TT)}_{\rho\lambda,00} + h^{(TT)}_{\rho 0,0\lambda} - h^{(TT)}_{\rho 0,0\lambda}\right] \simeq \frac{1}{2}\eta^{\alpha\rho}h^{(TT)}_{\rho\lambda,00} \tag{11.64}$$

as $h^{(TT)}_{0\tau} = 0$. Thus, for slowing moving particles ($ds \simeq cdt$), Eq.(11.62) can be written in the following form

$$\boxed{\frac{1}{c^2}\frac{d^2\Delta L^\alpha}{dt^2} = \frac{1}{2}\eta^{\alpha\rho}h^{(TT)}_{\rho\lambda,00}L^\lambda} \tag{11.65}$$

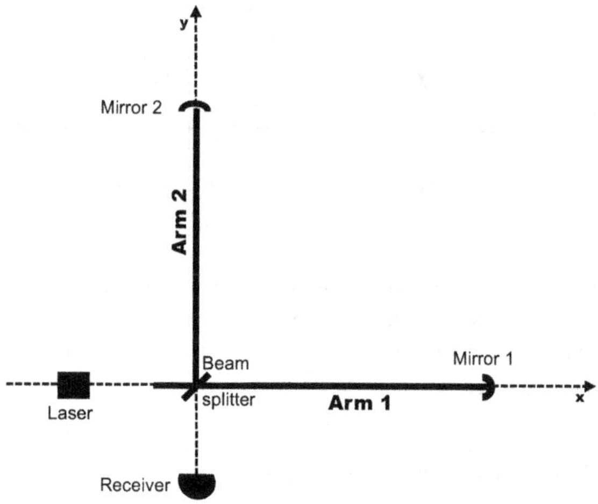

Figure 11.1 A schematic view of an interferometer for gravitational wave detection, see text.

11.10 INTERFEROMETERS

We can apply Eq.(11.65) to the case of interferometric detectors of gravitational waves. Such detectors consist of two orthogonal arms with mirrors at their ends. A laser (almost monochromatic) signal is sent toward a beam splitter and from it to the two mirrors. The reflected signals are then recombined at the beam splitter and received by a photodetector at the readout point [see Figure 11.1]. The interference pattern is used to measure the relative phase of one beam *w.r.t.* the other.

For the sake of simplicity, let's choose a coordinate system with the x- and y-axis aligned with the arms' directions. So, for a gravitational wave propagating along the z-axis, we can still consider valid Eq.(11.35). Consider a particle A at the origin of the reference frame, where the beam splitter is located, and a particle B (C) at the position of Mirror 1 (2) [see Figure 11.1]. Eq.(11.65) shows that a gravitational wave affects the optical path lengths of the arms. So, the interference pattern at the readout point should change when a gravitational wave is passing through the detector. With this configuration, Eq.(11.35) yields[2]

$$\begin{aligned}\frac{d^2\Delta L_x}{dt^2} &= -\frac{1}{2}\frac{\partial^2 h_+}{\partial t^2}L_x - \frac{1}{2}\frac{\partial^2 h_\times}{\partial t^2}L_y \\ \frac{d^2\Delta L_y}{dt^2} &= -\frac{1}{2}\frac{\partial^2 h_\times}{\partial t^2}L_x + \frac{1}{2}\frac{\partial^2 h_+}{\partial t^2}L_y\end{aligned} \tag{11.66}$$

[2] Let's simplify the notation by writing $L^1 = L_x$, $L^2 = L_y$, $L = L_x = L_y$ for the zeroth order quantities, and $\Delta L^1 = \Delta L_x$, $\Delta L^2 = \Delta L_y$ for the perturbed ones.

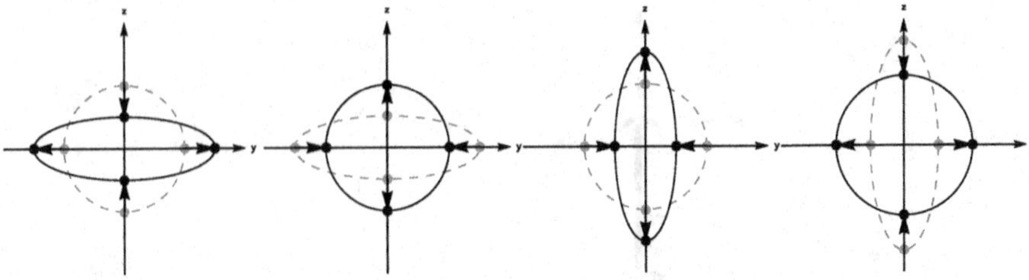

Figure 11.2 Test particles are initially disposed around a (dashed) circle. Because of the transit of a "+" polarized GW, particles move parallel to the coordinate axis. The continuous lines identify the patterns of the test particles at $t_n = nT/4$, with $n = 1,4$ (from left to right), T being the period of the wave.

A direct integration of this equation provides

$$\begin{aligned} \Delta L_x(t) &= -\frac{1}{2}h_+(t)L_x - \frac{1}{2}h_\times(t)L_y \\ \Delta L_y(t) &= -\frac{1}{2}h_\times(t)L_x + \frac{1}{2}h_+(t)L_y \end{aligned} \tag{11.67}$$

Let's assume that the gravitational wave has only a "+" polarization status. On the interferometer plane, we can write $h_+ = A_+ \cos(\omega t + \pi/2)$, implying that the transit of the wave starts at $t = 0$. Then, the fractional change in the comoving distance between the A and B particles is given by

$$\frac{\Delta L_x}{L} = -\frac{1}{2}A_+ \cos\left(\omega t + \frac{\pi}{2}\right) \tag{11.68}$$

whereas the fractional change in the comoving distance between the C particle, at the position of Mirror 2, and A (at the beam splitter) results to be

$$\frac{\Delta L_y}{L} = \frac{1}{2}A_+ \cos\left(\omega t + \frac{\pi}{2}\right) \tag{11.69}$$

To show the overall effect of a "+" polarized gravitational wave, let's imagine to have a number of test particles at the same distance from the origin, disposed along a circumference. Then, by using Eq.(11.67), we can see what happens at different times. As shown in Figure 12.2, the effect of a "+" polarized gravitational wave is to force each of the particle to move parallel to the coordinate axes. Thus, the effect of a h_+ polarized wave is to modulate the lengths of the interferometer arms around their zeroth order value, L. Note that an elongation along the x-axis corresponds to a contraction along the y-axis, and vice versa. These fractional changes in the length of the interferometer arms induce a variation in the interferometric pattern at the readout point. This in turn allows to accurately measure $\Delta L/L$ and, then, $h_{ij}^{(TT)}$.

If the wave is "cross" polarized, then $h_\times = A_\times \cos(\omega t + \pi/2)$. In this case, Eq.(11.67) provides

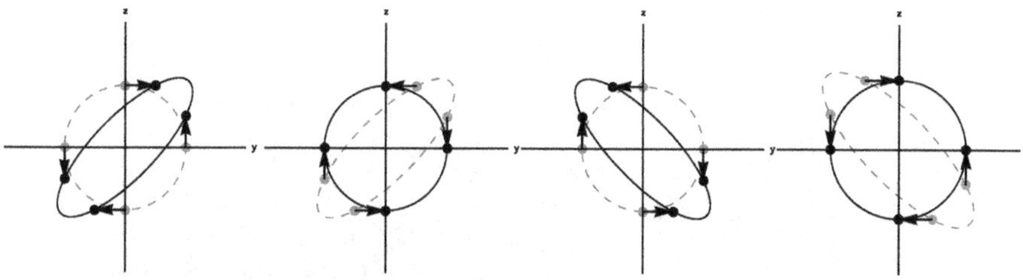

Figure 11.3 Test particles are initially disposed around a (dashed) circle. Because of the transit of a "×" polarized GW, particles move orthogonally to the coordinate axis. The continuous lines identify the patterns of the test particles at $t_n = nT/4$, with $n = 1,4$ (from left to right), T being the period of the wave.

$$\begin{aligned}\frac{\Delta L_x}{L}(t) &= -\frac{1}{2}A_\times \cos(\omega t + \pi/2)\\ \frac{\Delta L_y}{L}(t) &= -\frac{1}{2}A_\times \cos(\omega t + \pi/2)\end{aligned} \tag{11.70}$$

As shown in Figure 11.3, the effect of the "×" polarized gravitational wave forces the points on the circumference to move not parallel to the coordinate axes, but rather orthogonally to them. So, the evolution of the polarization pattern is the same as for the + polarized wave, but rotated by 45°. In real life, the interferometers are sensitive to a weighted combination of both the "+" and "×" polarization, the weights being defined by the direction of the wave *w.r.t.* the interferometer plane and orientation.

11.11 THE *DIRECT* DETECTION OF GRAVITATIONAL WAVES

The *direct* detection of gravitational waves requires to build ground-based facilities able to exploit what was discussed in the previous Section. The suggestion of using interferometers to detect gravitational waves dates back to the 60s of the last century [37, 40], but it is only in the first decade of this century that a number of interferometers became operational. The effort to push, year after year, the current technologies to their limits led in 2015 to the Advanced *Laser Interferometer Gravitational-Wave Observatory-LIGO* [2]. Advanced-*LIGO* is a very sophisticated, high sensitivity experiment. It is composed by two identical laser interferometers with 4*km* long arms. They are both US based, one in Hanford (Washington) and the other in Livingston (Lousiana). They are 3000*km* away to be sure that the detected signals are not of local origin, and to identify only those signal that are identical in the two interferometers. Working with two interferometers is essential to confirm each other's detections of *non-local* signals—that is, of gravitational waves. There is another *plus* in using two or even more interferometers. The study of the difference in the signal arrival time at each detector allows (and will allow) to (better) identify the direction on the sky from which a gravitational wave arrives.

In 2015, immediately after they were turned on, the two detectors of Advanced-*LIGO* observed a transient gravitational-wave signal. The strength of a gravitational

Figure 11.4 Aerial view of the Hanford-LIGO interferometer, reaching 4 km into the desert. Credit: Caltech/MIT/LIGO Lab. See `https://www.ligo.caltech.edu/system/avm_image_sqls/binaries/32/jpg_original/ligo-hanford-aerial-04.jpg?1447108890`

wave is quantified by a dimensionless number, $h = \Delta L_x/L_X - \Delta L_y/L_y$, the so-called *gravitational-wave strain*. The peak of the gravitational-wave strain observed by Advance-LIGO was $h \simeq 10^{-21}$. It is worth noting that such a strain corresponds to $\Delta L \simeq 10^{-16} cm$, almost three orders of magnitude smaller that the diameter of the proton. The form of the detected signal matches the waveform predicted by General Relativity for the inspiral and merger of a pair of black holes. The analysis of the Advanced-*LIGO* data provided the masses of the two black-holes, $M_1 = (36 \pm 5)M_\odot$ and $M_2 = (29 \pm 4)M_\odot$, their distance, $(410 \pm 180$ Mpc, and the energy radiated in the form of gravitational wave, $(3 \pm 0.5)M_\odot c^2$ [3]. This has been the first *direct* detection of gravitational waves and the first evidence of binary black-hole merger.

Direct observation of gravitational waves has open a new filed, the gravitational wave astronomy. Now the network of interferometers comprises also the Advanced-*Virgo* interferometer. This is based in Italy and has two perpendicular, $3km$ long arms. LIGO and Virgo detected 35 new events in their latest observation run, bring to 90 the gravitational waves detected by the global three-interferometer network. In addition to LIGO and Virgo, there will also be the Japanese KAGRA detector. The LIGO, Virgo, and KAGRA scientific collaborations are closely coordinating in order to start their common run of observations in mid-December, 2022.

11.12 THE *INDIRECT* EVIDENCE FOR GRAVITATIONAL WAVES

Let's conclude this chapter by discussing the *indirect* evidence for the existence of gravitational obtained through astronomical observations. This starts in 1975 with the discovery by Hulse and Taylor of the binary pulsar PSR B1913+16 [46]. The

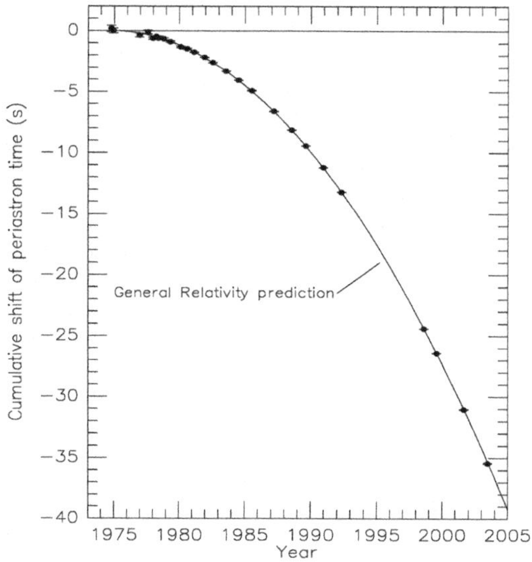

Figure 11.5 This is Figure 3 of the paper by J. M. Weisberg and Y. Huang [98], showing the rate of period decrease of PSR B1913+16 in 30 years of observations. The observational data points indicate the observed period change at periastron. The parabola shows the GR predictions for the rate of period decrease of a binary system emitting gravitational radiation.

pulsation period of this source, $\simeq 59$ ms, was observed to vary systematically by $39\,\mu$s over a cycle of roughly 8 hours. This periodic variation is consistent with having PSR B1913+16 in a binary orbit around a compact object of similar size. In the abstract of their paper, Hulse and Taylor state that their discovery "makes feasible a number of studies involving the physics of compact objects, the astrophysics of close binary systems, and special- and general-relativistic effects". In 1993, Hulse and Taylor were awarded by Nobel Prize in Physics "for the discovery of a new type of pulsar, a discovery that has opened up new possibilities for the study of gravitation". In fact, according to General Relativity, we do expect that a binary system emits energy in the form of gravitational waves. As a result, the orbit shrink and the period decreases. Such an evidence was indeed supported by the observations [89], and further confirmed on the basis of more data. In particular, the excellent agreement between the theoretical predictions for the rate of period decreases [$\dot{T}_{GR} = (-2.403 \pm 0.005)10^{-12}$] and the observations [$\dot{T}_{GR} = (-2.30 \pm 0.22)10^{-12}$] provides compelling evidence for the existence of gravitational radiation [89,90]. Thirty years of observations and analysis of the relativistic binary pulsar B1913+16 provide the plot shown in Figure 11.5 [98].

Part III

From Singularities to Cosmological Scales

12 Schwarzschild Black Holes

12.1 INTRODUCTION

The Schwarzschild solution proved to be extremely effective in providing observational tests that confirmed the validity of the field equations of General Relativity. These classical tests were done on planetary scales—that is, on scales much larger than the Schwarzschild radius of the Sun. A complete new phenomenology arises if we explore, still in the "vacuum", scales comparable or even smaller than the Schwarzschild radius of the considered structure [30, 53]. The goal of this chapter is to develop the tools needed for discussing specific predictions of General Relativity in this regime.

12.2 SINGULARITIES OF THE SCHWARZSCHILD METRIC

As already discussed, the Schwarzschild metric [*c.f.* Eq.(9.1)]

$$ds^2 = (1 - \mathscr{R}_S/r)\,dx^{0^2} - \frac{dr^2}{1 - \mathscr{R}_S/r} - r^2(d\theta^2 + \sin^2\theta d\phi^2) \tag{12.1}$$

describes the geometry of the space-time in the presence of a point-mass at the origin of the chosen coordinate system. The curvature of the Schwarzschild space-time cannot be described in terms of the Ricci scalar [*c.f.* Eq.(8.16)]. In fact, the Schwarzschild solution has been derived in the "vacuum", using field equations that require the Ricci tensor (and then its trace) to be identically equal to zero [*c.f.* Eq.(8.13]. For this reason, we have to describe the curvature of the Schwarzschild space-time by considering the scale-invariant, *Kretschmann scalar* $K \equiv R_{\alpha\beta\mu\nu}R^{\alpha\beta\mu\nu}$. In the Schwarzschild case, we have

$$K = 12\frac{\mathscr{R}_S^2}{r^6} \tag{12.2}$$

The Schwarzschild metric clearly shows two pathological points, at $r = 0$ and $r = \mathscr{R}_S$. In these points, the relation between proper and coordinate time is lost [*c.f.* Eq.(7.1)], as well as the relation between proper and coordinate radial lengths [*c.f.* Eq.(3.48)]. However, the Kretschmann scalar confirms the pathological behavior at $r = 0$, but *not* at $r = \mathscr{R}_S$. This is an important point. In fact, being a tensor with rank 0, K has values that, point by point, are independent on the choice of the coordinate system.

All this opens up the discussion on whether $r = \mathscr{R}_S$ is a *physical singularity*, or rather an artifact of the chosen coordinate system, a so-called *coordinate singularity*. On very general grounds, the latter highlights the fact that a given coordinate system is unable to properly describe the geometry of a space or space-time in some specific regions. For example, the geometry of a 2D sphere is usually described in terms of

DOI: 10.1201/9781003141259-12

polar coordinates. In this representation, the North and South poles are clear coordinate singularities. In fact, each of the two poles is identified by a single value of the latitude ($\theta = 0$ or $\theta = \pi$), but by any arbitrary value of the longitudinal angle: $0 \le \phi \le 2\pi$. Just to give another example, the metric given in Eq.(3.37) shows a singularity at $r = c/\omega$. In Section 3.6, we concluded that this singularity reflected the impossibility of building a physical reference frame, with rods and clocks, when rotation velocities become larger than c. In other words, the chosen coordinate system can't be used to cover the entire space-time.

It is worth remembering that the Schwarzschild geometry was derived with very specific assumptions about the time and radial coordinates: the coordinate time is the proper time of an observer at infinity, whereas the radial coordinate is chosen to render the angular part of the metric equal (apart from the minus sign) to the Euclidean one. Since t and r are just space-time coordinates, the interpretation of their physical role always requires great attention.

Note that forcing $r < \mathscr{R}_S$ in the Schwarzschild solution results in a change of sign in both g_{00} and g_{11} [*c.f.* Eq.(12.1)]. This is an interesting clue. Consider, for example, a free-falling test-particle that in a given dt changes its position by dr:

$$ds^2 = (1 - \mathscr{R}_S/r)dx^{0^2} - (1 - \mathscr{R}_S/r)^{-1}dr^2 \tag{12.3}$$

If $r > \mathscr{R}_S$, $ds^2 > 0$: as expected, the considered interval is time-like. However, if $r < \mathscr{R}_S$, for the same dt and dr, $ds^2 < 0$: the interval becomes now space-like. This has an interesting consequence on the orientation of the light-cones in a Schwarzschild space-time. The condition $ds = 0$ provides [*c.f.* Eq.(12.1)]

$$\frac{dr}{dx^0} = \pm\left(1 - \frac{\mathscr{R}_S}{r}\right) \tag{12.4}$$

Here the plus (minus) sign refers to an outgoing (ingoing) light signal, as dr increases (decreases) with time. Integration of Eq.(12.4) provides

$$\begin{aligned} \tau_{out} &= +\frac{1}{2}(\rho - \rho_0) + \frac{1}{2}\log\left(\left|\frac{\rho - 1}{\rho_0 - 1}\right|\right) \\ \tau_{in} &= -\frac{1}{2}(\rho - \rho_0) - \frac{1}{2}\log\left(\left|\frac{\rho - 1}{\rho_0 - 1}\right|\right) \end{aligned} \tag{12.5}$$

where $\tau = ct/(2\mathscr{R}_S)$, $\rho = r/\mathscr{R}_S$, and $\tau = 0$ for $\rho = \rho_0$.

Null geodesics are shown in Figure 12.1. For $\rho >> 1$, the light-cones have an opening angle of $\simeq 90°$, as expected in an almost Minkowski space-time. However, when ρ tends to 1 from the right, the opening angle decreases, keeping the light-cone axes still parallel to the τ-axis. Consider a source of pulsed light in free-fall toward the central mass. The narrowing of the light cones implies that an observer at infinity receives these pulses at a decreasing rate. This is consistent with the discussion done in Section 9.7: for an observer at infinity, the time required for a test particle to reach the Schwarzschild radius is infinite. On the other hand, as discussed above, when $\rho < 1$ time-like intervals become space-like. This forces the light-cones to rotate by

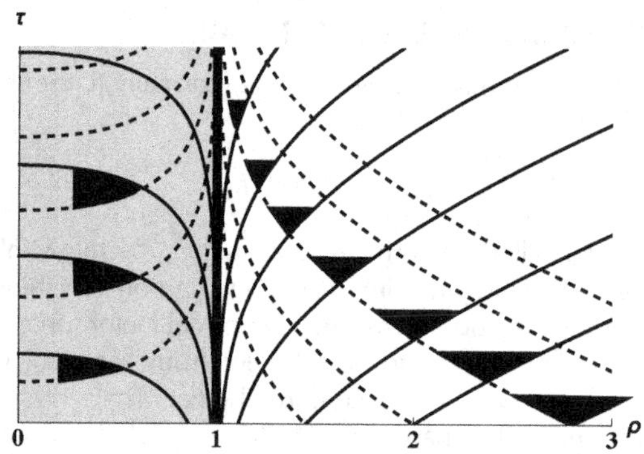

Figure 12.1 The Schwarzschild space-time. Here $\tau = ct/(2\mathscr{R}_S)$ and $\rho = r/\mathscr{R}_S$. Continuous (dashed) lines describe outgoing (ingoing) light signals. The black triangles show the light-cones of observer at different radial positions. Note the opening of the light-cones is reduced while approaching $\rho = 1$ from the right. Note also the 90° rotation of the light-cone axes when $\rho < 1$ (the light-gray area on the left of the Figure).

90°: now the axes of the light-cones are parallel to the ρ-axis [see Figure 12.1]. This has two important consequences: i) there is no way to communicate with an observer at infinity and ii) there is no way of avoiding the singularity at $r = 0$.

Let's refer to the *Schwarzschild sphere* as to the sphere of radius $\mathscr{R}_S$. Figure 12.1 shows that the Schwarzschild sphere acts as a "boundary" between the outer and the inner worlds: once one passed it, there is no way to communicate with the outer world, nor to be back to it. The Schwarzschild sphere is then called the *event horizon*: it separates events that can be seen by an observer at infinity, from those that cannot be seen. The region inside the Schwarzschild sphere is called *black hole*: it is a hole because matter is falling into it; it is black because even light is unable to escape from it. Whatever is in the inner world, it accretes mass without emitting any signal to the outer world. It is worth remembering that the Schwarzschild solution describes the geometry of the space-time produced by a point mass—that is, in the "vacuum". In reality, a black hole can exist only if the mass responsible for the curvature of the space-time is entirely contained inside the Schwarzschild sphere. This can occur only for very massive, but extremely compact astrophysical objects.

12.3 CONFORMALLY-FLAT COORDINATES

As discussed in the previous Section, intervals that are time-like (when $\rho > 1$) become space-like when ρ turns to be smaller than unity. Thus, inside the Schwarzschild sphere the "radial" and "time" coordinates must be reinterpreted as

BOX 12.1 CONFORMALLY FLAT SPACE-TIME

A pseudo-Riemannian space-time is said to be conformally flat if its metric can be written in the following way

$$g_{\mu\nu}(x^{\tau}) = H^2(x^{\tau})\eta_{\mu\nu} \tag{B12.1.a}$$

where $H(x^{\tau})$ is the so-called conformal-factor and $\eta_{\mu\nu}$ is the Minkowski metric. Note that a conformallly-flat space-time has indeed a non-vanishing curvature. In fact, being a function of the coordinates, the conformal factor forces the Christoffel symbols and the Riemann tensor to be different from zero [*c.f.* Eq.(5.39)]. Any 2D pseudo-Riemannian space-time is conformally flat. The terrestrial globe is an example of a conformally flat space. In fact, it is possible to use a stereographic projection and to map the globe in a flat $x-y$ space. In this case, the metric can be written as follows

$$d\ell^2 = d\theta^2 + \sin^2\theta d\phi^2 = \frac{4}{(1+r^2)^2}(dx^2+dy^2) \tag{B12.1.b}$$

Here θ e ϕ are the colatitude and longitude angles, where $r = \sqrt{x^2+y^2}$ is the distance from the origin of the (flat) $x-y$ coordinate system. The conformal-factor results then to be $2/(1+r^2)$.

"*time*" and "*space*" markers, respectively. This strongly suggests to use conformally flat coordinates, to write the Schwarzschild metric given in Eq.(12.4) in a somehow more abstract way

$$ds^2 = f^2(u,v)\left(dv^2 - du^2\right) \tag{12.6}$$

Note that we can write the Schwarzschild 2D metric of Eq.(12.4) in this form because any 2D pseudo-Riemannian space-time is conformally flat [see Box 12.1].

There are two clear advantages in writing the metric in the form given in Eq.(12.6). First, we don't need to assign *a priori* neither to u nor to v the role of "*time*" or "*space*" coordinates. So, they can play the role of a "*time*" or "*space*" markers depending on the physical situation we want to investigate—that is, scales larger or smaller than $\mathscr{R}_S$. Secondly, the conformal factor does not affect the shape of the light cones that will be equal to those of a Minkowski space-time. As a consequence, the coordinate speed of light is unity, as $du/dv = dv/du = 1$. This is clearly not the case in the Schwarzschild metric, where the coordinate speed of light becomes smaller and smaller when r is approaching the Schwarzschild radius: $dr/dt = c(1-\mathscr{R}_S/r)$ [*c.f.* Eq.(9.1)].

The coordinates v and u used in Eq.(12.6) are known as the *Kruskal-Szekeres* coordinates [53,88]. We have now to connect them to the Schwarzschild coordinates, t and r. To do so, let's start from the Jacobian of the transformation

$$J^{\alpha}{}_{\mu} \equiv \partial\bar{x}^{\alpha}/\partial x^{\mu} = \begin{pmatrix} \partial v/\partial x^0 & \partial v/\partial r \\ \partial u/\partial x^0 & \partial u/\partial r \end{pmatrix} \tag{12.7}$$

Here barred and unbarred coordinates refer to the Kruskal-Szekeres and Schwarzschild coordinates, respectively. Given the transformation law of the metric tensor, $g_{\mu\nu} = J^{\alpha}{}_{\mu} J^{\beta}{}_{\nu} \bar{g}_{\alpha\beta}$ [see *e.g.* Eq.(4.15)], it is immediate to derive the following relations

$$(1 - \mathscr{R}_S/r) = f^2(u,v)\left[\left(\partial v/\partial x^0\right)^2 - \left(\partial u/\partial x^0\right)^2\right] \quad (12.8a)$$

$$-(1 - \mathscr{R}_S/r)^{-1} = f^2(u,v)\left[(\partial v/\partial r)^2 - (\partial u/\partial r)^2\right] \quad (12.8b)$$

$$0 = (\partial v/\partial x^0)(\partial v/\partial r) - (\partial u/\partial x^0)(\partial u/\partial r) \quad (12.8c)$$

Let's simplify the notation by introducing a new function

$$\mathscr{F}(r) = \frac{1 - \mathscr{R}_S/r}{f^2(r)} \quad (12.9)$$

where we made the *ansatz* that f depends only on r. It is also convenient to introduce a new radial coordinate

$$\xi = r + \mathscr{R}_S \times \ln\left|\frac{r}{\mathscr{R}_S} - 1\right| \quad (12.10)$$

With the definitions given in Eq.(12.9) and Eq.(12.10), we can rewrite Eq.(12.8) in a more compact way

$$\left[\left(\partial v/\partial x^0\right)^2 - \left(\partial u/\partial x^0\right)^2\right] = \mathscr{F}(\xi) \quad (12.11a)$$

$$\left[(\partial v/\partial \xi)^2 - (\partial u/\partial \xi)^2\right] = -\mathscr{F}(\xi) \quad (12.11b)$$

$$(\partial v/\partial x^0)(\partial v/\partial \xi) = (\partial u/\partial x^0)(\partial u/\partial \xi) \quad (12.11c)$$

We can eliminate the still unknown function $\mathscr{F}$ by adding Eq.(12.11a) and Eq.(12.11b). Then, let's use the resulting expression two times: the first to add and the second to subtract twice Eq.(12.11c). We then obtain

$$\left(\partial v/\partial x^0 + \partial v/\partial \xi\right)^2 = \left(\partial u/\partial x^0 + \partial u/\partial \xi\right)^2 \quad (12.12a)$$

$$\left(\partial v/\partial x^0 - \partial v/\partial \xi\right)^2 = \left(\partial u/\partial x^0 - \partial u/\partial \xi\right)^2 \quad (12.12b)$$

The square roots of these expressions yield

$$\partial v/\partial x^0 + \partial v/\partial \xi = +\left(\partial u/\partial x^0 + \partial u/\partial \xi\right) \quad (12.13a)$$

$$\partial v/\partial x^0 - \partial v/\partial \xi = -\left(\partial u/\partial x^0 - \partial u/\partial \xi\right) \quad (12.13b)$$

Note that the signs of the roots on the *lhs* and *rhs* of Eq.(12.13a) are the same, unlike what happens for the root of Eq.(12.12b), where we have chosen the plus and the minus sign for the *lhs* and *rhs*, respectively. This is a necessary choice if we want the determinant of the Jacobian matrix different from zero. In fact, by adding and subtracting Eq.(12.13a) Eq.(12.13b), we get

$$\partial v/\partial x^0 = \partial u/\partial \xi; \qquad \partial v/\partial \xi = \partial u/\partial x^0 \quad (12.14)$$

that indeed imply $\det(J^{\alpha}{}_{\mu}) \neq 0$ [*c.f.* Eq.(12.7)]. Let's now differentiate each of the previous equations first *w.r.t.* x^0 and then *w.r.t.* ξ. After combining the results, we can write two wave equations for both v and u

$$\partial^2 v/\partial x^{02} - \partial^2 v/\partial \xi^2 = 0; \qquad \partial^2 u/\partial x^{02} - \partial^2 u/\partial \xi^2 = 0 \tag{12.15}$$

with solutions

$$v = h(\xi + x^0) + g(\xi - x^0); \qquad u = h(\xi + x^0) - g(\xi - x^0) \tag{12.16}$$

In order to find the unknown h and g functions, let's substitute Eq.(12.16) back in Eq.(12.11a). Indicating with a prime (') a derivative *w.r.t.* the whole argument of the h and g functions, we obtain

$$\left[h'(\xi + x^0) - g'(\xi - x^0)\right]^2 - \left[h'(\xi + x^0) + g'(\xi - x^0)\right]^2 = \mathscr{F}(\xi) \tag{12.17}$$

that is

$$\boxed{-4h'(\xi + x^0)g'(\xi - x^0) = \mathscr{F}(\xi)} \tag{12.18}$$

Note that we would have obtained exactly the same results if we had substituted Eq.(12.16) in Eq.(12.11b). Note also that Eq.(12.11c) is automatically satisfied by Eq.(12.16). Let's now differentiate Eq.(12.18) *w.r.t.* ξ and divide by the same equation: we get

$$\frac{h''(\xi + x^0)}{h'(\xi + x^0)} + \frac{g''(\xi - x^0)}{g'(\xi - x^0)} = \frac{\mathscr{F}'(\xi)}{\mathscr{F}(\xi)} \tag{12.19}$$

Likewise, let's differentiate Eq.(12.18) *w.r.t.* x_0 and divide by the same equation: we now obtain

$$\frac{h''(\xi + x^0)}{h'(\xi + x^0)} - \frac{g''(\xi - x^0)}{g'(\xi - x^0)} = 0 \tag{12.20}$$

Eq.(12.19) and Eq.(12.20) yield

$$\boxed{\left[\ln \mathscr{F}(\xi)\right]' = 2\left[\ln h'(\xi + x^0)\right]'} \tag{12.21}$$

12.4 KRUSKAL-SZEKERES VS. SCHWARZSCHILD COORDINATES

Eq.(12.18) and Eq.(12.21) allow us to find the explicit dependence of the Kruskal-Szekeres coordinates from the Schwarzschild ones. To do so, let's discuss separately what happens on scales larger and smaller than the Schwarzschild radius, $\mathscr{R}_S$.

• OUTSIDE THE SCHWARZSCHILD SPHERE

For $r > \mathscr{R}_S$, the function $\mathscr{F}$ is positive definite [*c.f.* Eq.(12.9)]. Then, because of Eq.(12.18), h' and g' must have opposite sign. The ' (prime) sign indicates a derivative *w.r.t.* the whole argument of the unknown functions h, g, and $\mathscr{F}$. So, we can consider ξ and $\xi + x^0$ (or $\xi - x^0$) as two independent variables. If this the case,

Eq.(12.21) is satisfied *if and only if* the *lhs* and the *rhs* are equal to the same constant, η say. Moreover, Eq.(12.21) suggests to look for exponential solutions. This is indeed the case, as the solutions of Eq.(12.18) can be written in the following way

$$h(\xi + x^0) = \pm \frac{e^{\eta(\xi + x^0)}}{2}; \quad g(\xi - x^0) = \mp \frac{e^{\eta(\xi - x^0)}}{2}; \quad \mathscr{F}(\xi) = \eta^2 e^{2\eta\xi}; \quad (12.22)$$

The notation of Eq.(12.22) highlights that the h and g functions have opposite signs. It is easy to verify that Eq.(12.22) provides solutions also for Eq.(12.20) and Eq.(12.21). For $r > \mathscr{R}_S$, Eq.(12.10) yields $e^{\xi} = (r/\mathscr{R}_S - 1)^{\mathscr{R}_S} e^r$. Then, by using Eq.(12.9) and Eq.(12.16), we can now write

$$v = \pm \left(\frac{r}{\mathscr{R}_S} - 1\right)^{\mathscr{R}_S \eta} e^{\eta r} \sinh(\eta x^0); \qquad u = \pm \left(\frac{r}{\mathscr{R}_S} - 1\right)^{\mathscr{R}_S \eta} e^{\eta r} \cosh(\eta x^0) \quad (12.23)$$

This shows that *both* the v and u functions must be chosen with the *same* sign, either positive or negative. Also, the last of Eq.(12.22) provides

$$f^2(r) = \frac{\mathscr{R}_S}{\eta^2 r} \left(\frac{r}{\mathscr{R}_S} - 1\right)^{1 - 2\mathscr{R}_S \eta} e^{-2\eta r} \quad (12.24)$$

The constant η is chosen to ensure the regularity of the conformal-factor f [*c.f.* Eq.(12.24)] in $r = \mathscr{R}_S$. This clearly requires $\eta = (2\mathscr{R}_S)^{-1}$. Given this, we can finally write

$$\boxed{v = \pm\sqrt{\rho - 1}\, e^{\rho/2} \sinh\tau;} \quad \boxed{u = \pm\sqrt{\rho - 1}\, e^{\rho/2} \cosh\tau;} \quad \boxed{f^2 = \frac{4\mathscr{R}_S^2}{\rho} e^{-\rho}} \quad (12.25)$$

where let's repeat it again, $\rho = r/\mathscr{R}_S$ and $\tau = ct/(2\mathscr{R}_S)$. These equations provide the explicit dependence of the v, u, and f functions on the Schwarzschild coordinates x^0 and r. Note that the conformal-factor is divergent for $\rho = 0$, consistently with what already found with the Kretschmann scalar [*c.f.* Eq.(12.2)].

• INSIDE THE SCHWARZSCHILD SPHERE

For $r < \mathscr{R}_S$, the function $\mathscr{F}$ is negative defined [*c.f.* Eq.(12.9)]. Then, h' and g' must have the same sign and be either both positive or both negative [*c.f.* Eq.(12.18)]. Again, we can consider ξ and $\xi + x^0$ (or $\xi - x^0$) as independent variables. Then, Eq.(12.21) is satisfied *if and only if* its *lhs* and *rhs* are equal to the same constant, call it η again. Eq.(12.21) suggests to look again for exponential solutions. Then, it is easy to verify that the following expressions

$$h(\xi + x^0) = \pm \frac{e^{\eta(\xi + x^0)}}{2}; \quad g(\xi - x^0) = \pm \frac{e^{\eta(\xi - x^0)}}{2}; \quad F(\xi) = -\eta^2 e^{2\eta\xi}; \quad (12.26)$$

are solutions of Eq.(12.19), Eq.(12.20), and Eq.(12.21). The notation of Eq.(12.26) requires to take for *both* the h and g functions either the plus or the minus sign. When

$r < \mathscr{R}_S$, Eq.(12.10) yields $e^{\xi} = (1 - r/\mathscr{R}_S)^{\mathscr{R}_S} e^{r}$. Then,

$$v = \pm\left(1 - \frac{r}{\mathscr{R}_S}\right)^{\mathscr{R}_S \eta} e^{\eta r} \cosh(\eta x^0); \tag{12.27a}$$

$$u = \pm\left(1 - \frac{r}{\mathscr{R}_S}\right)^{\mathscr{R}_S \eta} e^{\eta r} \sinh(\eta x^0) \tag{12.27b}$$

$$f^2 = \frac{\mathscr{R}_S}{\eta^2 r}\left(1 - \frac{r}{\mathscr{R}_S}\right)^{1-2\mathscr{R}_S \eta} e^{2\eta r} \tag{12.27c}$$

Let's stress once more that the v and u functions must have the same sign, either positive or negative. Again, the constant η must be chosen by requiring the regularity of the conformal factor at $r = \mathscr{R}_S$. Note that this yields the same value used in Section 12.4, $\eta = (2\mathscr{R}_S)^{-1}$. Then, for $r < \mathscr{R}_S$, the Schwarzschild coordinates map into the Kruskal-Szekeres coordinates as indicated in the following expressions

$$\boxed{v = \pm\sqrt{1-\rho}\, e^{\rho/2} \cosh\tau;} \quad \boxed{u = \pm\sqrt{1-\rho}\, e^{\rho/2} \sinh\tau;} \quad \boxed{f^2 = \frac{4\mathscr{R}_S^2}{\rho} e^{-\rho}} \tag{12.28}$$

where again, $\rho = r/\mathscr{R}_S$ and $\tau = ct/(2\mathscr{R}_S)$.

• CONCLUSION

Eq.(12.25) and Eq.(12.28) show that the conformal factor has the same expression for both $\rho > 1$ and $\rho < 1$. Thus, on the basis of the above consideration and discussion, we can now write Eq.(12.6) in its final form

$$ds^2 = \frac{4\mathscr{R}_S^2}{\rho} e^{-\rho} \left(dv^2 - du^2\right) - r^2 \left(d\theta^2 + \sin^2\theta d\phi^2\right) \tag{12.29}$$

This clearly shows that the singularity of the Schwarzschild metric at $r = \mathscr{R}_S$ is just a *coordinate singularity*. Thus, the Kruskal-Szekeres coordinate system has the advantage of providing a metric that well-behaves everywhere except for the only *physical singularity* of the Schwarzschild solution at $r = 0$.

12.5 THE KRUSKAL-SZEKERES PLANE

Consider now the Kruskal-Szekeres $u - v$ plane and write both Eq.(12.25) and Eq.(12.28) in the following form:

$$\rho > 1 \quad \Rightarrow \quad \boxed{u^2 - v^2 = (\rho - 1)e^{\rho};} \quad \boxed{v = u \tanh\tau} \tag{12.30a}$$

$$\rho < 1 \quad \Rightarrow \quad \boxed{v^2 - u^2 = (1 - \rho)e^{\rho};} \quad \boxed{u = v \tanh\tau} \tag{12.30b}$$

This clearly shows the consistency of these two equations once the roles of v and u have been swapped. As we will see, this will ensure continuity in passing from the exterior to the interior of the Schwarzschild sphere.

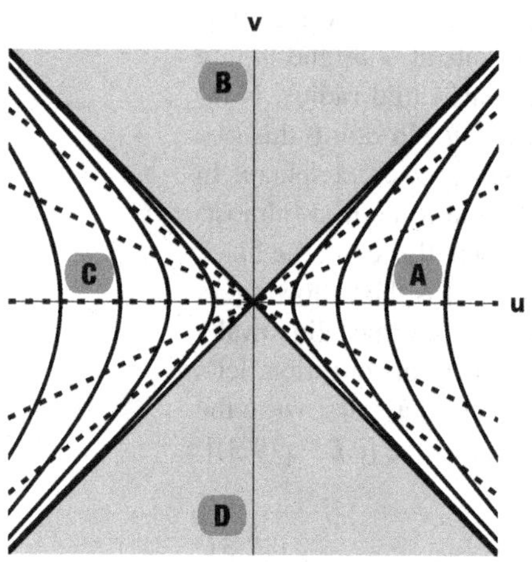

The Kruskal-Szekeres plane. For $\rho > 1$, the gray (B and D) regions are forbid-
th the same Schwarzschild radial and time coordinate, ρ and τ, are hyperbolae
es) and straight (dashed) lines, see text.

THE SCHWARZSCHILD SPHERE

e, $\rho > 1$, Eq.(12.30a)] holds. Then, we must have either $u > |v|$ or
se two regions of the Kruskal-Szekeres plane are labeled A and C in
The two gray areas, labeled B and D in the same Figure, do not fulfill
equirements and then are forbidden regions when $\rho > 1$.

)] also shows that events with the same Schwarzschild radial coordi-
apped into hyperbolae that intersect the u-axis. As shown in Figure
ing ρ pushes the hyperbolae toward their asymptotes, the bisectors
plane, while their vertexes get nearer and nearer to the origin of the
eres coordinate system. Thus, events with $\rho \simeq 1$ are mapped into the
of the bisectors, $v = \pm u$, delimiting the A and C regions. Note that for
perbola degenerates to the origin of the $u - v$ plane [*c.f.* Eq.(12.25)]
h the same value of the Schwarzschild time coordinate τ are mapped
nes, with angular coefficient $\tanh\tau$ [*c.f.* Eq.(12.30a)]. It follows that the
s all the events with a Schwarzschild time coordinate $\tau = 0$. For pos-
e) t values, the angular coefficient increases (decreases) progressively
2.2]. Note that for $t \to \infty$ ($t \to -\infty$), $v = u$ ($v = -u$). Thus, the two bi-
iting the A and C regions of the $u - v$ plane are lines corresponding to
$= \infty$.

vent of Schwarzschild coordinates τ and ρ is then mapped in the A or
e $u - v$ plane at the intersection point between the hyperbola of given
ight line of given τ [*c.f.* Figure 12.2].

EDDING" PROCEDURE

essing that in spite of their obvious symmetry, the A and C regions are
intrinsic and fundamental sense. To discuss this point, let's consider

vents that occur at $v = 0$ in $-|u|$ and $|u|$. By construction, $v = 0$ corresponds Schwarzschild coordinate $\tau = 0$, whereas both the events at $-|u|$ and $|u|$ are cterized by being at the same distance from the Schwarzschild radius. Now, an imagine to keep $v = 0$ and to progressively increase $|u|$. In doing this, we lentifying events further and further away from the Schwarzschild sphere. In ular, for $u \to -\infty$, we are far enough from the black hole to be in a (almost) owski space-time. The same occurs when $u \to \infty$ as, also in this case, the black is at an infinite distance. Now the question to be answered is the following: is Iinkowski space-time of an observer with $v = 0$ and $u = \infty$ the same Minkowsky -time of an observer with $v = 0$ but $u = -\infty$? To answer to this question, let's v the same "embedding" procedure used in Section 9.3. So, let's first write the rface ($t = const$ and $\theta = \pi/2$) of the Schwarzschild space-time [*c.f.* Eq.(9.3)]

$$d\sigma^2 = \left(1 - \frac{\mathcal{R}_S}{r}\right)^{-1} dr^2 + r^2 d\phi^2 \tag{12.31}$$

embed" it into a 3D Euclidean space with cylindrical coordinates

$$d\ell^2 = d\bar{r}^2 + \bar{r}^2 d\bar{\phi}^2 + d\bar{z}^2 \tag{12.32}$$

one in Section 9.3, we equate Eq.(12.31) and Eq.(12.32) to conclude that $\bar{r} = r$ $\bar{\phi} = \phi$. Then, by using again Eq.(9.5), we get

$$\left(\frac{\partial \bar{z}}{\partial r}\right) = \pm\sqrt{\left(1 - \frac{\mathcal{R}_S}{r}\right)^{-1} - 1} \tag{12.33}$$

rating Eq.(12.34) leads to explicitly write the function $\bar{z}(r)$

$$\boxed{\bar{z}(r) = \pm\sqrt{4\mathcal{R}_S(r - \mathcal{R}_S)}} \tag{12.34}$$

that we are considering here both the positive and negative values of $\bar{z}(r)$, un- what was done in Eq.(9.7). The 2D surface described by the metric given in 12.31) is shown in Figure 12.3. The embedding procedure shows that the geom- of the 2D equatorial ($\theta = \pi/2$) slice of the 3D Schwarzschild hypersurface at consists of *two*, distinct, aymptotically flat universes connected by a bridge or *hole*. Such a bridge is the so-called *Einstein-Rosen bridge*.

), the condition of being outside the Schwarzschild sphere provides two differ- olutions for the Kruskal-Szekeres coordinates, one with the plus and the other the minus sign [*c.f.* Eq.(12.34)], corresponding, *e.g.*, to the A and C portions of $-v$ plane [see Figure 12.2]. This is an important point: the Kruskal-Szekeres dinates provide the maximal analytical extension of the Schwarzschild solution, atter covering only one half of a broader domain.

must also be stressed that the surface shown Figure 12.3 is a sort of snap- at Schwarzschild coordinate $\tau = const$. However, on the basis of the Jebsen- hoff theorem, we can say that the geometry outside the Schwarzschild radius

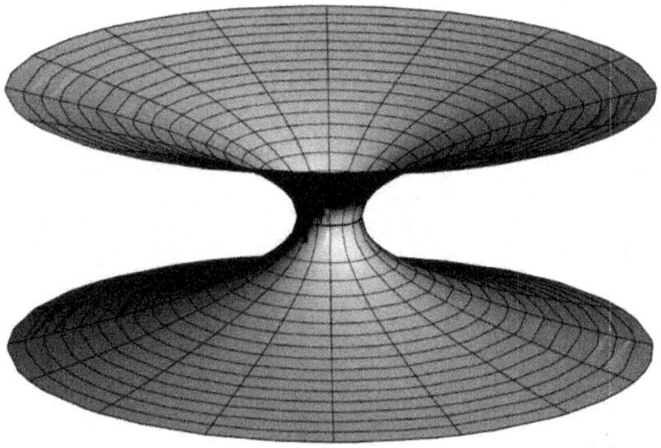

The 2D surface at $\tau = const$ and $\theta = \pi/2$ is embedded in a 3D Euclidean space. ...er) portion of this surface corresponds to $\bar{z} > 0$ ($\bar{z} < 0$) and to the A (C) region So, the A and C regions have two distinct asymptotic Minkowski space-times.

...t on time. On the other hand, the existence and the dimension of the ...n bridge is subjected on how this surface evolve on scales less than the ...d radius.

...E SCHWARZSCHILD SPHERE

...(12.30b) requires $v > |u|$ and $v < -|u|$. This excludes the A and C re... ...ed above [see Figure 12.4]. In the allowed regions, B and D, events with ... still mapped into hyperbolae that now intersect the v-axis. Note that ...d singularity at $\rho = 0$ is mapped into two hyperbolae, $v = \sqrt{u^2+1}$ and ... It follows that the two dark gray regions of Figure 12.4—those above ...ese two (dashed) hyperbolae—are forbidden region of the Kruskal-...e. Then, when $\rho < 1$, the only allowed regions of the Kruskal-Szekeres ...e the white portion of the B and D regions [*c.f.* Figure 12.4]. In spite ...netry, the B and D regions are again profoundly different. First, the D ...valent to the B region but with the time inverted. So, the singularity in ... $v = -\sqrt{u^2+1}$ occurred in the past. This is the so-called *white hole*. ...erence between a white and a black hole is that matter escape from the ...ut being trapped as in the latter. Also, the event horizon of a white hole ...ry rather than the exit of matter (we cannot enter into something the ... past), while for a black hole is exactly the contrary.

...eases from 0 to unity, the hyperbolae get nearer and nearer to their ...while the vertexes progressively approach the origin of the $u - v$ plane. ...vents with $\rho \simeq 1$ are again mapped into the two bisectors, $v = \pm u$, de... ... and D regions. Note that the functional form of the hyperbolae given ... is the same for ρ greater or smaller than unity. This is an important fea... ...uskal-Szekeres coordinate system. In fact, it is this feature that ensures ...nd, most of all, continuity in the treatment of the two regimes: $\rho < 1$ ...deed, as $\rho \to 1$, the bisectors $v = u$ and $v = -u$ are approached by both ...given in Eq.(12.30), although from opposite sides.

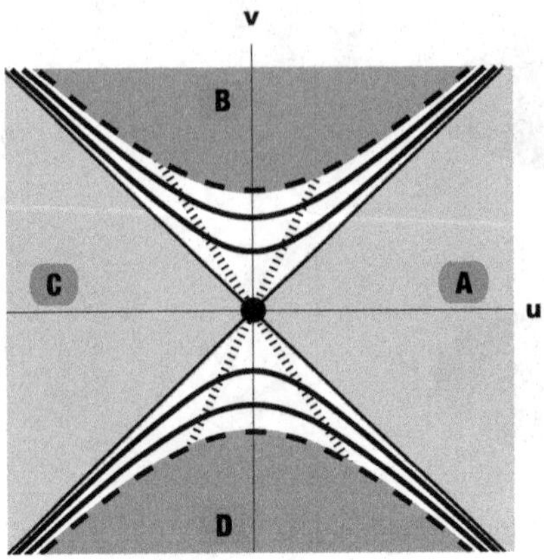

e 12.4 For $\rho < 1$, the light gray, A and C, regions are forbidden. The hyperbolae (con-
ıs lines) identify events with the same radial coordinate $\rho < 1$. The straight (dotted) lines
fy events with the same time coordinate τ. The dark gray region, above and below the
dashed) hyperbolae representing $\rho = 0$, is also forbidden regions. The white portion is
allowed for $\rho < 1$.

given value of the Schwarzschild time coordinate τ identifies straight lines of
$- v$ plane with angular coefficient $1/\tanh\tau$. For positive (negative) values of
e angular coefficient of the corresponding straight lines decreases (increases)
essively. Note that for $t \to \pm\infty$, these straight lines become bisectors of the
plane. In particular, the bisector corresponding to $\tau = +\infty$ separates the A and
gions, as well as the C and D ones. Likewise, the bisector corresponding to
$-\infty$ separates the B and C regions, as well as the D and A ones [see Figure 12.4]

THE SCHWARZSCHILD BLACK HOLE

important to stress that the metric given in Eq.(12.29) is an exact mathemat-
solution of the Einstein field equations. The question about the physical rele-
e of white holes and wormholes remain open. In any case, remember that the
varzschild solution is valid in the "vacuum". So, in real life, the world line of
int on the surface of a collapsing star can give the boundary of the meaning-
ortion of the Kruskal-Szekeres plane. The region on the right of this world line
ifies the "vacuum", whereas the region on the left of the worldline requires the
ion of the field equation in the presence of matter. This point, addressed by
varzschild in 1916, will be discussed later in the book, in Section 14.2. So, it
out that the A and B regions are of more direct physical interest to discuss the
varzschild black hole.

ote that, by construction, the bisectors of the $u - v$ plane describe the light cone.
g conformally flat coordinates allows to draw a strict analogy with Special Rel-
ty. In particular, it makes sense to assign to the Kruskal-Szekeres u and v coordi-

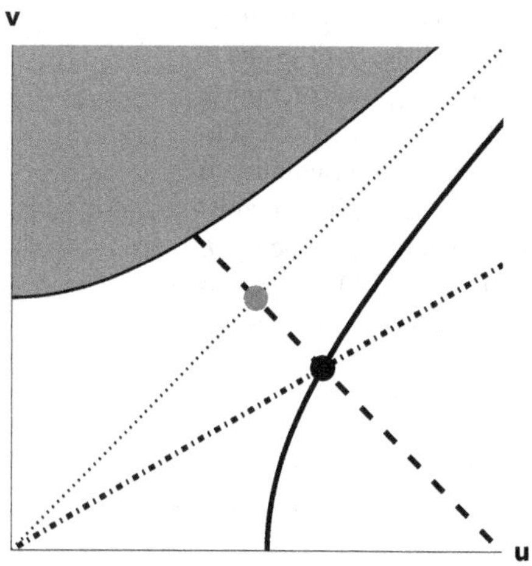

-ight traveling inwards moves along the dashed line, a bisector of the Kruskal-
:. The black dot is at the intersection of the hyperbola (continuous black
g the position, ρ, with the straight (dot-dashed line) identifying the time, τ.
reaches $\rho = 1$, the (dotted) bisector of the Kruskal-Szekeres plane, enters the
surface and is captured by the central body.

of "*radial*" and "*time*" markers. Note that because of Eq.(12.29), both aintain their sign when we cross the horizon. So, the v and u directions ne- and space-like. This is in contrast with the case of Schwarzschild vhere ρ and τ are space- and time-like (time- and space-like) outside orizon.

nother point that must be stressed. The metric given in Eq.(12.6) is *lat*. It follows that the light cones have an opening of 90° with their arallel to the v-axis. It follows that the world-line of test particles must : inside these cones. In other words, free-falling test-particles in the A rced to move vertically and not horizontally. In so doing, they cannot ılarity in the B region. Let's consider light traveling inward toward the metry of the Schwarzschild geometry. As seen in Section 9.7, light is ıe central body if the impact parameter b is less than $3\sqrt{3}\mathscr{R}_S/2$. This ived on the basis of the Schwarzschild solution and, then, for radial dis-han $\mathscr{R}_S$. However, the use of Kruskal-Szekeres coordinates allows us to ısiderations of Section 9.7 to the case $\rho < 1$. Light coming from infin-ıching the central body is described, for example, by the dashed line of n this specific case, light ray arrives in $r = 1.3\mathscr{R}_S$ at a coordinate time : this event is marked with a black bullet point (see Figure 12.5). When ntersect the bisector $v = u$, the Schwarzschild coordinate time is $t = \infty$ ɔullet point in Figure 12.5). Note that when we move further along the oward $r = 0$), both the Schwarzschild radial and the time coordinates Section 12.5]. This underlines the inadequacy of the Schwarzschild te to play the role of time marker inside the Schwarzschild sphere

a very analogous way, we can study the complementary situation of light mov-
adially outwards *w.r.t.* to central body. As light moves outwards, both ρ and τ
ase. When light intersects the bisector $v = u$ (the gray bullet point of Figure
, the Schwarzschild coordinate time is formally $t = \infty$. On the basis of the dis-
on of Section 7.5, we know that an observer at infinity measures a gravitational
ift of the emitted radiation. Since $g_{00}(\mathscr{R}_S) = 0$ and $g_{00}(\infty) = 1$, a photon emit-
om the Schwarzschild sphere will be observed at infinity at zero frequency [*c.f.*
.9)]—that is, the photon cannot be observed. So, the Schwarzschild sphere ab-
the radially incoming radiation without emitting none: this is why, as already
pated, a black hole is black.

13 Field Equations in Non-"Empty" Space-Times

13.1 INTRODUCTION

The field equations used up to now have provided the tool for studying the geometry of the space-time in the "vacuum". It was possible to resolve them analytically, to find the Schwarzschild solution, and, most importantly, to observationally test the General Relativity predictions on planetary scales. We want now to extend this line of reasoning to find the geometry of space-time not *outside*, but rather *within* a given matter/energy distribution. The goal of this chapter is then to write the field equations of General Relativity not in the "vacuum", but rather in non-*empty* region of the space.

13.2 FIELD EQUATIONS: REQUIREMENTS

As discussed in Section 8.4, the field equations in the "vacuum" are provided by the vanishing of the Einstein tensor. In the weak field limit, these equations contain the Laplace equation. In the same limit, but in the presence of matter, we have rather to deal with the Poisson equation, $\Delta U = 4\pi G\rho$. This suggests that the field equations should be obtained by equating the Einstein tensor, describing the geometry of space-time, to a rank-two tensor, describing the properties of the matter/radiation energy.

To find how to covariantly describe these properties, consider the simple case of a Minkowski space-time, and let's choose two inertial reference frames: the lab. frame $\mathcal{K}$, and a frame $\mathcal{K}'$ uniformly sliding *w.r.t.* $\mathcal{K}$ along its x-axis at constant velocity V. Consider now a cloud of particles at rest in $\mathcal{K}'$. If m_0 and n_0 are their mass and number density, then the cloud proper mass density is given by $\rho_0 = m_0 n_0$. However, an observer in the lab. frame would measure a mass density

$$\rho = \rho_0 \gamma^2 \tag{13.1}$$

The γ^2 appears because in the lab. frame: i) the mass of the particles increases ($m_0 \to m_0\gamma$); ii) the volume containing the cloud decreases because of the length contraction along the x-axis ($V = V_0/\gamma$). Thus, Eq.(13.1) clearly shows that the mass density is neither an invariant nor the component of a four-vector. In the last case, Eq.(B2.1.a) and Eq.(2.16) would only yield a linear dependence on γ. So, to have a quadratic dependence on the Lorentz factor, we have to conclude that the mass density must necessarily be the component of a rank-two tensor. Let's restrict to the case of a perfect fluid, with no viscosity or heat conduction. It follows that the quantities useful to construct a rank-two tensor are i) the fluid proper density and pressure, ρ_0 and p_0; ii) the fluid four-velocity, u^μ; iii) the metric tensor, $\eta_{\alpha\beta}$. Let's then consider the

DOI: 10.1201/9781003141259-13

following combination, which clearly provides a rank two, symmetric tensor

$$T^{\mu\nu} = \rho_0 c^2 u^\mu u^\nu + p_0 \left[u^\mu u^\nu - \eta^{\mu\nu}\right] \tag{13.2}$$

For a fluid with a vanishing pressure, Eq.(13.2) yields

$$T^{00} = \rho_0 c^2 \gamma^2 = \rho c^2 \tag{13.3}$$

Thus, the time-time component of the $T^{\mu\nu}$ tensor provides the relativistic energy density of the fluid, containing the contributions coming from both the rest energy and the motion of the fluid, in full agreement with Eq.(13.1).

13.3 CONSERVATION LAWS FOR A RELATIVISTIC FLUID

To verify that the whole combination given in Eq.(13.2) makes sense at all, let's consider its non-relativistic limit: $p_0 \ll \rho_0 c^2$; $u^\mu \simeq \{1, \mathrm{v}^k/c\}$; $(\mathrm{v}^k/c)^2 \ll 1$. In this limit,

$$T^{00} \simeq \rho_0 c^2; \qquad T^{0i} \simeq \rho_0 c \mathrm{v}^i; \qquad T^{ik} \simeq \rho_0 \mathrm{v}^i \mathrm{v}^k + p_0 \delta^{ik} \tag{13.4}$$

where δ^{ik} is the Kronecker symbol. Let's now impose the vanishing of the four-divergence of the energy-momentum tensor: $T^{\mu\nu}{}_{,\nu} = 0$. The time-component of this equation provides

$$\frac{\partial}{\partial x^0}(\rho_0 c^2) + \frac{\partial}{\partial x^i}(\rho_0 c \mathrm{v}^i) = 0 \tag{13.5}$$

that is, the familiar continuity equation of classical mechanics. On a similar line, $T^{i\nu}{}_{,\nu} = 0$ provides

$$\rho_0 c \frac{\partial}{\partial t} \mathrm{v}^i + \rho_0 c \left(\mathrm{v}^k \frac{\partial}{\partial x^k}\right) \mathrm{v}^i + \frac{\partial p_0}{\partial x^i} = 0 \tag{13.6}$$

that is, the classical Euler equation of fluid dynamics. Thus, we state the following.

THE ENERGY-MOMENTUM TENSOR IN SPECIAL RELATIVITY

The properties of a perfect fluid are described by its *energy-momentum tensor*

$$\boxed{T^{\mu\nu} = \left(\rho_0 c^2 + p_0\right) u^\mu u^\nu - p_0 \eta^{\mu\nu}} \tag{13.7}$$

The vanishing of its divergence

$$\boxed{T^{\mu\nu}{}_{,\nu} = 0} \tag{13.8}$$

provides equations that contain the classical continuity and Euler equations.

13.4 THE MATTER ENERGY-MOMENTUM TENSOR

Let's now extend the discussion of the previous section to the more general case of pseudo-Riemannian, curved space-times. On the basis of the Principle of General

Covariance [*c.f.* Section 4.7], all the physical laws must have the same form in all reference frames. This is why we want to write them in terms of tensor equations. Reversing the order of reasoning, we can state that if a tensor equation is valid in the absence of gravity, *e.g.* in the local inertial reference frame, then it must also be valid in the presence of *any* arbitrary gravitational field. Now, Eq.(13.7) and Eq.(13.8) were derived in the framework of Special Relativity or, equivalently, in the local inertial reference frame where the metric reduces to the Minkowski one and the covariant derivatives are just ordinary ones. However, in a generic frame, $\eta_{\mu\nu} \to g_{\mu\nu}(x^\tau)$ and ordinary derivatives become covariant derivatives. So, on the basis of general covariance principle, it is natural to generalize Eq.(13.7) and Eq.(13.8) as follows.

THE MATTER ENERGY-MOMENTUM TENSOR

The covariant expression of the *energy-momentum tensor* is given by

$$\boxed{T^{\mu\nu} = \left(\rho_0 c^2 + p\right) u^\mu u^\nu - p g^{\mu\nu}} \tag{13.9}$$

The covariant expression of its four-divergence must be written as follows

$$\boxed{T^{\mu\nu}{}_{;\nu} = 0} \tag{13.10}$$

Eq.(13.10) can be conveniently rewritten in the following way [see Exercise A.38],

$$\frac{1}{\sqrt{-g}} \frac{\partial}{\partial x^\sigma} \left(\sqrt{-g} T^{\mu\sigma}\right) + \Gamma^\mu_{\sigma\nu} T^{\nu\sigma} = 0 \tag{13.11}$$

Because of the last term, Eq.(13.11) cannot be technically called a conservation equation. So, it could be more appropriate to refer to Eq.(13.11) as to the equations describing the properties and the motion of matter that moves in a curved space-time.

To better see this point, let's consider a "cold" cloud of particles that in the interval $t_1 < t < t_2$ remain confined inside a four-box of proper volume $V_p^{(4)}$ [see Figure 13.1]. By construction, there is not a net flux of particles through the "lateral" hypersurfaces of the four-box. Thus, on those hypersurfaces, the energy-momentum tensor can be safely be set to zero.

The infinitesimal proper and coordinate four-volume elements are related: $dV_p^{(4)} = \sqrt{-g}\, dV^{(4)}$ [*c.f.* Eq.(8.22)]. It follows that the integral of Eq.(13.11) over an arbitrary proper four-volume can be written as follows:

$$\int_V \frac{1}{\sqrt{-g}} \frac{\partial}{\partial x^\sigma} \left(\sqrt{-g} T^{\mu\sigma}\right) \sqrt{-g}\, dV^{(4)} + \int_V \Gamma^\mu_{\sigma\nu} T^{\nu\sigma} \sqrt{-g}\, dV^{(4)} = 0 \tag{13.12}$$

The first integral can be easily evaluated by applying the divergence theorem. In fact, the only non-vanishing flux of particles has to do with the entrance and exit from "below" and from "above", *i.e.* through the lower and upper spatial hypersurface of the four-box at t_1 and t_2 [see Figure 13.1]. Then,

$$\int \frac{\partial}{\partial x^\sigma} \left(\sqrt{-g} T^{\mu\sigma}\right) d^4V = \int \sqrt{-g} T^{\mu 0} d^3x \Big|_{ct_2} - \int \sqrt{-g} T^{\mu 0} d^3x \Big|_{ct_1} \tag{13.13}$$

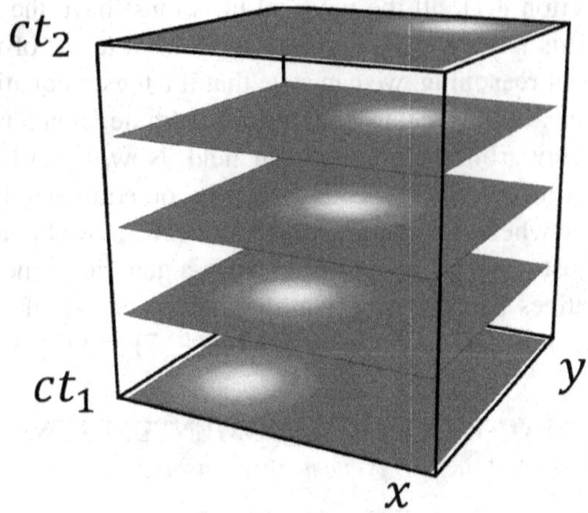

Figure 13.1 The motion of a "cold" cloud of particles in space-time. The density of the cloud is shown as contour plots in the hyperplanes at constant t: from white (highest density) to dark gray (zero density). For $t_1 < t < t_2$, the cloud is always inside the chosen four-box.

Note that for a cold cloud ($p \simeq 0$) $T^{\mu 0} = \rho_0 c^2 (dx^\mu/ds)(dx^0/ds)$. Also, proper and coordinate four-volumes are related: $dV_p^{(4)} = c\,d\tau dV_p^{(3)} \equiv ds\,dV_p^{(3)} = \sqrt{-g}dx^0 d^3x$, where $d\tau$ and $dV_p^{(3)}$ are the infinitesimal proper time interval and proper volume element. Taking all this into account, we can rewrite Eq.(13.13) as follows

$$\int_{x_0=ct_2} dV_p^{(3)} \rho_0 c^2 \frac{dx^\mu}{ds} - \int_{x_0=ct_1} dV_p^{(3)} \rho_0 c^2 \frac{dx^\mu}{ds} \tag{13.14}$$

For a "cold" cloud, the individual particle four-velocities basically coincide with the bulk four-velocity of the cloud itself. Then, they are not expected to vary on the two spatial hypersurfaces at t_1 and t_2. If this is the case, the integral over $dV_p^{(3)}$ involves only the proper density and not the four-velocities. If we define the proper mass of the cloud in the standard way, $M = \int \rho_0 dV_p^{(3)}$, Eq.(13.14) provides

$$Mc^2 \left[\left.\frac{dx^\mu}{ds}\right|_{ct_2} - \left.\frac{dx^\mu}{ds}\right|_{ct_1} \right] = Mc^2 \int_{x_0=ct_1}^{x_0=ct_2} \frac{d^2x^\mu}{ds^2} ds \tag{13.15}$$

On the basis of similar arguments, the second term of Eq.(13.8) becomes:

$$\int \Gamma^\mu_{\sigma\nu} \left(\rho_0 c^2 \frac{dx^\sigma}{ds} \frac{dx^\nu}{ds} \right) ds\,dV_p^{(3)} = Mc^2 \int \Gamma^\mu_{\sigma\nu} \frac{dx^\sigma}{ds} \frac{dx^\nu}{ds} ds \tag{13.16}$$

Combining Eq.(13.15) and Eq.(13.16) shows that Eq.(13.11) yields, in this simple pressureless ($p \simeq 0$) configuration, the geodesic equations:

$$\frac{d^2x^\mu}{ds^2} + \Gamma^\mu_{\sigma\nu} \frac{dx^\sigma}{ds} \frac{dx^\nu}{ds} = 0 \tag{13.17}$$

This is not unexpected: a fluid element subjected only to an external gravitational field moves along geodesics of the space-time.

13.5 THE EM ENERGY-MOMENTUM TENSOR

It is of interest to extend the considerations done in Section 13.3 to find an explicit expression for the energy-momentum tensor of an electromagnetic field. To do so, let's consider a charged fluid element moving in an electric and/or magnetic field. As it is well known, that element will move according to the following equation

$$\rho_0 \left[\frac{\partial \vec{v}}{\partial t} + \left(\vec{v} \cdot \vec{\nabla}\right) \vec{v}\right] = \sigma_0 \left[\vec{E} + \frac{1}{c}\left(\vec{v} \times B\right)\right] \tag{13.18}$$

Here ρ_0 and σ_0 are the matter and electric charge proper densities, whereas $\vec{v}$ is the velocity of the fluid element. The term on the *lhs* can be interpreted as the four-divergence of the energy-momentum tensor of a perfect, pressure-less fluid [*c.f.* Eq.(13.6)]. The term on the *rhs* can be conveniently written in terms of the electromagnetic tensor, $F^{k\beta}$, and of the four-current density J_β [see Box 13.1]. In fact, Eq.(13.18) can be written in a very compact way:

$$T^{k\beta}{}_{,\beta} = -\frac{1}{c} F^{k\beta} J_\beta \tag{13.19}$$

To keep contact with the formalism of Section 13.3, let's introduce a radiation energy-momentum tensor, $S^{\alpha\beta}$ say, and write

$$\left(T^{\alpha\beta} + S^{\alpha\beta}\right)_{,\beta} = 0 \tag{13.20}$$

Consistency with Eq.(13.19) implies

$$S^{\alpha\sigma}{}_{,\sigma} = \frac{1}{c} F^{\alpha\sigma} J_\sigma = \frac{1}{4\pi} F^{\alpha\sigma} F_\sigma{}^{\mu}{}_{,\mu} = \frac{1}{4\pi} F^{\alpha\sigma} \eta_{\sigma\nu} F^{\nu\mu}{}_{,\mu} \tag{13.21}$$

where we used Eq.(B13.1.cb). Note that Eq.(13.21) clearly shows that $S^{\alpha\sigma}$ is a quadratic function of the electromagnetic tensor. Let's then consider the following expression, quadratic in $F_{\mu\nu}$

$$S^{\alpha\sigma} = \mathscr{C}_1 \left[\eta_{\mu\nu} F^{\alpha\mu} F^{\sigma\nu} + \mathscr{C}_2 \eta^{\alpha\sigma} F_{\mu\nu} F^{\mu\nu}\right] \tag{13.22}$$

BOX 13.1 EM TENSOR AND MAXWELL EQUATIONS

The electric and magnetic fields are conveniently written in terms of the components of a four-potential vector $A^\mu \equiv \{\phi, \vec{A}\}$, ϕ and $\vec{A}$ being the scalar and vector potentials: $\vec{E} = -\vec{A}_{,0} - \vec{\nabla}\phi$; $\vec{B} = \vec{\nabla} \times \vec{A}$. The rank-two, antisymmetric *electromagnetic tensor* is defined as follows:

$$F_{\mu\nu} = A_{\nu,\mu} - A_{\mu,\nu} \tag{B13.1.a}$$

In terms of the E and B components,

$$F_{\alpha\beta} = \begin{pmatrix} 0 & -E_x & -E_y & -E_z \\ E_x & 0 & B_z & -B_y \\ E_y & -B_z & 0 & B_x \\ E_z & B_y & -B_x & 0 \end{pmatrix} \tag{B13.1.b}$$

It is easy to verify that Eq.(B13.1.b) yields $F_{\mu\nu}F^{\mu\nu} = 2\left(B^2 - E^2\right)$ and $F_{\alpha\beta,\gamma} + F_{\gamma\alpha,\beta} + F_{\beta\gamma,\alpha} = 0$. The *four-current* is defined as follows: $J^\alpha = \{c\sigma, \vec{j}\}$, where σ is the density of electric charges and $\vec{j}$ is the 3D current density. The Maxwell equations can then be conveniently expressed in terms of the electromagnetic tensor and of the four-current

$$\left.\begin{array}{l} \vec{\nabla}\cdot\vec{B} = 0 \\ \vec{\nabla}\cdot B + \dfrac{1}{c}\dfrac{\partial \vec{B}}{\partial t} = 0 \end{array}\right\} \Leftrightarrow \quad F_{\alpha\beta,\gamma} + F_{\gamma\alpha,\beta} + F_{\beta\gamma,\alpha} = 0 \tag{B13.1.ca}$$

$$\left.\begin{array}{l} \vec{\nabla}\cdot\vec{E} = 4\pi\sigma \\ \vec{\nabla}\times\vec{B} = \dfrac{4\pi}{c}\vec{j} + \dfrac{1}{c}\dfrac{\partial \vec{E}}{\partial t} \end{array}\right\} \Leftrightarrow \quad F^{\alpha\beta}{}_{,\beta} = \frac{4\pi}{c}J^\alpha \tag{B13.1.cb}$$

To extend all this to curved space-time, let's apply the principle of general covariance: i) $\eta_{\alpha\beta} \to g_{\alpha\beta}(x^\tau)$; ordinary $\to$ covariant derivatives; $d^4\xi \to \sqrt{-g}d^4x$. It is easy to verify that all these operations still lead to

$$F_{\alpha\beta} \equiv A_{\beta;\alpha} - A_{\alpha;\beta} = A_{\beta,\alpha} - A_{\alpha,\beta} \tag{B13.1.da}$$

$$F_{\alpha\beta;\gamma} + F_{\gamma\alpha;\beta} + F_{\beta\gamma;\alpha} = F_{\alpha\beta,\gamma} + F_{\gamma\alpha,\beta} + F_{\beta\gamma,\alpha} = 0 \tag{B13.1.db}$$

whereas Eq.(B13.1.cb) becomes

$$F^{\alpha\beta}{}_{;\beta} \equiv \frac{1}{\sqrt{-g}}\frac{\partial(\sqrt{-g}F^{\alpha\sigma})}{\partial x^\sigma} = \frac{4\pi}{c}J^\alpha \tag{B13.1.e}$$

The two constants, $\mathscr{C}_1$ and $\mathscr{C}_2$, are determined by imposing that the four-divergence of Eq.(13.22) is consistent with Eq.(13.21). So, let's first write

$$S^{\alpha\sigma}{}_{,\sigma} = -\mathscr{C}_1\eta_{\mu\nu}F^{\alpha\mu}F^{\nu\sigma}{}_{,\sigma} + \mathscr{C}_1\left[\eta_{\mu\nu}F^{\alpha\mu}{}_{,\sigma}F^{\sigma\nu} + 2\mathscr{C}_2\eta^{\alpha\sigma}F_{\mu\nu,\sigma}F^{\mu\nu}\right] \tag{13.23}$$

The factor of two in the last term on the *rhs* comes from being $F_{\mu\nu}F^{\mu\nu}{}_{,\sigma} = F^{\mu\nu}F_{\mu\nu,\sigma}$. Let's now rewrite the squared parenthesis in Eq.(13.23)

$$\eta_{\mu\nu}\eta^{\alpha\sigma}\eta^{\mu\tau}F_{\sigma\tau,\rho}F^{\rho\nu} + 2\mathscr{C}_2\eta^{\alpha\sigma}F_{\mu\nu,\sigma}F^{\mu\nu} = \frac{1}{2}\eta^{\alpha\sigma}F^{\beta\nu}\left(2F_{\sigma\nu,\beta} + 4\mathscr{C}_2F_{\beta\nu,\sigma}\right) \tag{13.24}$$

Note that $F^{\beta\nu}F_{\sigma\nu,\beta} = -F^{\beta\nu}F_{\sigma\beta,\nu}$. Then, for $\mathscr{C}_2 = 1/4$, the round parenthesis in Eq.(13.24) vanishes because of Eq.(B13.1.ca). Thus, if $\mathscr{C}_1 = -1/(4\pi)$, Eq.(13.23) reduces to Eq.(13.21), as we wanted. At last, by exploiting again the principle of general covariance, we can state the following.

ENERGY-MOMENTUM TENSOR OF AN EM FIELD

The covariant expression of the energy-momentum tensor of an EM field is given by

$$S^{\alpha\beta} = \frac{1}{4\pi}\left[-g_{\mu\nu}F^{\alpha\mu}F^{\beta\nu} + \frac{1}{4}g^{\alpha\beta}F_{\mu\nu}F^{\mu\nu}\right] \tag{13.25}$$

where $F^{\beta\nu}$ is the EM tensor and $g_{\mu\nu}$ is the metric tensor.

13.6 AN ISOTROPIC RADIATION FIELD

Given Eq.(13.25), we can now evaluate the single components of the radiation energy-momentum tensor. We first find $F_{\mu\nu}F^{\mu\nu} = 2\left(B^2 - E^2\right)$. It follows that the quantity $B^2 - E^2$ is a Lorentz invariant. Then, for $\{\alpha,\beta\}$ equal to $\{0,\beta\}$ we get

$$\begin{aligned} S^{00} &= \frac{1}{8\pi}\left(B^2+E^2\right) = \varepsilon \\ S^{01} &= \frac{1}{4\pi}\left(B_zE_y - B_yE_z\right) = \frac{1}{c}P_x \\ S^{02} &= \frac{1}{4\pi}\left(B_xE_z - B_zE_x\right) = \frac{1}{c}P_y \\ S^{03} &= \frac{1}{4\pi}\left(B_yE_x - B_xE_y\right) = \frac{1}{c}P_z \end{aligned} \tag{13.26}$$

where ε is the energy density and $\vec{P} = c(\vec{E}\times\vec{B})/(4\pi)$ is the Poynting vector. It follows that the time component of the equation $S^{\alpha\beta}{}_{,\beta} = 0$ provides

$$\frac{\partial\varepsilon}{\partial t} + \vec{\nabla}\cdot\vec{P} = 0 \tag{13.27}$$

stating the conservations of energy. For the other components, we have, for example, for $\{\alpha,\beta\}$ equal to $\{1,k\}$

$$S^{11} = \frac{-B_x^2+B_y^2+B_z^2-E_x^2+E_y^2+E_z^2}{8\pi}; \qquad S^{12} = -\frac{B_xB_y+E_xE_y}{4\pi}$$
$$S^{13} = -\frac{B_xB_z+E_xE_z}{4\pi} \tag{13.28}$$

Let's consider an isotropic radiation field obtained by superposing incoherent plane waves that propagate in all directions with equal intensities. Let's take time averages of all the $S^{\alpha\beta}$ components over a sufficient long period of time. We will have $\langle E_k\rangle = 0$, $\langle B_k\rangle = 0$, $\langle |E_k|^2\rangle = \langle |E|^2\rangle/3$ and $\langle |B_k|^2\rangle = \langle |B|^2\rangle/3$, where $k = 1,3$ identifies the spatial components of the electric and magnetic vectors. This implies that the radiation momentum tensor becomes diagonal

$$S^{\alpha\beta} = \text{diag}\left[\varepsilon, \frac{\varepsilon}{3}, \frac{\varepsilon}{3}, \frac{\varepsilon}{3}\right] \tag{13.29}$$

On the other hand, a perfect fluid not subjected to internal forces ($\vec{\nabla} p = 0$) will be at rest *w.r.t.* a comoving frame [*c.f.* Box 11.2]. The energy-momentum tensor becomes diagonal: $T^{\alpha\beta} = \mathrm{diag}\{\varepsilon, p, p, p\}$. By comparing $T^{\alpha\beta}$ with $S^{\alpha\beta}$, we concude that an isotropic radiation field can be described by a photon gas with equation of state $p = \varepsilon/3$.

13.7 FIELD EQUATIONS INSIDE A MATTER/ENERGY DISTRIBUTION

Having now the Poisson equation in the back of our mind [*c.f.* Section 13.2], it is quite natural to ask what happens if we equate, unless of a constant, the Einstein and the energy-momentum tensors.

$$G^{\alpha\beta} = \chi T^{\alpha\beta} \tag{13.30}$$

This seems reasonable also because the covariant four-divergence of both these tensors vanishes [*c.f.* Eq.(8.21) and Eq.(13.5)]. Now, in force of Eq.(8.15), we can rewrite Eq.(13.30) in terms of the Ricci tensor, for example, in a mixed form

$$R^{\alpha}{}_{\beta} - \frac{1}{2}\delta^{\alpha}{}_{\beta} R = \chi T^{\alpha}{}_{\beta} \tag{13.31}$$

The contraction of indices provides a relation between the traces of the Ricci and the energy-momentum tensor: $R = -\chi T$. Then, Eq.(13.31) yields

$$R^{\alpha}{}_{\beta} = \chi \left[T^{\alpha}{}_{\beta} - \frac{1}{2}\delta^{\alpha}{}_{\beta} T \right] \tag{13.32}$$

Now, we know that in the weak field limit $R^{0}{}_{0} = \nabla^2 U/c^2$ [*c.f.* Eq.(8.12)]. In the same limit, $p_0 \ll \rho_0 c^2 \simeq \rho c^2$ and $u^{\mu} u_{\mu} = 1 \simeq u^0 u_0$. It follows that $T^{0}{}_{0} = T = \rho c^2$. Thus, the time-time component of Eq.(13.31) provides

$$\frac{1}{c^2}\nabla^2 U = \frac{1}{2}\chi \rho c^2 \tag{13.33}$$

So, the weak field limit of Eq.(13.32) contains the Poisson equation, provided that $\chi = 8\pi G/c^4$. We can then generalize the results of Section 8.5 to the case of non-*empty* space, including, for sake of generality, the contribution of a non-vanishing Cosmological Constant.

FIELD EQUATIONS INSIDE A MATTER/ENERGY DISTRIBUTION

The Einstein field equations in non-"empty" space can be written either in terms of the Einstein tensor

$$\boxed{G^{\alpha\beta} - \Lambda g^{\alpha\beta} = \frac{8\pi G}{c^4} T^{\alpha\beta}} \tag{13.34}$$

or, equivalently, in terms of the Ricci tensor

$$\boxed{R^{\alpha\beta} + \Lambda g^{\alpha\beta} = \frac{8\pi G}{c^4}\left[T^{\alpha\beta} - \frac{1}{2} g^{\alpha\beta} T \right]} \tag{13.35}$$

Note that we can always rewrite Eq.(13.34) as $G_{\alpha\beta} = 8\pi G\widetilde{T}_{\alpha\beta}/c^4$, defining

$$\widetilde{T}_{\alpha\beta} \equiv T_{\alpha\beta} + T^{(\Lambda)}_{\alpha\beta} = (\tilde{\rho}c^2 + \tilde{p})u_\alpha u_\beta - \tilde{p}g_{\alpha\beta} \tag{13.36}$$

Here

$$T^{(\Lambda)}_{\alpha\beta} \equiv \frac{\Lambda c^4}{8\pi G} g_{\alpha\beta} \tag{13.37}$$

whereas $\tilde{\rho}c^2 = \rho c^2 + \Lambda c^4/(8\pi G)$ and $\tilde{p} = p - \Lambda c^4/(8\pi G)$. This is to show that we can associate with the Cosmological Constant an energy density [*c.f.* Eq.(8.48)] and a pressure

$$\boxed{\rho_\Lambda c^2 = \frac{\Lambda c^4}{8\pi G};} \qquad \boxed{p_\Lambda = -\frac{\Lambda c^4}{8\pi G}} \tag{13.38}$$

It follows that the Cosmological Constant acts as perfect fluid with an unconventional equation of state: $p_\Lambda = -\rho_\Lambda c^2$.

13.8 THE REISSNER-NORDSTROM SOLUTION

The field equations given in Eq.(13.34) allow us to extend the Schwarzschild solution to the case of a point mass with a net electric charge. This leads to the so-called *Reissner-Nordstrom solution.* We will again assume spherical symmetry and a polar coordinate system. As a consequence, the metric can be still written in a stationary form and in terms of only two unknown functions [*c.f.* Eq.(8.57]:

$$ds^2 = e^{\nu(r)}dx^{0\,2} - e^{\lambda(r)}dr^2 - r^2 d\Omega^2 \tag{13.39}$$

• THE EINSTEIN TENSOR

Given the Ricci tensor and Ricci scalar [see Exercise A.39], we can evaluate the components of the Einstein tensor [*c.f.* Eq.(8.15)]

$$G_{00} = \frac{e^\nu}{r^2}\left[1 - \left(re^{-\lambda}\right)'\right] \tag{13.40a}$$

$$G_{11} = -\frac{e^\lambda}{r^2} + \frac{1}{r^2} + \frac{\nu_r}{r} \tag{13.40b}$$

$$G_{22} = e^{-\lambda} r\left[\left(\frac{\nu_r}{2} - \frac{\lambda_r}{2}\right) + r\left(-\frac{\lambda_r\nu_r}{4} + \frac{\nu_r^2}{4} + \frac{\nu_{rr}}{2}\right)\right] \tag{13.40c}$$

$$G_{33} = e^{-\lambda} r\sin^2(\theta)\left[\left(\frac{\nu_r}{2} - \frac{\lambda_r}{2}\right) + r\left(-\frac{\lambda_r\nu_r}{4} + \frac{\nu_r^2}{4} + \frac{\nu_{rr}}{2}\right)\right] \tag{13.40d}$$

• THE EM TENSOR

Because of the symmetry of the problem, the electromagnetic tensor has only one non-vanishing independent component[1]

$$F_{01} = -F_{10} = -E(r); \qquad F^{01} = -F^{10} = e^{-(\nu+\lambda)}E(r) \tag{13.41}$$

[1] We do not consider here the possible existence of magnetic monopoles. So, we set $\vec{B} = 0$.

We are interested in the geometry of the space-time outside the central body—that is, in the "vacuum". So, by ignoring the current density in Eq.(B13.1.cb), we have $\partial(\sqrt{-g}F^{01})/\partial r = 0$. This yields

$$r^2 \mathrm{e}^{-(\nu+\lambda)/2} E(r) = \mathscr{C} \tag{13.42}$$

where $\mathscr{C}$ is an integration constant we have to determine. To do so, remember that at large distance from the point mass, the space-time must be (almost) flat: $\lim_{r\to\infty} \nu(r) = \lambda(r) = 0$. So, $\lim_{r\to\infty} E(r) = \mathscr{C}/r^2$. This Coulomb's inverse-square law behavior suggests to identify the constant $\mathscr{C}$ with the net electric charge of the central body, Q say. As a result,

$$E(r) = e^{(\nu+\lambda)/2} \frac{Q}{r^2} \tag{13.43}$$

and Eq.(13.41) becomes

$$F_{01} = -F_{10} = -e^{(\nu+\lambda)/2} \frac{Q}{r^2}; \qquad F^{01} = -F^{10} = \mathrm{e}^{-(\nu+\lambda)/2} \frac{Q}{r^2} \tag{13.44}$$

• THE ENERGY-MOMENTUM TENSOR

Written in a total covariant form, Eq.(13.25) yields

$$S_{\alpha\beta} = \frac{1}{4\pi} \left[-g^{\mu\nu} F_{\alpha\mu} F_{\beta\nu} + \frac{1}{4} g_{\alpha\beta} F_{\mu\nu} F^{\mu\nu} \right] \tag{13.45}$$

The first term on the *rhs* provides

$$-g^{00} F_{\alpha 0} F_{\beta 0} - g^{11} F_{\alpha 1} F_{\beta 1} = \mathrm{diag}\{\mathrm{e}^{\nu}, -\mathrm{e}^{\lambda}, 0, 0\} \left(\frac{Q}{r^2} \right)^2 \tag{13.46}$$

whereas the second term, with $F_{\mu\nu} F^{\mu\nu} = -2\left(Q/r^2\right)^2$ [*c.f.* Eq.(13.44)], yields

$$\frac{1}{4} g_{\alpha\beta} F_{\mu\nu} F^{\mu\nu} = -\frac{1}{2} \mathrm{diag}\{\mathrm{e}^{\nu}, -\mathrm{e}^{\lambda}, -r^2, -r^2 \sin^2\theta\} \left(\frac{Q}{r^2} \right)^2 \tag{13.47}$$

Thus, the electromagnetic energy-momentum tensor can be written as follows

$$\boxed{S_{\alpha\beta} = \frac{1}{8\pi} \mathrm{diag}\{\mathrm{e}^{\nu}, -\mathrm{e}^{\lambda}, r^2, r^2 \sin^2\theta\} \left(\frac{Q}{r^2} \right)^2} \tag{13.48}$$

• THE FIELD EQUATIONS

Let's assume, again, a vanishing Cosmological Constant. The field equation in the "vacuum" can then be written as follows [*c.f.* Eq.(13.34)]

$$G_{\alpha\beta} = \frac{8\pi G}{c^4} S_{\alpha\beta} \tag{13.49}$$

These equations provide

$$\alpha = \beta = 0) \qquad \frac{1}{r^2}\left[1-\left(re^{-\lambda}\right)'\right] = \frac{G}{c^4}\left(\frac{Q}{r^2}\right)^2 \tag{13.50a}$$

$$\alpha = \beta = 1) \qquad e^{-\lambda}\left[\frac{\nu'}{r}+\frac{1}{r^2}\right] - \frac{1}{r^2} = -\frac{G}{c^4}\left(\frac{Q}{r^2}\right)^2 \tag{13.50b}$$

where the prime indicates a derivative *w.r.t.* the radial coordinate r. Adding these two equations yields

$$-\frac{\left(re^{-\lambda}\right)'}{r^2} + e^{-\lambda}\left[\frac{\nu'}{r}+\frac{1}{r^2}\right] = 0 \tag{13.51}$$

This implies $\nu' + \lambda' = 0$ and, then, $e^{\nu}e^{\lambda} = 1$ because of our boundary condition of an asymptotically flat space-time. Then, integration of Eq.(13.50a) provides

$$re^{-\lambda} = r + \frac{G}{c^4}\frac{Q^2}{r} + \mathscr{C} \tag{13.52}$$

The integration constant $\mathscr{C}$ is set by requiring that for $Q = 0$ we recover the weak field limit of Eq.(6.16). Then,

$$e^{\nu} = e^{-\lambda} = 1 - \frac{2m}{r} + \frac{q^2}{r^2} \tag{13.53}$$

where $m = GM/c^2$ and $q^2 = GQ^2/c^4$ are the mass and the electric charge, both in geometrical units ($G = c = 1$). Thus, the *Reissner-Nordstrom metric* finally writes as follows

$$\boxed{ds^2 = \left(1 - \frac{2m}{r} + \frac{q^2}{r^2}\right) - \left(1 - \frac{2m}{r} + \frac{q^2}{r^2}\right)^{-1} dr^2 - r^2(d\theta^2 + \sin^2\theta d\phi^2)} \tag{13.54}$$

• HORIZON VS. HORIZONS

As in the Schwarzschild case, the metric shows a true, *physical* singularity at $r = 0$, and *coordinate* singularities given by the vanishing of the round parentheses in Eq.(13.54). There are three cases of interest

• If $q < m$, we have two distinct horizons at $r_\pm = m \pm \sqrt{m^2 - q^2}$. Here r_+ and r_- corresponds to the outer and inner horizon, respectively. Note that $\lim_{q\to 0} r_+ = 2m$ and $\lim_{q\to 0} r_- = 0$ [see Figure 13.2], consistently with the Schwarzschild solution.

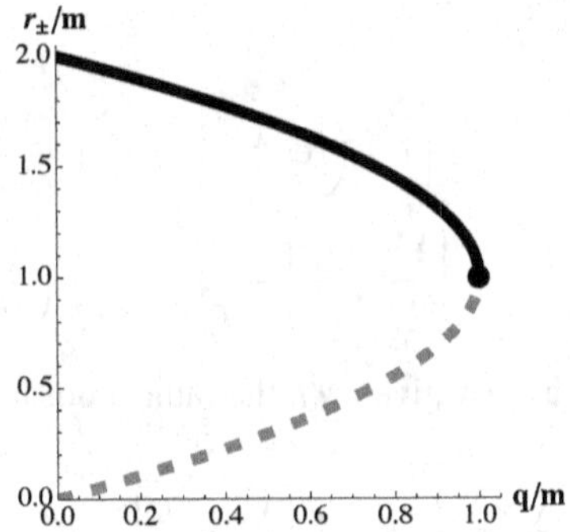

Figure 13.2 The outer (black, continuous line) and the inner (gray, dashed line) horizons are plotted as a function of q/m. Increasing q makes the two horizons to become more similar until they coincide for $q = m$, the extremal black hole solution.

- If $q = m$, the outer and inner horizon coincides: $r_+ = r_- = m$. This is the case of the so-called *extremal black hole*: it has the minimum mass compatible with a given electric charge.

- If $q > m$, the metric well-behaves: both g_{00} and g_{11} are positive defined for any value of $r > 0$

13.9 ORBITS IN A REISSNER-NORDSTROM GEOMETRY

The possible orbits in a Reissner-Nordstrom geometry can be studied by generalizing the considerations of Section 9.8 to take into account the metric given in Eq.(13.54). Thus, Eq.(9.16) now provides

$$\frac{1}{\sqrt{1-\beta^2}} = \frac{E}{mc^2}\frac{1}{\sqrt{1-\mathscr{R}_S/r+q^2/r^2}} \tag{13.55}$$

leading to [*c.f.* Eq.(9.26)]

$$h = r^2\frac{d\phi}{cd\tau}\frac{E}{mc^2}\frac{1}{\sqrt{1-\mathscr{R}_S/r+q^2/r^2}} \tag{13.56}$$

It follows that Eq.(9.27) now writes as follows

$$\frac{1}{c^2}\left(\frac{d\ell}{d\tau}\right)^2 = 1-\left(\frac{mc^2}{E}\right)^2\left(1-\frac{\mathscr{R}_S}{r}+\frac{q^2}{r^2}\right)\left[1+\frac{h^2}{r^2}\right] \tag{13.57}$$

providing

$$\left(\frac{E}{mc^2}\right)^2 \geq f(x) \equiv 1-\frac{1}{x}+\frac{\alpha^2+\beta^2}{x^2}-\frac{\alpha^2}{x^3}+\frac{\alpha^2\beta^2}{x^4} \tag{13.58}$$

where $x = r/\mathscr{R}_S$, $\alpha = h/\mathscr{R}_S$, $\beta = q/\mathscr{R}_S$, and $\mathscr{R}_S = 2m$.
To simplify the discussion, let's limit to the case of an extremal black hole, that is $q = m$ or $\beta = 0.5$. To find where $f'(x) = 0$, we have to solve a cubic algebraic equation

$$\frac{1}{x^2} - 2\left(\alpha^2 + \frac{1}{4}\right)\frac{1}{x^3} + 3\frac{\alpha^2}{x^4} - \frac{\alpha^2}{x^5} = 0 \tag{13.59}$$

which provides three solutions:

$$x_1 = 0.5; \qquad x_2 = \alpha^2 - \alpha\sqrt{\alpha^2 - 2}; \qquad x_3 = \alpha^2 + \alpha\sqrt{\alpha^2 - 2} \tag{13.60}$$

There are then two interesting cases to be discussed.

- $\alpha < \sqrt{2}$: There is only one real solution at $x_1 = 0.5$. This is a minimum, as the second derivative of the effective potential is positive defined for $0 \leq \alpha < \sqrt{2}$: $f''(x_1) = 8 + 32\alpha^2$. As in the Schwarzschild case, a particle coming from infinity is *not* gravitationally captured by the central body [see Figure 13.3].

- $\alpha \geq \sqrt{2}$: We have three different real solutions [see Figure 13.4]. $f(x)$ has a maximum at x_2 and two minima at x_1 and x_3 [see Figure 13.5]. So, as in the Schwarzschild case, stable orbits occur outside the horizon at radii $\simeq x_3$. For $\alpha > 2$, there are unstable circular orbits at a radius x_2.

13.10 FREE-FALL ON THE REISSNER-NORDSTROM BLACK HOLE

Following what was done in Section 9.7, let's study the radial infall of a neutral test-particle into a Reissner-Nordstrom black hole. For the sake of simplicity, let's assume that the particle is initially at rest at infinity. Then, Eq.(9.16) provides

$$\sqrt{1 - \mathscr{R}_S/r + q^2/r^2} \simeq \sqrt{1 - \mathrm{v}^2(r)/c^2} \tag{13.61}$$

This leads to

$$\mathrm{v}(r) = -c\sqrt{\frac{\mathscr{R}_S}{r} - \frac{q^2}{r^2}} \tag{13.62}$$

where, as in Eq.(9.17), the minus sign is chosen because of the inward motion of the test particle. The proper velocity is also given by [*c.f.* Eq.(9.18)]

$$\mathrm{v}(r) = \frac{1}{1 - \mathscr{R}_S/r + q^2/r^2}\frac{dr}{dt} \tag{13.63}$$

By combining Eq.(13.62) and Eq.(13.63), one gets

$$\int_{t_{in}}^{t} \frac{cdt}{\mathscr{R}_S} = \int_{r}^{r_{in}} \frac{(\mathscr{R}_S/r - q^2/r^2)^{-1/2}}{1 - \mathscr{R}_S/r + q^2/r^2}\frac{dr}{\mathscr{R}_S} = \int_{x}^{x_{in}} \frac{x^3(x - \beta^2)^{-1/2}}{x^2 - x + \beta^2}dx \tag{13.64}$$

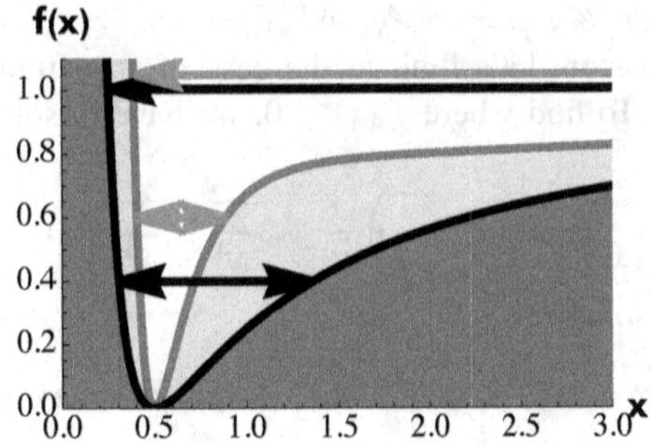

Figure 13.3 Orbits in a Reissner-Nordstrom extremal black hole for $\alpha \leq \sqrt{2}$. The continuous, black and gray lines refer to $\alpha = 0$ and $\alpha = 1.3$. The shadowed areas are the forbidden regions of the parameter space.

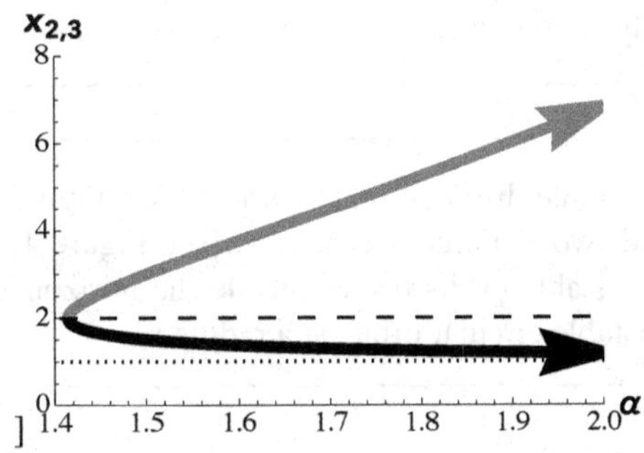

Figure 13.4 The x_2 (black line) and x_3 (gray line) values are plotted as a function of α. For $\alpha = \sqrt{2}$, $x_2 = x_3 = 2$, whereas $\lim_{\alpha \to \infty} x_2(\alpha) = 1$ and $\lim_{\alpha \to \infty} x_3(\alpha) = \infty$, see text.

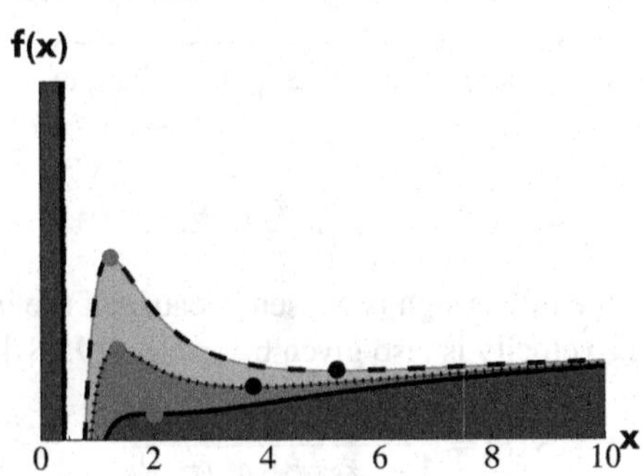

Figure 13.5 The function $f(x)$ is plotted for $\alpha = \sqrt{2}$ (continuous line), $\alpha = 1.6$ (dotted line) and $\alpha = 1.8$ (dashed line). There is a minimum at x_1, a maximum at x_2 (gray dots), and a minimum at x_3 (black dots).

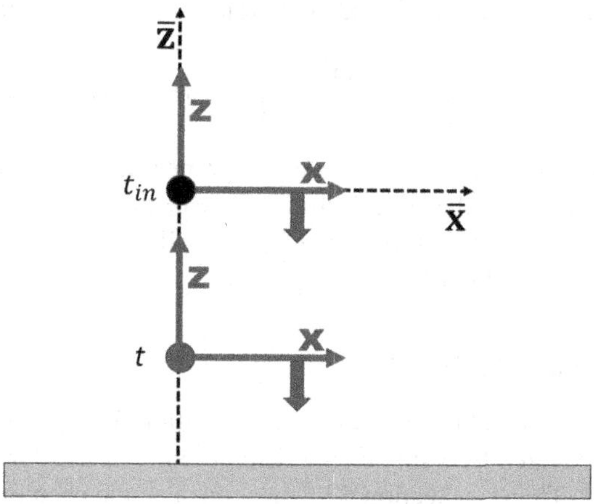

Figure 13.6 A test-particle is free-falling ($\alpha = 0$) onto a Reissner-Nordstrom extremal black hole. For an observer at infinity, the proper time needed for the particle to reach x_1 is infinite (black line). On the contrary, for an observer in free-fall with the test-particle, the proper time needed to reach x_1 is finite (gray line). Here $x_1 = r_1/\mathscr{R}_S$ is derived in Eq.(13.60).

having defined, as above, $x = r/\mathscr{R}_S$ and $\beta = q/\mathscr{R}_S$. In the case of an extremal black-hole, $\beta = 0.5$ and the primitive of the last integral in Eq.(13.64) provides

$$f(x) = \sqrt{4x-1}\frac{(2x^2+6x-5)}{6x-3} + \log\left(\frac{1-\sqrt{4x-1}}{1+\sqrt{4x-1}}\right) \tag{13.65}$$

showing a clear divergence at the event horizon—that is, for $x \to 0.5$. The solution

$$T_\infty \equiv \frac{c(t-t_{in})}{2\mathscr{R}_S} = \frac{1}{2}\left[f(x_{in}) - f(x)\right] \tag{13.66}$$

is plotted in Figure 13.6. As in the case of the Schwarzschild solution, an observer at infinity concludes that a test particle will never reach the horizon of an extremal black hole—equivalently, it will take an infinite time to do so. This is not the case for an observer in free-fall with the test particle. For this observer, the proper time necessary to cross the event horizon of an extremal black hole is finite. Note that $d\tau/dt = ds/dx^0 = g_{00}$ Then, by using Eq.(13.62) and Eq.(13.63), we get

$$d\tau = \frac{d\tau}{dt}\frac{dt}{dr}dr = -\frac{1}{c}\frac{dr}{\sqrt{\mathscr{R}_S/r - q^2/r^2}} \tag{13.67}$$

The indefinite integral, $\int x dx/\sqrt{x-1/4}$, provides

$$g(x) = \frac{2}{3}\sqrt{x-\frac{1}{4}}\left(x+\frac{1}{2}\right) \tag{13.68}$$

Thus, the observer in free-fall will reach the radial position x at a proper time

$$T_{ff}(x) \equiv \frac{c\tau_{ff}(x)}{2\mathscr{R}_S} = \frac{1}{2}\left[g(x_{in}) - g(x)\right] \tag{13.69}$$

This is also shown in Figure 13.6, where the proper times of two observers, one at rest at infinity and the other one in free-fall with the test-particle, are properly compared. As anticipated, for an observer in free-fall, the proper time necessary to cross the event horizon is finite.

13.11 THE KERR SOLUTION

In Section 9.2 we discussed the Schwarzschild solution for a point mass. This solution is fully characterized by a single parameter, the geometrical mass of the central body $m = GM/c^2$. Then, in Section 13.8 we consider the case of a charged point mass and discussed the Reissner-Nordstrom solution. This solution is fully characterized by two parameters, the geometrical mass of the central body, m, and its geometrical charge $q = GQ^2/c^4$. However, real astronomical objects show very often a non-vanishing angular momentum. So, it makes sense to generalize our previous discussions to include a third physical quantity, the angular momentum J of the central body. It is then convenient to define the parameter $a = GJ/(c^3 M)$ that measures the angular momentum per unit mass of the source. We can refer to the quantity Ma as to the geometric angular momentum, setting $G = c = 1$. To simplify the discussion and to make a better contact with the Reissner-Nordstrom solution, we will limit here to the case $q = 0$, so to have again only two parameters, m and a. The metric describing the geometry of the space-time outside a body with mass *and* angular momentum was derived in 1963 by Kerr [52]. It can be written as follows

$$\begin{aligned} ds^2 =& dx^{02} - dx^2 - dy^2 - dz^2 \\ & - \frac{2mr}{r^4 + a^2 z^2}\left[dx^0 + \frac{r(xdx + ydy)}{a^2 + r^2} + \frac{a(ydx - xdy)}{a^2 + r^2} + \frac{z}{r}dz\right]^2 \end{aligned} \tag{13.70}$$

Note that the coordinates used in Eq.(13.70), the so-called Kerr-Schild coordinates, are not Cartesian coordinates. Indeed, they are related to polar coordinates by the following transformations

$$\begin{aligned} x &= \sqrt{r^2 + a^2}\sin\theta\cos\phi \\ y &= \sqrt{r^2 + a^2}\sin\theta\sin\phi \\ z &= r\cos\theta \end{aligned} \tag{13.71}$$

which imply

$$\frac{x^2 + y^2}{r^2 + a^2} + \frac{z^2}{r^2} = 1 \tag{13.72}$$

This clearly shows that the correct Cartesian relation is recovered only if $a = 0$. Note also that the metric given in Eq.(13.70) shows a physical singularity at $r = 0$ and

$z = 0$, that is at [*c.f.* Eq.(13.72)]

$$x^2 + y^2 = a^2 \tag{13.73}$$

We can then conclude that Kerr black hole exhibit, independently on the values of m and a, a physical singularity which, unlike what happens for Schwarzschild solution, is not point-like: it is a ring of radius a in the equatorial plane [remember that $z = 0$ implies $\theta = \pi/2$,*c.f.* Eq.(13.71)]. For $a \to 0$, the ring collapse to a point in the origin and we recover the physical, point-like singularity of the Schwarzschild metric. We can use Eq.(13.71) to re-express Eq.(13.70) in terms of the Boyer-Linquist coordinates $\{x^0, r, \theta, \phi\}$. After some algebra, we get

$$\begin{aligned} ds^2 = & \left(1 - \frac{2mr}{\rho^2}\right) dx^{02} + \frac{4mra\sin^2\theta}{\rho^2} dx^0 d\phi - \frac{\rho^2}{\Delta} dr^2 - \rho^2 d\theta^2 \\ & - \left[\left(r^2 + a^2\right)\sin^2\theta + \frac{2mra^2\sin^4\theta}{\rho^2}\right] d\phi^2 \end{aligned} \tag{13.74}$$

Here, we defined

$$\rho^2 = r^2 + a^2\cos^2\theta \tag{13.75}$$

and

$$\Delta = r^2 - 2mr + a^2 \tag{13.76}$$

Note that for $r \to \infty$, both ρ and Δ tend to r. It turns out that in this limit Eq.(13.74) reduces to the Minkowski metric, as we should have expected: the Kerr space-time is asymptotically flat. Also, the metric is stationary, as the metric coefficient does not explicitly depend on time. However, because of the off diagonal term proportional to $dx^0 d\phi$, the metric is not invariant under time inversion, $t \to -t$. So, we don't expect the metric to be static. Last, but not least, the metric coefficients don't depend on ϕ as required by the axial symmetry of the problem.

The metric given in Eq.(13.74) becomes singular when $\rho = 0$, that is when $r = 0$ and $\theta = \pi/2$ (corresponding to $z = 0$). This is the same physical singularity given by Eq.(13.73): a ring of radius a in the equatorial plane. There are, however, other *coordinate* singularities that didn't appear in Eq.(13.70), but that it is important to discuss. Let's consider first the g_{11} metric coefficient. It doesn't well behave if $\Delta = 0$. This might happen depending on the a/m ratio. If $a/m < 1$, we have two distinct horizons

$$r_\pm = m \pm \sqrt{m^2 - a^2} \tag{13.77}$$

These horizons coincide if $a/m = 1$, but disappear if $a/m > 1$: in this case, the equation $\Delta = 0$ doesn't admit real solutions. Note that when $a \to 0$, as it could have been expected, $r_- \to 0$ and $r_+ \to 2m = \mathscr{R}_S$, the event horizon of the Schwarzschild space-time. The outer and inner horizons are plotted in Figure 13.7. Note the similarity with the Reissner-Nordstrom results shown in Figure 13.2. The solutions given in Eq.(13.77) actually identify two closed surfaces, Σ_+ and Σ_-, the latter being always contained inside the former. So, it turns out that Σ_+ is the relevant surface dividing the region accessible from the external region (that is, from r much larger than m and

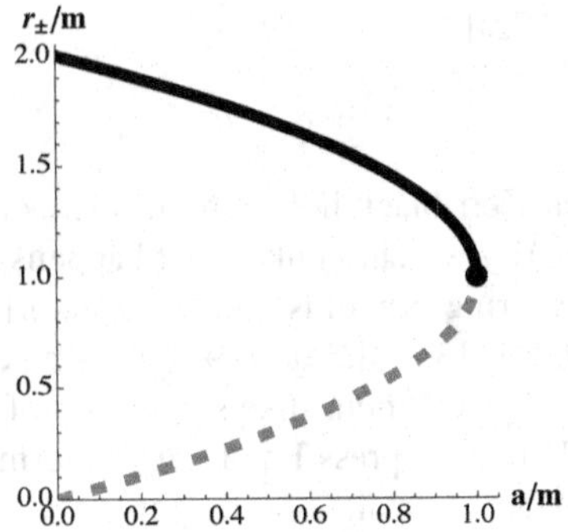

Figure 13.7 The outer (black, continuous line) and the inner (gray, dashed line) Kerr horizons are plotted as a function of a/m. Increasing a makes the two horizons to become more similar until they coincide for $a = m$, the extremal Kerr black hole case.

a), playing the same role of the Schwarzschild surface or of the external horizon in the Reissner-Nordstrom solution.

In the Schwarzschild and Reissner-Nordstrom solutions, the Schwarzschild surface and the external Reissner-Nordstrom horizon represent also the surfaces of infinity redshift, because of the simultaneous vanishing of g_{00} and g_{11}. This is not the case for the Kerr black hole. In fact, the condition $g_{00} = 0$ implies $2mr = \rho^2$ [*c.f.* Eq.(13.74)]. For $a < m$, we get the following solutions

$$s_{\pm} = m \pm \sqrt{m^2 - a^2 \cos^2 \theta} \tag{13.78}$$

corresponding to two closed surfaces, S_+ and S_-. Now it turns out that the surface S_+ always contains the surface Σ_+, while the reverse is true for S_- and Σ_- [see Figure 13.8].

The nature of S_+ and Σ_+ is very different indeed. The infinite redshift surface, S_+, does not represent a "barrier": both massive test-particles and photons can cross it in either directions, with the only exception of the poles ($\theta = 0$ and $\theta = \pi$) where $r_+ = s_+$. On the contrary, Σ_+ must be regarded as the critical surface below which a body becomes a black hole, analogous to the Schwarzschild or to the Reissner-Nordstrom black hole. The region between S_+ and Σ_+ is called *ergosphere*. There are several interesting effects due to the not-coincidence of the infinite redshift surface with the external horizon. For example, in the ergosphere the motion of test-particles is skewed in the same direction as the black hole rotation, even if they were initially in free-fall with zero angular momentum. Another interesting consequence of the existence of an ergosphere is the so-called Penrose effect [64]. Imagine a particle entering the ergosphere and, while there, decaying in two particles. If one of the decay product crosses the event horizon, the other can get out of the ergosphere. The point is that, under appropriate circumstances, the latter can leave the ergosphere with a greater energy than the initial particle had when it entered. So, the Penrose effects predict the possibility of extracting energy from a black hole, at the expense of

Figure 13.8 The surfaces discussed in the context of the Kerr black hole for $a/m = 0.95$. Going from the outside to the inside we have: i) the surface of infinite redshift, S_+; ii) the outer event horizon, Σ_+; iii) the inner event horizon, Σ_-; iv) the surface of infinite redshift, S_-. See text.

its rotational energy. We will refer to other text books for a more detailed discussion of all these effects.

14 Further Applications of the Field Equations

14.1 INTRODUCTION

We discussed the field equations needed to study the geometry of the space-time within a given matter/energy distribution, fully described by a suitable energy-momentum tensor. In that framework, we derived the Reissner-Nordstrom solution, generalizing the Schwarzschild solution to the case of a charged point mass. The goal of this chapter is to apply the field equations to different matter/energy configurations, relaxing the point mass approximation used up to now.

14.2 THE INNER SCHWARZSCHILD SPACE-TIME

Consider an isotropic and incompressible mass distribution. Let's assume that the gravitational attraction is completely counterbalanced by the pressure gradients, *i.e.*, the considered structure is in *hydrostatic equilibrium*. The question to be asked is the following: which is the geometry of the space-time *inside* this matter distribution? This problem was addressed for the first time by Karl Schwarzschild in 1916 [80,83], who found an analytical solution of the field equations. We refer to this solution as to the *inner Schwarzschild solution*.

Note that the symmetries of the problem are the same as those discussed in Section 8.8. In addition, because of the assumed equilibrium configuration, we don't expect the metric to be time dependent. Thus, we can write the square of the line element as in Eq.(8.57), but ignoring the time dependence of the ν and λ unknown functions:

$$ds^2 = e_{int}^{\nu(r)} dx^{0\,2} - e_{int}^{\lambda(r)} dr^2 - r^2 d\Omega^2 \tag{14.1}$$

It follows that the components of the Einstein tensor are those given in Eq.(13.40). The subscript *int* indicates that the values of the radial coordinate are less or equal to the coordinate radius of the considered structure: $r \leq R$. To find the two unknown functions, $\nu(r)$ and $\lambda(r)$, we will use the equations of motion [*c.f.* Eq.(13.10)] and the field equations with a vanishing Cosmological Constant [*c.f.* Eq.(13.34)].

• HYDROSTATIC EQUILIBRIUM

Since we want to deal with an incompressible fluid, we can safely assume that the mass density is a constant: $\rho = const$. Also, hydrostatic equilibrium implies that the four-velocities of the fluid elements have vanishing spatial components: $u^\alpha(r) \equiv \{e^{-\nu(r)/2}, 0, 0, 0\}$. Thus, the energy momentum tensor results to be [*c.f.* Eq.(13.9)]

$$\boxed{T^{\alpha\beta} = \text{diag}\left[\rho c^2 e^{-\nu(r)}, p(r)e^{-\lambda(r)}, \frac{p(r)}{r^2}, \frac{p(r)}{r^2 \sin^2\theta}\right]} \tag{14.2}$$

DOI: 10.1201/9781003141259-14

Let's now consider the spatial component of the equations of motion [*c.f.* Eq.(13.5)].

$$T_1{}^{\mu}{}_{;\mu} = \frac{1}{\sqrt{-g}}\frac{\partial\sqrt{-g}T_1{}^{\sigma}}{\partial x^{\sigma}} - \Gamma^{\sigma}_{1\mu}T_{\sigma}{}^{\mu} = 0 \tag{14.3}$$

It is easy to show that $\Gamma^{\sigma}_{1\mu}T_{\sigma}{}^{\mu} = g_{\mu\rho,1}T^{\rho\mu}/2$ [see Exercise A.40]. Then, Eq.(14.3) yields [see Exercise A.41]

$$\boxed{\frac{dp}{dr} + \frac{1}{2}(\rho c^2 + p)\frac{d\nu}{dr} = 0} \tag{14.4}$$

In the weak field limit, $p \ll \rho c^2$, $e^{\nu} \simeq 1 + \nu$ and $\nu \simeq 2U/c^2$ [see, *e.g.*, Eq.(6.18)]. Then, Eq.(14.4) reduces to the classical hydrostatic equilibrium condition

$$\frac{dp}{dr} = \rho\frac{dU}{dr} \tag{14.5}$$

where gravity is completely counterbalanced by the pressure gradients. Eq.(14.4) shows that similar concepts are also found in General Relativity, provided that we add to the classical term a new term, proportional to the pressure. It follows that in General Relativity pressure not only fights against gravity, but it also contributes to it. This is why the resolution of the hydrostatic equilibrium problem requires to know, besides the geometry of the space-time, also the behavior of the matter density and pressure as a function of the radial coordinate. Under the hypothesis of a uniform density distribution, the integration of Eq.(14.4) provides

$$\boxed{(\rho c^2 + p) = \frac{c^4}{8\pi G}\mathscr{C}_3 e^{-\nu/2}} \tag{14.6}$$

where, just for the sake of convenience, we wrote the exponential of the integration constant as $c^4\mathscr{C}_3/(8\pi G)$. The constant $\mathscr{C}_3$ will be determined below.

• THE λ FUNCTION

The *covariant* time-time component of the field equations, $G_{00} = \chi T_{00}$, provides

$$\frac{d}{dr}\left(r e^{-\lambda}_{int}\right) = 1 - \frac{8\pi G}{c^4}\rho c^2 r^2 \tag{14.7}$$

and, then,

$$\boxed{e^{-\lambda(r)}_{int} = 1 - \frac{r^2}{r_0^2} = 1 - \frac{2GM_{int}(r)}{c^2 r}} \tag{14.8}$$

Here

$$r_0^2 \equiv \frac{3c^2}{8\pi G\rho} \tag{14.9}$$

whereas

$$M_{int}(r) \equiv \frac{4}{3}\pi\rho r^3 \tag{14.10}$$

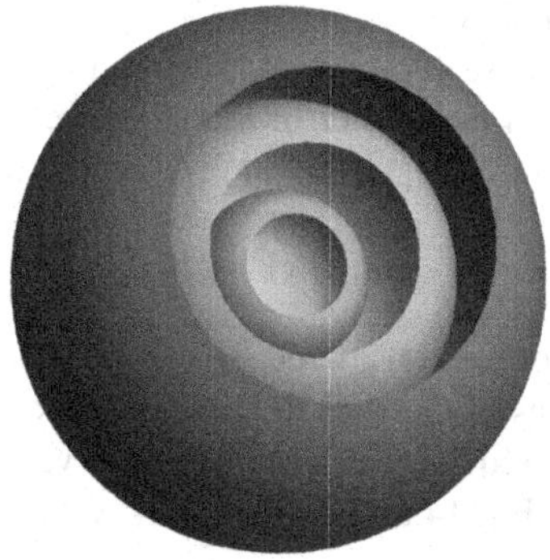

Figure 14.1 An isotropic mass distribution. Three shells are shown, of coordinate radius $r_1 = 2R/3$, $r_2 = R/3$ and $r_3 = R$. Here R is the coordinate radius of the considered structure.

is the mass contained inside a sphere of coordinate radius r. The form of Eq.(14.8) is not unexpected. In a way, it shows an extension to non-"empty" space of the Jebsen-Birkhoff theorem [*c.f.* Section 9.4]: the space-time geometry at a coordinate distance r from the center of symmetry is determined only by $M_{int}(r)$, the mass inside a sphere of radius r [see Figure 14.1]. We will discuss this point again in Section 14.5.

• THE ν FUNCTION

Let's write the field equations in a mixed form, and let's consider the following combination $G_0^0 - G_1^1 = 8\pi G(T_0^0 - T_1^1)/c^4$. We can use Eq.(13.40) and Eq.(14.2) to find

$$\frac{e^{-\lambda}}{r}(\lambda_r + \nu_r) = \frac{8\pi G}{c^4}(\rho c^2 + p) \tag{14.11}$$

Substituting Eq.(14.6) in Eq.(14.11) yields

$$-\frac{de^{-\lambda}}{dr} + \nu_r e^{-\lambda} = \mathscr{C}_3 e^{-\nu/2} r \tag{14.12}$$

Then, using Eq.(14.8), we find

$$\left(1 - \frac{r^2}{r_0^2}\right)\frac{de^{\nu(r)/2}}{dr} + \frac{r}{r_0^2}e^{\nu(r)/2} = \frac{1}{2}\mathscr{C}_3 r \tag{14.13}$$

This equation can be integrated analytically and provides

$$\boxed{e^{\nu/2} = \mathscr{C}_1 - \mathscr{C}_2\sqrt{1 - r^2/r_0^2}} \tag{14.14}$$

where both

$$\mathscr{C}_1 \equiv \frac{1}{2}\mathscr{C}_3 r_0^2 \tag{14.15}$$

and $\mathscr{C}_2$ are constants to be determined.

• BOUNDARY CONDITIONS

We clearly have to impose continuity between the *internal* and *external* Schwarzschild solutions. On the border of the structure, $r = R$. Then, we must require that $\mathrm{e}_{int}^{-\lambda(R)} = \mathrm{e}_{ext}^{-\lambda(R)}$ [*c.f.* Eq.(14.8) and Eq.(9.1)]. It turns out that, as expected,

$$M_{int}(R) = M \tag{14.16}$$

where M is the total mass of the considered structure, indeed the mass used in Chapter 10 to test the predictions of General Relativity on planetary scales. We can also exploit another constraint, requiring that the pressure vanishes on the border of the structure: $p(R) = 0$. Then, Eq.(14.6) and Eq.(14.14) yield

$$\frac{8\pi G}{c^2}\left(\mathscr{C}_1 - \mathscr{C}_2\sqrt{1 - R^2/r_0^2}\right)\rho = \mathscr{C}_3 \tag{14.17}$$

Remember that ρ has assumed to be constant. Then, we can use Eq.(14.9) and Eq.(14.15) to write $\rho = 3c^2/(8\pi G r_0^2)$ and $\mathscr{C}_3 \equiv 2\mathscr{C}_1/r_0^2$. Thus, Eq.(14.17) provides

$$\mathscr{C}_1 = 3\mathscr{C}_2\sqrt{1 - R^2/r_0^2} \tag{14.18}$$

It follows that Eq.(14.14) writes

$$\mathrm{e}_{int}^{\nu(r)/2} = \mathscr{C}_2\left\{3\sqrt{1 - R^2/r_0^2} - \sqrt{1 - r^2/r_0^2}\right\} \tag{14.19}$$

To find $\mathscr{C}_2$, let's exploit again boundary conditions: $\mathrm{e}_{int}^{\nu(R)/2} = \mathrm{e}_{ext}^{\nu(R)/2}$ [*c.f.* Eq.(14.19) and Eq.(9.1)]. It follows that $2\mathscr{C}_2\sqrt{1 - R^2/r_0^2} = \sqrt{1 - \mathscr{R}_S/R}$. Given Eq.(14.9), this provides

$$\mathscr{C}_2 = 1/2 \tag{14.20}$$

Thus, the time-time component of the metric coefficients can be written as follows:

$$\boxed{\mathrm{e}_{int}^{\nu(r)/2} = \frac{3}{2}\sqrt{1 - R^2/r_0^2} - \frac{1}{2}\sqrt{1 - r^2/r_0^2}} \tag{14.21}$$

• THE PRESSURE PROFILE

Remember that $\mathscr{C}_3 = 2\mathscr{C}_1/r_0^2$ [*c.f.* Eq.(14.15)], $r_0^{-2} = 8\pi G\rho/(3c^2)$ [*c.f.* Eq.(14.9)] and $\mathscr{C}_1 = (3/2)\sqrt{1 - R^2/r_0^2}$ [*c.f.* Eq.(14.18) and Eq.(14.20)]. Then, to find the pressure profile needed to keep a homogeneous and isotropic mass distribution in hydrostatic equilibrium, we can use Eq.(14.6) to obtain

$$\boxed{p(r) = \rho c^2 \frac{\sqrt{1 - r^2/r_0^2} - \sqrt{1 - R^2/r_0^2}}{3\sqrt{1 - R^2/r_0^2} - \sqrt{1 - r^2/r_0^2}}} \tag{14.22}$$

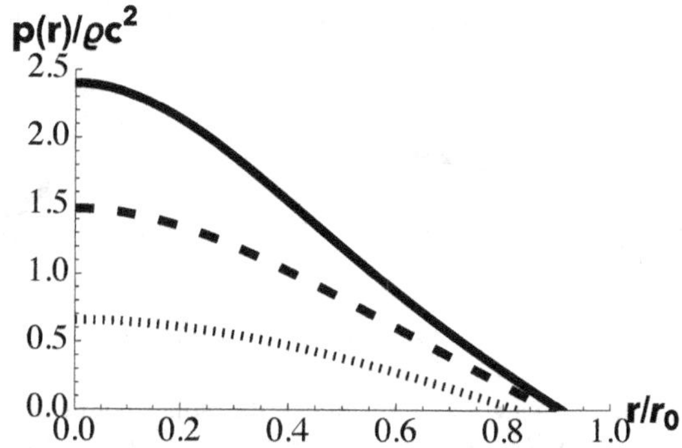

Figure 14.2 Pressure profiles of an isotropic, incompressible mass distribution in hydrostatic equilibrium. The function $p(r)$ is plotted in units of ρc^2 against the radial coordinate r in units of r_0. We assume that the coordinate radius R of the structure in units of r_0 is 0.91 (black line), 0.89 (dashed line) and 0.83 (dotted line), respectively.

Clearly, we have to impose that the pressure is positive for all the possible values of r, from the center to the border of the structure. When this condition is enforced at the origin, we get

$$\boxed{R < \sqrt{\frac{8}{9}}r_0} \tag{14.23}$$

or, equivalently [*c.f.* Eq.(14.9)],

$$\boxed{R > \frac{9}{8}\mathscr{R}_S} \tag{14.24}$$

• THE INNER SCHWARZSCHILD SOLUTON

Eq.(14.8) and Eq.(14.21) allow to write the following expression for the metric of the space-time *inside* a uniform, isotropic matter distribution

$$\boxed{ds^2 = \left(\frac{3}{2}\sqrt{1-R^2/r_0^2} - \frac{1}{2}\sqrt{1-r^2/r_0^2}\right)^2 dx^{0\,2} - \frac{dr^2}{1-r^2/r_0^2} - r^2 d\Omega^2} \tag{14.25}$$

where $d\Omega^2 = d\theta^2 + \sin^2\theta d\phi^2$. The constraints given in Eq.(14.23) and Eq.(14.24) force us to conclude that a homogeneous and isotropic mass distribution in hydrostatic equilibrium can exist *if and only if* its radius is *larger* than the Schwarzschild radius by at least a factor of 9/8.[1] Because of this constraint, the metric does not show any pathology. The g_{00} and g_{11} metric coefficients are shown in Figure 14.3. Note that g_{11} remain finite, although discontinuous at the border of the structure. This is not unexpected, as there both ρ and $p(R)$ vanish.

[1] Note that a slighter looser condition can also be derived by noting that $e_{int}^{-\lambda(r)}$ is positive defined. Thus, Eq.(14.8) yields $R < r_0$ or $R > \mathscr{R}_S$.

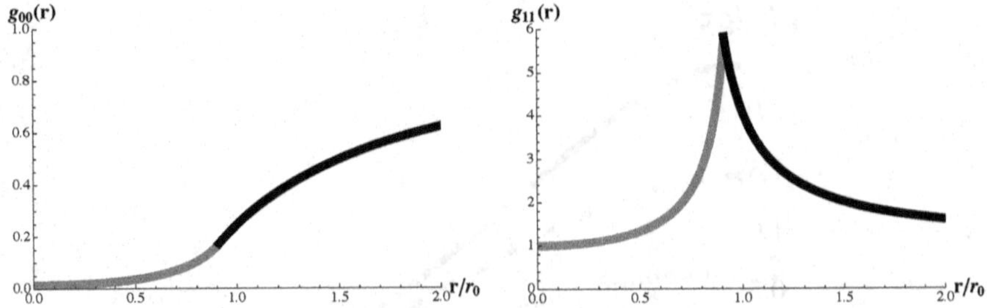

Figure 14.3 The metric coefficients as a function of the radial coordinate r in units of r_0 [*c.f.* Eq.(14.9)]. We show $g_{00}(r)$ (left) and $g_{11}(r)$ (right). The gray and black lines refer to the inner and outer solutions, respectively.

• "EMBEDDING"

We can now follow the same line of reasoning of Section 9.3, to embed the 2D, equatorial ($\theta = \pi/2$) surface at $t = const$

$$d\sigma^2 = \frac{dr^2}{1 - r^2/r_0^2} + r^2 d\phi^2 \tag{14.26}$$

into a 3D Euclidean space with metric $d\ell^2 = \left[1 + (\partial z/\partial r)^2\right] dr^2 + r^2 d\phi^2$ [*c.f.* Eq.(9.5)]. Comparing the radial parts yields

$$\frac{\partial z}{\partial r} = \frac{r}{r_0}\left[1 - \left(\frac{r}{r_0}\right)^2\right]^{-1/2} \tag{14.27}$$

Integrating this equation provides

$$z(r) = -r_0\sqrt{1 - r^2/r_0^2} + \mathscr{C} \tag{14.28}$$

where the constant $\mathscr{C}$ is fixed by requiring that at the boundary, $r = R$, the internal and external solution coincide. This function is shown in Figure 14.4a for $r \leq R$. In Figure 14.4b this solution is shown together with what was obtained for the external Schwarzschild solution [see, *e.g.*, Eq.(9.7)]. It is clear the smooth transition from the inside ($r \leq R$) and the outside ($r \geq R$) of the considered structure, as long as $R > \mathscr{R}_S$.

14.3 PROPER VS. OBSERVABLE MASS

Joining the inner and outer Schwarzschild solutions led us to Eq.(14.16), where M is the total, *observable* mass of the central body. As already mentioned, it is this mass that can be measured at large distances with a number of tests, like those discussed in Chapter 10. The observable mass M can then be interpreted as the body *active* mass, the one responsible for the curvature of the space-time at the exterior of the

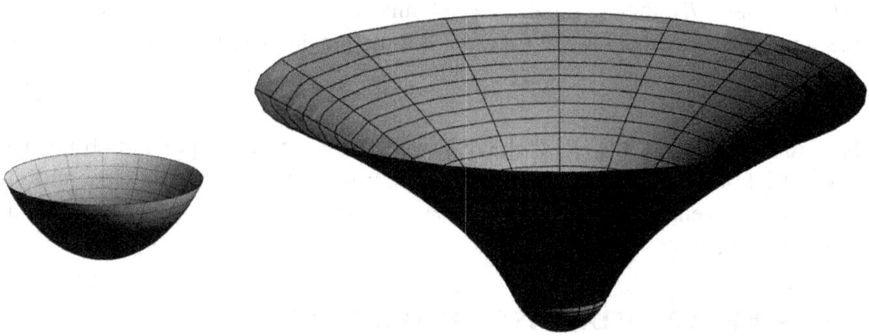

Figure 14.4 The "embedding" of the equatorial plane ($\theta = \pi/2$) of the inner Schwarzschild solution is shown in the left panel for $R > \mathscr{R}_S$. The (smooth) transition from the inner to the outer Schwarzschild solution is shown in the right panel.

central body [*c.f.* Eq.(9.1)]. To the observable mass, contribute the rest mass, as well as the equivalent mass due to any other form of energies, including the gravitational energy. To see this, let's use Eq.(14.10) and write $M_{int}(R)$ in a slightly different form

$$M_{int}(R) = \int_{V_p} \rho dV_p + \left[\int_V \rho dV - \int_{V_p} \rho dV_p\right] \tag{14.29}$$

where $dV = 4\pi r^2 dr$ and

$$dV_p = 4\pi \frac{r^2}{\sqrt{1-(r/r_0)^2}} dr \underset{r/r0 \ll 1}{\simeq} 4\pi r^2 \left[1 + \frac{1}{2}\left(\frac{r}{r_0}\right)^2\right] dr \tag{14.30}$$

are the coordinate and proper volume elements, respectively. The first integral in Eq.(14.29) defines the proper mass M_p of the central body:

$$M_p = 4\pi \int_0^R \rho e^{\lambda(r)/2} r^2 dr \tag{14.31}$$

For a uniform density distribution (and in the weak field approximation), the squared parenthesis in Eq.(14.29) provides

$$-2\pi\rho \int \frac{r^4}{r_0^2} dr = -\frac{16\pi G \rho^2}{3} \int r^4 dr = -\frac{G}{c^2} \int \frac{M_{int}(r)}{r} dM_{int}(r) = -\frac{|W|}{c^2} \tag{14.32}$$

that is, the classical expression for the gravitational binding energy. This energy contributes additively (and with its sign) to the proper mass of the central body. In the weak field limit, Eq.(14.29) provides

$$M_{int}(R) = M_p(R) - \frac{|W|}{c^2} \tag{14.33}$$

If $M_p(R) > M_{int}(R)$, then to break apart a bound structure it is necessary to do some work on it. Thus, we can rewrite Eq.(14.33) to define the binding energy as follows:

$$|W| = \Delta M c^2 \tag{14.34}$$

where $\Delta M = M_p(R) - M_{int}(R)$ is the difference between the proper and observable masses. Eq.(14.34) should sound more than plausible, showing how in General Relativity the mass-energy equivalence can be extended also to the case of the gravitational binding energy.

14.4 SPHERICAL RELATIVISTIC HYDRODYNAMICS

We want to extend the considerations of Section 14.2 by relaxing two assumptions: the uniform density distribution and the hydrostatic equilibrium. We will still consider an isotropic mass distribution, as this greatly helps in simplifying the writing of the metric. Without loss of generality, we can choose *comoving* Lagrangian rather than Eulerian coordinates. Thus, r, θ and ϕ must be considered as (constant) flags that identify a given fluid element. In this case, the metric can be written as follows:

$$\boxed{ds^2 = \mathrm{e}^{\nu(r,t)} dt^2 - \mathrm{e}^{\lambda(r,t)} dr^2 - R^2(r,t)\left[d\theta^2 + \sin^2\theta d\phi^2\right]} \tag{14.35}$$

Note that unlike what has been done in Eq.(8.57) and Eq.(14.1), we have now to consider a new function $R(r,t)$. From a geometrical point of view, $R(r,t)$ acts as a scale factor. From a physical point of view, $R(r,t)$ plays the role of an angular diameter distance[2] of a shell of comoving radius r at a given time t. The surface and the volume of a sphere of coordinate radius r are then given by $4\pi R^2(r,t)$ and $4\pi R^3(r,t)/3$, respectively.

For what follows, it is useful to define two new operators, $\mathscr{D}_t$ and $\mathscr{D}_r$. The first indicates a partial derivative *w.r.t. proper time*, $\mathscr{D}_t \equiv \mathrm{e}^{-\nu/2}(\partial/\partial t)$, while the second a partial derivative *w.r.t.* the *proper radial length*, $\mathscr{D}_r \equiv \mathrm{e}^{-\lambda/2}(\partial/\partial r)$. Since R is the (proper) angular distance of a shell of comoving coordinate r, it is reasonable to interpret the quantity

$$V(r,t) \equiv \mathscr{D}_t R(r,t) \tag{14.36}$$

as the proper (expansion or collapsing) velocity of that shell. To find the unknowns of our problems, the functions $\nu(r,t)$, $\lambda(r,t)$ and $R(r,t)$, we will use the equations of motion [*c.f.* Eq.(13.10)] and the field equations with a vanishing Cosmological Constant [*c.f.* Eq.(13.34)].

• EQUATION OF MOTION

For comoving coordinates, the spatial components of the four-velocity are zero, even in the absence of hydrostatic equilibrium. Then, $u^0 u_0 = 1$. It follows that the energy momentum tensor of a perfect fluid can be written in a very simple form

$$T^{\alpha}{}_{\beta} = \mathrm{diag}\{\varepsilon, -p, -p, -p\} \tag{14.37}$$

[2] See, *e.g.*, Section 10.5.

The *time* component of Eq.(13.10) provides

$$\varepsilon_t + \frac{\lambda_t}{2}(\varepsilon + p) + 2\frac{R_t}{R}(\varepsilon + p) = 0 \qquad (14.38)$$

Multiply this equation by $R^2 e^{\lambda/2}$ to write

$$\boxed{\frac{\partial}{\partial t}\left(\varepsilon R^2 e^{\lambda/2}\right) + p\frac{\partial}{\partial t}\left(R^2 e^{\lambda/2}\right) = 0} \qquad (14.39)$$

The proper volume of a shell of (comoving) coordinate r and thickness Δr is given by $4\pi R^2 e^{\lambda/2}\Delta r$. Then, Eq.(14.39) is nothing else that the First Principle of Thermodynamics written for an adiabatic transformation: it describes how the energy of a shell changes due to the work done by pressure forces.

The *space* component of Eq.(13.10) yields

$$-\frac{\nu_r + \lambda_r}{2}p - 2\frac{R_r}{R}p - p_r - \frac{1}{2}\nu_r\varepsilon + \frac{1}{2}\lambda_r p + 2\frac{R_r}{R}p = 0 \qquad (14.40)$$

which reduces to Eq.(14.4)

$$\boxed{p_r + \frac{\nu_r}{2}(\varepsilon + p) = 0} \qquad (14.41)$$

This seems a contradictory result, as we are now studying a configuration out of the hydrostatic equilibrium. The point is that we are using a coordinate system *comoving* with the fluid. In this frame, the fluid elements are and remain at rest.

• *THE λ FUNCTION*

The time-space component of the field equations [*c.f.* Eq.(13.34)] provides $G^0{}_1 = 0$, that is,

$$\lambda_t = 2\frac{R_{tr}}{R_r} - \nu_r\frac{R_t}{R_r} \qquad (14.42)$$

Let's multiply both sides of this equation by $e^{-\nu/2}/2$ to get

$$\frac{1}{2}e^{-\nu/2}\lambda_t = \frac{\partial(e^{-\nu/2}R_t)/\partial r}{\partial R/\partial r} \qquad (14.43)$$

or, equivalently,

$$\boxed{\frac{1}{2}\mathscr{D}_t\lambda = \frac{\partial V}{\partial R}} \qquad (14.44)$$

On a similar line, the *time-time* component of the field equations, $G^0{}_0 = \chi T^0{}_0$, provides

$$-2\frac{R_{rr}}{R}e^{-\lambda} + \frac{\lambda_t}{c^2}\frac{R_t}{R}e^{-\nu} + \lambda_r\frac{R_r}{R}e^{-\lambda} + \frac{e^{-\nu}}{c^2}\frac{R_t^2}{R^2} + \frac{R_r^2}{R^2}e^{-\lambda} + \frac{1}{R^2} = \frac{8\pi G}{c^4}\varepsilon \qquad (14.45)$$

BOX 14.1 THE OBSERVABLE MASS FUNCTION $M(r,t)$

The function

$$M(r,t) = \frac{4\pi}{c^2}\int_0^r \varepsilon R^2 R_{r'} dr' \tag{B14.1.a}$$

is the mass of a sphere of radius r. The time derivative of Eq.(B14.1.a) yields

$$\frac{\partial M(r,t)}{\partial t} = \frac{4\pi}{c^2}\int_0^r \left(\varepsilon_t R^2 R_{r'} + 2\varepsilon R R_t R_{r'} + \varepsilon R^2 R_{r't}\right) dr' \tag{B14.1.b}$$

Eliminate ε_t, λ_t and ν_r by using Eq.(14.38), Eq.(14.42) and Eq.(14.41). Then,

$$\frac{\partial M(r,t)}{\partial t} = -\frac{4\pi}{c^2}\int \frac{\partial}{\partial r}(pR^2 R_t)dr = -\frac{4\pi}{c^2} pR^2 R_t \tag{B14.1.c}$$

The integration constant vanishes, as $M(0,t) = 0$. Multiply by $e^{-\nu/2}c^2$ both sides of Eq.(B14.1.c) to get

$$\boxed{\mathscr{D}_t \left[M(r,t)c^2\right] = -4\pi pR^2(r,t)V(r,t)} \tag{B14.1.d}$$

Thus,

- $M(r,t)c^2$ varies with time because of the work done by the pressure forces.
- $M(r,t)c^2$ is an energy.
- $M(r,t)$ plays the role of the observable mass of a sphere of radius r.
- If $p = 0$, $M(r,t)$ depends only on the radial coordinate.

Let's multiply both sides of this equation by $R^2 R_r$ and use Eq.(14.42) to eliminate λ_t. We get

$$\frac{\partial}{\partial r}\left(R - e^{-\lambda}R_r^2 R\right) + \frac{1}{c^2}\left(2\frac{R_{tr}}{R_r} - \nu_r \frac{R_t}{R_r}\right) R_t R R_r e^{-\nu} + \frac{e^{-\nu}}{c^2} R_t^2 R_r = \frac{8\pi G}{c^4}\varepsilon R^2 R_r \tag{14.46}$$

We can further simplify the writing of this equation by using the definition of mass function given in Eq.(B14.1.a). Then, Eq.(14.46) can be written as follows:

$$\frac{\partial}{\partial r}\left(R - e^{-\lambda}R_r^2 R + \frac{1}{c^2}e^{-\nu}R_t^2 R\right) = \frac{2G}{c^2}\frac{\partial M(r,t)}{\partial r} \tag{14.47}$$

providing

$$R - e^{-\lambda}R_r^2 R + \frac{1}{c^2}e^{-\nu}R_t^2 R = \frac{2GM(r,t)}{c^2} \tag{14.48}$$

Here the integration constant has been set to zero, as both $R(r,t)$ and $M(r,t)$ vanish for $r = 0$. Let's define a new function

$$\boxed{\mathscr{E} = \mathscr{D}_r R} \tag{14.49}$$

This new function plays the role of an energy measured in unit of the rest energy of a test-particle [see Box 14.2]. Let's divide Eq.(14.48) by R and use the definition given in Eq.(14.49), to get

$$\boxed{\mathscr{E}^2(r,t) = 1 + \frac{V^2(r,t)}{c^2} - \frac{2GM(r,t)}{c^2 R(r,t)}} \tag{14.50}$$

BOX 14.2 THE FUNCTION $\mathscr{E}$

Given Eq.(14.49),we can write

$$\mathscr{E} = \frac{1}{2\pi} \frac{\partial(2\pi R)}{e^{\lambda/2}\partial r} \tag{B14.2.a}$$

Thus, $\mathscr{E}$ is proportional to the derivative of the proper circumference length *w.r.t.* to its proper radius.

- In Euclidean spaces, this ratio has to be 2π and, then, $\mathscr{E}$ has to be unity.
- In Special Relativity, we expect a Lorentz contraction in the radial direction. So, we have $\mathscr{E} = \gamma = \sqrt{1+p^2/(m_p c)^2}$. This shows that $V \equiv \mathscr{D}_t R$ is a momentum per unit mass [*c.f.* Eq.(14.50) with $M(r,t) = 0$].
- For the internal Schwarzschild solution [*c.f.* Eq.(14.25)], $R \to r$ and $R_r \to 1$. Then, $\mathscr{E} = \sqrt{1 - 2GM/(c^2 r)}$, showing that now curvature comes into play.
- In the weak field limit, Eq.(14.50) yields $\mathscr{E} = 1 + V^2/(2c^2) - GM(r,t)/(c^2 R)$.

In the absence of pressure gradients, $\mathscr{E}$ depends only on the radial coordinate. To see this point, let's first derive Eq.(14.49) *w.r.t.* to proper time and, then, divide the result by $\mathscr{E}$. We get

$$\frac{1}{\mathscr{E}} D_t \mathscr{E} = -\frac{1}{2} D_t \lambda + \frac{e^{-\nu/2}}{R_r} R_{tr} \tag{14.51}$$

The last term can be conveniently written in the following form:

$$\frac{1}{R_r}\frac{\partial}{\partial r}\left(e^{-\nu/2} R_t\right) + \frac{\nu_r}{2}\frac{e^{-\nu/2}}{R_r} R_t = \frac{V_r}{R_r} - \frac{V}{\varepsilon + p}\frac{\partial p}{\partial R} \tag{14.52}$$

where we used $V \equiv \mathscr{D}_t R$ and Eq.(14.41). Since $\mathscr{D}_t \lambda = 2V_r/R_r$ [*c.f.* Eq.(14.44)], then Eq.(14.51) provides

$$\boxed{\frac{1}{\mathscr{E}} D_t \mathscr{E} = -\frac{V}{\varepsilon + p}\frac{\partial p}{\partial R}} \tag{14.53}$$

As anticipated, in the absence of pressure gradients, $\mathscr{E}$ depends only on the radial coordinate. This is particularly so, for a pressureless configuration. Note that the definition given in Eq.(14.49) implies

$$\boxed{g_{11} \equiv e^{\lambda} = \frac{R_r}{\mathscr{E}}} \tag{14.54}$$

- THE DYNAMICS OF A SHELL OF MATTER

Consider the *space-space* components of the field equations, $G^1{}_1 = \chi T^1{}_1$. In this case, we have [*c.f.* Eq.(14.37)]

$$\frac{2}{c^2}\frac{R_{tt}}{R} e^{-\nu} - \frac{\nu_t}{c^2}\frac{R_t}{R} e^{-\nu} - \frac{R_r}{R}\nu_r e^{-\lambda} + \frac{e^{-\nu}}{c^2}\frac{R_t^2}{R^2} - \frac{R_r^2}{R^2} e^{-\lambda} + \frac{1}{R^2} = -\frac{8\pi G}{c^4} p \tag{14.55}$$

Multiply both sizes of the previous equation by $R/2$ and use Eq.(14.41) to substitute V_r. We get

$$\frac{e^{-\nu/2}}{c^2}\frac{\partial}{\partial t}\left(e^{-\nu/2}R_t\right) = -\frac{4\pi G}{c^4}pR - \frac{e^{-\lambda}R_r^2}{\varepsilon+p}\frac{\partial p}{\partial R} - \frac{1}{2R^2}\left[\frac{e^{-\nu}R_t^2 R}{c^2} - Re^{-\lambda}R_r^2 + R\right] \tag{14.56}$$

Remembering the relations given in Eq.(14.48) and Eq.(14.49), we can write Eq.(14.56) in a very compact form.

$$\boxed{\mathscr{D}_t V(r,t) = -\frac{\mathscr{E}^2(r,t)}{\varepsilon(r,t)+p(r,t)}\frac{\partial p(r,t)}{\partial R} - \frac{G}{R^2(r,t)}\left[M(r,t) + \frac{4\pi}{c^2}p(r,t)R^3(r,t)\right]} \tag{14.57}$$

Thus, the derivative *w.r.t.* the proper time of the proper velocity of a shell of (co-moving) coordinate radius r is given by the sum of two terms: the first is proportional to the pressure gradient; the second generalizes the Newtonian gravitational term, $\propto M(r,t)$, by adding the general relativistic effect associated to the "self-regeneration" of the pressure [see the discussion following, *e.g.*, Eq.(14.4)].

• THE ν FUNCTION

With the introduction of the operator $\mathscr{D}_t = e^{-\nu/2}(\partial/\partial t)$, the element g_{00} of the metric tensor formally disappeared from all the equations we have derived so far. To provide its explicit expression, let's consider the particle number density, $n(r,t)$, of a shell of comoving radius r and width Δr. The number of particles in the shell is conserved if $\partial(nR^2e^{\lambda/2})/\partial t = 0$, that is, if

$$\frac{1}{R^2e^{\lambda/2}}\frac{\partial}{\partial t}\left(R^2e^{\lambda/2}\right) = -\frac{1}{n}\frac{\partial n}{\partial t} \tag{14.58}$$

On the other hand, Eq.(14.39) yields

$$\varepsilon_t R^2e^{\lambda/2} + \varepsilon\frac{\partial}{\partial t}\left(R^2e^{\lambda/2}\right) = -p\frac{\partial}{\partial t}\left(R^2e^{\lambda/2}\right) \tag{14.59}$$

The combination of these two equations provides

$$\frac{d\varepsilon}{\varepsilon+p} = \frac{dn}{n} \tag{14.60}$$

Eq.(14.41) can then be written as follows:

$$\frac{1}{2}d\nu = -\frac{d(\varepsilon+p)}{\varepsilon+p} + \frac{dn}{n} \tag{14.61}$$

with solution

$$\frac{1}{2}\nu = -\ln(\varepsilon+p) + \ln n + \mathscr{C}(t) \tag{14.62}$$

It follows that

$$e^{\nu/2} = \frac{n}{\varepsilon+p}F(t) \tag{14.63}$$

The function $F(t) = e^{\mathscr{C}(t)}$ can be eliminated by a change of the time coordinate. In fact, if m_p is the mass of the fluid particles, we can write $F(t) = m_p c^2 f(t)$. Now, define a new time coordinate: $dt' = f(t)dt$. We are left with $nm_p c^2/(\varepsilon + p)$, that is,

$$\boxed{g_{00} = \left(\frac{\varepsilon}{\varepsilon + p}\right)^2} \tag{14.64}$$

14.5 THE GRAVITATIONAL COLLAPSE

We want to use the results of the previous section to discuss the case of a spherical, gravitational collapse. For the sake of simplicity, we will limit ourselves to the case of a pressureless collapse. This approximation is certainly better suited to study the earlier phases of the collapse. Also, neglecting pressure in General Relativity is more critical than in the Newtonian case, because of the "regeneration" of pressure [*c.f.* Eq.(14.4)]. However, a proper treatment of the collapse with a non-vanishing pressure would require attacking the problem numerically, and this is beyond the scope of the present discussion. On the other hand, imposing a pressureless collapse allows us to derive simple, semi-analytical solutions, whose formalism can be very easily compared with the Newtonian one. In fact, if pressure vanishes, we have a number of interesting simplifications to discuss:

• $g_{00} = 1$ [*c.f.* Eq.(14.64)]
Proper and coordinate time coincide; $\mathscr{D}_t \to \partial_t$; the metric writes [*c.f.* Eq.(14.54)]

$$ds^2 = dx^{0\,2} - \frac{R_r^2(r,t)}{\mathscr{E}^2(r,t)} dr^2 - R^2(r,t)[d\theta^2 + \sin^2\theta d\phi^2] \tag{14.65}$$

• $M = M(r)$ [*c.f.* Eq.(B14.1.c)]
The mass contained in each shell doesn't change with time and can be used as a convenient (Lagrangian) comoving coordinate

$$ds^2 = dx^{0\,2} - \frac{(\partial R/\partial M)^2}{\mathscr{E}^2} dM^2 - R^2(M,t)[d\theta^2 + \sin^2\theta d\phi^2] \tag{14.66}$$

• $\mathscr{E} = \mathscr{E}(M)$ [*c.f.* Eq.(14.53)]
Rewrite Eq.(14.50) to obtain a differential equation for $R(M,t)$

$$\frac{R_t{}^2(M,t)}{c^2} - \frac{2GM}{c^2 R(M,t)} = \mathscr{E}^2(M) - 1 \tag{14.67}$$

Note that the Eq.(14.67) provides a quite strict analogy with the Newtonian formalism, where $R(M,t)$ is the radius of a sphere containing a mass M and $R_t(M,t)$

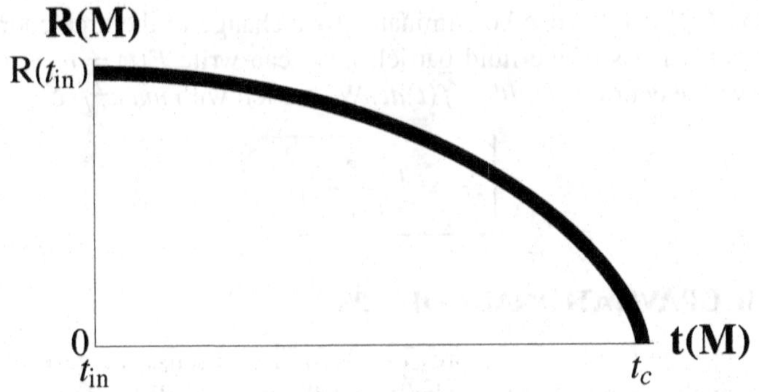

Figure 14.5 A shell containing a mass M is at rest at some initial time, t_{in}, and it collapses, in the absence of pressure, to a singularity at time t_c (black line). The collapse (proper) time as measured by an observer in free-fall with the shell is given by $t_c - t_{in}$ [*c.f.* Eq.(14.69)].

is the corresponding velocity. Then, according to Eq.(14.67) each shell is elliptic, parabolic or hyperbolic depending on whether $\mathscr{E}^2(M) - 1$ is less, equal or larger than zero. Note also that the dynamics of a shell enclosing a given mass M depends only on that mass. This is true for all the shells of the considered structure[3]. This sort of "onion" structure [see Figure 14.1] rests, in fact, on the following theorem: *the motions of particles inside any comoving sphere of symmetry are entirely determined by the matter inside the sphere.* This theorem was discussed by Bondi [12], it is a generalization of the Jebsen-Birkhoff theorem [see Section 9.4] and it is here formally stated by Eq.(14.67). Given these premises, it is not a surprise that for a bound structure, the solution of Eq.(14.67) can be written in a parametric form as follows [*c.f.* Exercise A.42]:

$$R(M,t) = \frac{GM}{c^2\mathscr{E}_-^2(M)}(1-\cos\eta) \tag{14.68a}$$

$$t = \frac{GM}{c^3\mathscr{E}_-^3(M)}(\eta - \sin\eta) \tag{14.68b}$$

where $\mathscr{E}_-^2(M) \equiv 1 - \mathscr{E}^2(M)$ and Eq.(14.68) is the classical cycloid equation. Let's now assume that at some initial time, t_{in}, $R_t(M,t_{in}) = 0$ for any value of M. Then, from Eq.(14.67), $\mathscr{E}_-^2(M) = 2GM/(c^2R_{in})$. As the mass is conserved, this also implies writing $\mathscr{E}_- = \sqrt{8\pi G\rho_{in}R_{in}^2/(3c^2)}$. According to Eq.(14.68), the collapse starts at $\eta = \pi$ and formally ends at $\eta = 2\pi$ (see Figure 14.5). Then, we can conclude that the collapse (proper) time for an observer comoving (that is, in free-fall) with the

[3] This statement is strictly true until different shells evolve independently. If there is shell crossing (an inner shell overtakes an outer shell, or an outer shell collapses faster than an inner one), this is not true anymore and the mass M cannot act as a Lagrangian coordinate anymore.

fluid is finite and equal to

$$\tau_c = \pi \frac{GM}{c^3 \mathscr{E}_-^3} = \sqrt{\frac{3\pi}{32G\rho_{in}}} \tag{14.69}$$

If the mass distribution of the structure under study is initially uniform, and initially at rest, the collapse time will be the same for all the shells, as Eq.(14.69) does not depend on the mass. The entire structure will then collapse to a singularity.

It is of interest to note that Eq.(9.23) and Eq.(14.69) yield very similar result. In fact, Eq.(9.23) evaluated for $x = 0$ provides

$$\tau_c = \frac{2}{3c} \frac{r_{in}^{3/2}}{\mathscr{R}_S^{1/2}} = \frac{1}{\sqrt{6\pi G\rho_{in}}} \tag{14.70}$$

where in this case we define ρ_{in} as the density obtained by spreading the mass M of the central point mass over a volume $4\pi r_{in}^3/3$. The difference in the numerical factors of Eq.(14.69) and Eq.(14.70) is due to the different initial conditions: in the former case the collapsing shell is bound ($\mathscr{E}^2 \equiv 1 - \mathscr{E}_-^2 \leq 0$), whereas in the latter case the motion of the test particle was assumed to be parabolic ($\mathscr{E}^2 = 1$).

15 Theoretical Cosmology

15.1 INTRODUCTION

The formalism provided by General Relativity can be used to study the global evolution of the universe. This task seems to be very complex, if not impossible. However, the problem substantially simplifies by assuming the *Cosmological Principle*: when observed on sufficiently large scales, the universe is homogeneous and isotropic. The goal of this chapter is to discuss those cosmological models that, based on the Cosmological Principle, are solutions of the field equations of General Relativity.

15.2 AN ISOTROPIC AND HOMOGENEOUS MATTER DISTRIBUTION

Let's start from the simplest case of a universe composed by an ideal, pressureless fluid. We can take advantage from the discussion of Section 14.5 and write the metric of the space-time filled by an *isotropic* matter distribution in the following form:

$$ds^2 = dx^{02} - \frac{R_r^2(r,t)}{\mathscr{E}^2(r)} dr^2 - R^2(r,t)[d\theta^2 + \sin^2\theta d\phi^2]; \qquad (15.1)$$

where r, θ and ϕ are comoving coordinates. Note that the scale-factor $R(r,t)$ has the dimension of a length, whereas r is an adimensional flag associated with a given fluid element. To be consistent with the Cosmological Principle, we have now to enforce *homogeneity*. To do so, let's consider two observers, A and B, say. The proper radial distances of B (A) from A (B) are given by the following expression:

$$\Delta\ell_{\|}\,|_A^B = \frac{R_r(r_A,t)}{\mathscr{E}^2(r_A)}\Delta r\,|_A^B\,; \qquad \Delta\ell_{\|}\,|_B^A = \frac{R_r(r_B,t)}{\mathscr{E}^2(r_B)}\Delta r\,|_B^A \qquad (15.2)$$

These relations imply

$$\frac{\partial \Delta\ell_{\|}\,|_A^B}{\partial t} = \frac{R_{rt}(r_A,t)}{R_r(r_A,t)}\Delta\ell_{\|}\,|_A^B\,; \qquad \frac{\partial \Delta\ell_{\|}\,|_B^A}{\partial t} = \frac{R_{rt}(r_B,t)}{R_r(r_B,t)}\Delta\ell_{\|}\,|_B^A \qquad (15.3)$$

As usual, the subscript r and t indicate a partial derivative *w.r.t.* r or t, respectively. Now, if the universe has to be *homogeneous*, the relative motion of A and B *cannot* depend on the position of the observer. So, in order to enforce homogeneity, we have to factorize the time and space dependence of the scale-factor given in Eq.(15.1), and write

$$\boxed{R(r,t) = \mathscr{R}(t)r} \qquad (15.4)$$

where the new scale-factor $\mathscr{R}(t)$ has still the dimension of a length. Under this condition, Eq.(15.3) yields

$$\frac{d\ell}{dt} = H(t)\ell \qquad (15.5)$$

DOI: 10.1201/9781003141259-15

This states that in a homogeneous universe, the rate of change of the proper distance between two, randomly chosen, cosmic observers is given by the rate of change of the scale-factor, the so-called *Hubble parameter*:

$$H(t) = \frac{\mathscr{R}_t(t)}{\mathscr{R}(t)} \tag{15.6}$$

In conclusion, the metric associated with an isotropic and homogeneous matter distribution writes as follows:

$$\boxed{ds^2 = dx^{0^2} - \mathscr{R}^2(t)\left[\frac{dr^2}{\mathscr{E}^2(r)} - r^2(d\theta^2 + \sin^2\theta d\phi^2\right]} \tag{15.7}$$

15.3 THE FLRW METRIC

The function $\mathscr{E}(r)$ was defined as the energy of a fluid element in units of its rest mass [*c.f.* Box 11.2]. Let's now use the field equations of General Relativity to simplify the writing of $\mathscr{E}(r)$ on the light of the Cosmological Principle.

For a pressureless fluid of energy density ε, we can write the energy momentum tensor and its trace as follows: $T^\alpha{}_\beta = \text{diag}(\varepsilon, 0, 0, 0)$; $T = \varepsilon$. Then, the field equations [*c.f.* Eq.(13.35)] provide

$$R_{\alpha\beta} = \text{diag}\left[\frac{\chi\varepsilon}{2} - \Lambda, -\left(\frac{\chi\varepsilon}{2} + \Lambda\right)g_{11}, -\left(\frac{\chi\varepsilon}{2} + \Lambda\right)g_{22}, -\left(\frac{\chi\varepsilon}{2} + \Lambda\right)g_{33}\right] \tag{15.8}$$

where $\chi = 8\pi G/c^4$. It follows that

$$R_{11} = \left(\frac{1}{2}\chi\varepsilon + \Lambda\right)\frac{R^2(t)}{\mathscr{E}^2(r)} \quad \text{and} \quad R^3_3 = -\left(\frac{1}{2}\chi\varepsilon + \Lambda\right) \tag{15.9}$$

implying that

$$\boxed{R_{11} = -\frac{\mathscr{R}^2(t)}{\mathscr{E}^2(r)}R^3{}_3} \tag{15.10}$$

On the other hand, from the definition of the Ricci tensor [*c.f.* Eq.(8.4) and Eq.(8.8)], we have [see Exercises A.43 and A.44]

$$\begin{aligned} R_{11} &= \frac{2\mathscr{R}_t^2}{c^2\mathscr{E}^2} + \frac{\mathscr{R}\mathscr{R}_{tt}}{c^2\mathscr{E}^2} - \frac{2\mathscr{E}_r}{r\mathscr{E}} \\ R^3{}_3 &= -\frac{2\mathscr{R}_t^2}{c^2\mathscr{R}^2} - \frac{\mathscr{R}_{tt}}{c^2\mathscr{R}} + \frac{\mathscr{E}^2}{r^2\mathscr{R}^2} - \frac{1}{r^2\mathscr{R}^2} + \frac{\mathscr{E}\mathscr{E}_r}{r\mathscr{R}^2} \end{aligned} \tag{15.11}$$

Thus, Eq.(15.10) implies

$$-\frac{1}{r^2\mathscr{E}^2} + \frac{1}{r^2} - \frac{\mathscr{E}_r}{r\mathscr{E}} = 0 \tag{15.12}$$

This equation can be integrated, yielding

$$\boxed{\mathscr{E}^2 = 1 - kr^2} \tag{15.13}$$

where k can assume the values of $+1$, 0 or -1. It follows that the metric given in Eq.(15.14) can be written in the following, probably more familiar form:

$$\boxed{ds^2 = dx^{0^2} - \mathscr{R}^2(t)\left[\frac{dr^2}{1-kr^2} + r^2(d\theta^2 + \sin^2\theta d\phi^2\right]} \tag{15.14}$$

This is the so-called *Friedmann-Lemaître-Robertson-Walker* (FLRW) metric describing the geometry of space-times with a matter distribution compliant with the requirements of the Cosmological Principle.

15.4 THE SPATIAL SECTOR OF THE FLRW SPACE-TIME

To have a better grasp on what Eq.(15.14) is telling us about the geometry of our universe—that is, the spatial sector of the space-time—let's use again the embedding technique. Consider first a 3D-hypersphere embedded in a 4D Euclidean space, whose metric is given by

$$d\Xi^2 = d\rho^2 + \rho^2 d\theta^2 + \rho^2 \sin^2\theta d\phi^2 + dw^2 \tag{15.15}$$

An hypersphere has a radius $\mathscr{R} = \rho^2 + w^2$. It follows that $dw^2 = \rho^2 d\rho^2/(\mathscr{R}^2 - \rho^2)$. Then, Eq.(15.15) can be rewritten as follows:

$$d\Sigma_+^2 = \mathscr{R}^2\left[\frac{d\nu^2}{1-\nu^2} + \nu^2\left(d\theta^2 + \sin^2\theta d\phi^2\right)\right] \tag{15.16}$$

Here $\nu = \rho/\mathscr{R}$ is an a dimensional (comoving) radial coordinate, whereas the subscript "+" indicates that the hypersphere has in every point a constant, positive-defined curvature [see Exercise A.45]. Embedding a 3D hyperplane, $w = const$, immediately provides

$$d\Sigma_0^2 = \mathscr{R}^2\left[d\nu^2 + \nu^2\left(d\theta^2 + \sin^2\theta d\phi^2\right)\right] \tag{15.17}$$

where now $\mathscr{R}$ is a suitable normalization constant, $\nu^2 = \rho/\mathscr{R}$ and the subscript "0" indicates that the hyperplane has in every point zero curvature. We might want to consider also the case of a 3D pseudosphere. Since its equation is given by $\rho^2 + w^2 = -\mathscr{R}^2$, it follows that $dw^2 = \rho^2 d\rho^2/(\mathscr{R}^2 + \rho^2)$. So, the metric of a 3D pseudosphere can be written as follows:

$$d\Sigma_-^2 = \mathscr{R}^2\left[\frac{d\nu^2}{1+\nu^2} + \nu^2\left(d\theta^2 + \sin^2\theta d\phi^2\right)\right] \tag{15.18}$$

Here, again, $\nu = \rho/\mathscr{R}$ is an adimensional (comoving) radial coordinate, whereas the subscript "−" indicates that the hypersphere has in every point the same, negative-defined curvature. We can conveniently write Eq.(15.16), Eq.(15.17) and Eq.(15.18) in a parametric form

$$\boxed{d\Sigma^2 = \mathscr{R}^2\left[\frac{d\nu^2}{1-k\nu^2} + \nu^2\left(d\theta^2 + \sin^2\theta d\phi^2\right)\right]} \tag{15.19}$$

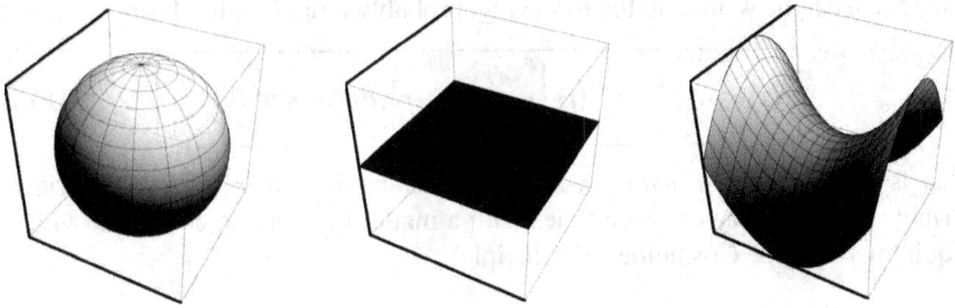

Figure 15.1 2D examples of closed ($k = +1$), flat ($k = 0$) and open ($k = -1$) surfaces. See text.

where k is the *sign* of the curvature: $+1$ for the hypersphere and 0 for the hyperplane and -1 for the 3D pseudosphere [see Figure 15.1]. On the basis of Eq.(3.48), the 3D metric of the purely spatial section (or spatial hypersurface) of the space-time can be written as $d\ell^2 = \gamma_{ik} dx^i dx^k$. In the case of Eq.(15.14), we have $\gamma_{ik} = -g_{ik}$ and then

$$\boxed{d\ell^2 = \mathscr{R}^2(t)\left[\frac{dr^2}{1-kr^2} + r^2\left(d\theta^2 + \sin^2\theta d\phi^2\right)\right]} \tag{15.20}$$

This is exactly the same expression of Eq.(15.19), a part from the time dependence of the "radius" $\mathscr{R}$. This should not be a surprise. In fact, from one hand, Eq.(15.19) describes the geometry of the only 3D hypersurfaces that are both *isotropic* and *homogeneous*—that is, with not preferred directions and with the same value of the curvature at any point. On the other hand, Eq.(15.20) is the outcome of the field equations [*c.f.* Eq.(15.10)], once we assume the homogeneity and isotropy of the space-time [*c.f.* Eq.(15.14)].

15.5 THE FRIEDMANN EQUATIONS

In the absence of pressure gradients, the function $\mathscr{E}(r)$ does not depend on time [*c.f.* Eq.(14.53)]. Also, if $g_{00} = 1$, then $V = R_t$ [*c.f.* Eq.(14.36)]. Moreover, for a homogenous matter distribution, $M = 4\pi\rho R^3/3$. It follows that Eq.(14.50) can be rewritten as follows:

$$1 - kr^2 = 1 + \frac{\mathscr{R}_t^2 r^2}{c^2} - \frac{8\pi G\rho}{3c^2}\mathscr{R}^2 r^2 \tag{15.21}$$

where we have used Eq.(15.4) and Eq.(15.13). After rearranging the various terms, we can write

$$\frac{\mathscr{R}_t^2}{\mathscr{R}^2} = \frac{8\pi G\rho}{3} - \frac{kc^2}{\mathscr{R}^2} \tag{15.22}$$

This equation can be generalized to the case of a non-vanishing Cosmological Constant by writing the vacuum energy density [*c.f.* Eq.(8.48)] as $\rho_\Lambda = 3c^2/(8\pi G)$ [*c.f.* Eq.(13.38)]. Thus, if $\rho \to \rho + \rho_\Lambda$, Eq.(15.22) provides the so-called *first Friedmann equation*

$$\boxed{\left(\frac{\mathscr{R}_t}{\mathscr{R}}\right)^2 + \frac{kc^2}{\mathscr{R}^2} = \frac{8\pi G}{3}\rho + \frac{1}{3}\Lambda c^2} \tag{15.23}$$

In the absence of pressure gradients, Eq.(14.57) simplifies in

$$R_{tt} = -\frac{G}{R^2(r,t)}\left[\frac{4\pi}{3}\rho R^3(r,t) + \frac{4\pi}{c^2}pR^3(r,t)\right] \tag{15.24}$$

that is

$$\frac{\mathscr{R}_{tt}}{\mathscr{R}} = -\frac{4\pi G}{3}\left[\rho + 3\frac{p}{c^2}\right] \tag{15.25}$$

where we used Eq.(15.4). We can again generalize the writing of Eq.(15.25) by exploiting the definition given in Eq.(13.38). Thus, if $\rho \to \rho + \rho_\Lambda$ and $p \to p + p_\Lambda$, Eq.(15.25) yields the so-called *second Friedmann equation*.

$$\boxed{\frac{\mathscr{R}_{tt}}{\mathscr{R}} = -\frac{4\pi G}{3}\left[\rho + 3\frac{p}{c^2}\right] + \frac{1}{3}\Lambda c^2} \tag{15.26}$$

15.6 EQUATION OF MOTIONS

To resolve the Friedmann equations, we have to know how the properties of the cosmic fluid evolve with time. Remember that the time component of the equations of motion [*c.f.* Eq.(13.8)] led to Eq.(14.39)

$$\boxed{\frac{\partial}{\partial t}\left(\varepsilon R^2 e^{\lambda/2}\right) + p\frac{\partial}{\partial t}\left(R^2 e^{\lambda/2}\right) = 0} \tag{15.27}$$

In the present context, $\varepsilon = \varepsilon(t)$ because of the assumed homogeneity. Moreover, $R = \mathscr{R}r$ [*c.f.* Eq.(15.4)] and $e^{\lambda/2} = \mathscr{R}/(1-kr^2)$ [*c.f.* Eq.(14.54) and Eq.(15.13)]. Thus, Eq.(15.27) can be written as follows:

$$\boxed{\frac{\partial}{\partial t}\left(\varepsilon\mathscr{R}^3\right) + p\frac{\partial}{\partial t}\left(\mathscr{R}^3\right) = 0} \tag{15.28}$$

Note that a spherical region of comoving radial coordinates r has a proper volume

$$V^{(p)} = 4\pi\mathscr{R}^3(t)\int_0^r \frac{r'^2}{\sqrt{1-kr'^2}}dr' \propto \mathscr{R}^3(t) \tag{15.29}$$

So, the variation of the energy content of this spherical region is determined by the work done by the pressure forces [*c.f.* Eq.(15.28)]. This is what is expected for an adiabatic transformation. On the other hand, adiabaticity is a necessary condition to

maintain over time both the homogeneity and the isotropy of space, as required by the Cosmological Principle. In fact, a net flux of energy would falsify the isotropy (if there is a preferential energy flow direction) or homogeneity (if the outward/inward flux is isotropic).
Note that the spatial components of Eq.(13.10) provide now an identity. In fact, for $g_{00} = 1$ and in the absence of pressure gradients, both v_r and p_r vanish [*c.f.* Eq.(14.41)].

15.7 COSMOLOGICAL PARAMETERS

The first Friedmann equation given in Eq.(15.23) shows that $\mathscr{R}(t)$ varies with time at a rate that is determined by the three key ingredients: i) the matter/energy density; ii) the space curvature; iii) the Cosmological Constant. To simplify the notation, let's work in terms of a normalized (or reduced) scale-factor

$$\boxed{a(t) \equiv \mathscr{R}(t)/\mathscr{R}_0} \tag{15.30}$$

where $\mathscr{R}_0 = \mathscr{R}(t_0)$ and t_0 is the present time. So, by construction, $a(t_0) = 1$ today. For a dust-filled universe, Eq.(15.28) implies mass conservation: $\rho(t) = \rho_0/a^3$, where ρ_0 is the matter density at the present time. Thus, Eq.(15.23) becomes

$$\frac{1}{H_0^2}\left(\frac{\dot{a}}{a}\right)^2 = \frac{8\pi G\rho_0}{3H_0^2}\frac{1}{a^3} - \frac{kc^2}{H_0^2\mathscr{R}_0^2}\frac{1}{a^2} + \frac{\Lambda c^2}{3H_0^2} \tag{15.31}$$

Here $H_0 \equiv \dot{a}(t_0)$ is the *Hubble constant*, the value of the Hubble parameter today. At the present, Eq.(15.31) writes

$$1 = \frac{8\pi G\rho_0}{3H_0^2} - \frac{kc^2}{H_0^2\mathscr{R}_0^2} + \frac{\Lambda c^2}{3H_0^2} \tag{15.32}$$

This suggests to define three cosmological parameters: the density, the curvature and the Cosmological Constant parameters

$$\boxed{\Omega_0 \equiv \frac{\rho_0}{\rho_{crit}};} \qquad \boxed{\Omega_k \equiv -\frac{kc^2}{H_0^2\mathscr{R}_0^2};} \qquad \boxed{\Omega_\Lambda \equiv \frac{\Lambda c^2}{3H_0^2}.} \tag{15.33}$$

These parameters are clearly not independent. In fact, from Eq.(15.32), we have

$$\boxed{\Omega_0 + \Omega_k + \Omega_\Lambda = 1} \tag{15.34}$$

The density parameter, Ω_0, measures the present density of the universe, ρ_0, in units of the critical density

$$\rho_{crit} \equiv \frac{3H_0^2}{8\pi G} = \left(1.88 \cdot 10^{-29} h^2\right) \text{g cm}^{-3} \tag{15.35}$$

Here $h = H_0/(100\ \mathrm{km\,s^{-1}/Mpc})$ is the normalized Hubble constant and G is the gravitational constant. Eq.(15.34) allows one to re-express Ω_k, a geometrical quantity connected with the curvature of the spatial hypersurface, in terms of two observable quantities: the matter density parameter and the Cosmological Constant term. In terms of the cosmological parameters introduced in Eq.(15.33), we can rewrite Eq.(15.23) and Eq.(15.26) as follows:

$$\frac{1}{H_0^2}\left(\frac{\dot{a}}{a}\right)^2 = \frac{\Omega_0}{a^3} + \frac{\Omega_k}{a^2} + \Omega_\Lambda \tag{15.36a}$$

$$\frac{1}{H_0^2}\frac{\ddot{a}}{a} = -\frac{1}{2}\frac{\Omega_0}{a^3} + \Omega_\Lambda \tag{15.36b}$$

We will discuss some solutions of these equations in the next sections.

15.8 THE DE SITTER MODEL

Consider the case of a flat, $\Omega_k = 0$, and empty $\Omega_0 = 0$ universe, with a non-vanishing, positive Cosmological Constant, $\Omega_\Lambda > 0$. In this case, Eq.(15.36a) provides an exponential solution

$$\mathscr{R}(t) = \mathscr{R}_* \exp\left[\sqrt{\frac{\Lambda c^2}{3}}(t - t_*)\right] \tag{15.37}$$

where $\mathscr{R}_*$ is the value of the scale-factor at some reference time t_*. It follows that the metric of this space-time is given by [*c.f.* Eq.(15.14)]

$$ds^2 = dx^{0^2} - \mathscr{R}_*^2 \exp\left[2\sqrt{\frac{\Lambda c^2}{3}}(t - t_*)\right]\left(dr^2 + r^2 d\Omega^2\right) \tag{15.38}$$

This metric doesn't have singularities (a part from the one at $t = -\infty$). So, a de Sitter universe has always existed, with a constant vacuum energy density, $\rho_\Lambda c^2 = \Lambda c^4/(8\pi G)$ [*c.f.* Eq.(13.38)], and a constant expansion rate, $\mathscr{R}_t/\mathscr{R} = \sqrt{\Lambda c^2/3}$. This model satisfies the so-called *Perfect Cosmological Principle*: the universe is homogeneous and isotropic *both* in space *and* in time. Note that it is possible to transform Eq.(15.38) in a static metric form, *à la* Schwarzschild [see Exercises A.46]

15.9 THE CLOSED FRIEDMANN UNIVERSE

In 1922, Friedmann discussed a model with $\Omega_0 > 1$, $\Omega_k = 1 - \Omega_0 < 0$ and a vanishing Cosmological Constant $\Omega_\Lambda = 0$ [35]. The spatial sector of this model has a positive curvature, $k = +1$. Note that in this case, the *rhs* of Eq.(15.36a) vanishes when the density, Ω_0/a^3, and the curvature, $(1 - \Omega_0)/a^2$, terms are equal. This happens for a specific value of the scale-factor $a_{max} = \Omega_0/(\Omega_0 - 1)$ that defines the epoch t_{max} when $\dot{a}(t_{max}) = 0$. After defining two new variables, $\xi = a/a_{max}$ and $\tau = H_0 t(\Omega_0 - 1)^{3/2}/\Omega_0$, Eq.(15.36a) becomes

$$\frac{d\xi}{d\tau} = \left(\frac{1}{\xi} - 1\right)^{1/2} \tag{15.39}$$

If we pose $\xi \equiv \sin^2(\eta/2)$, we obtain the cycloid solution

$$\boxed{\xi = \frac{1}{2}[1-\cos\eta];} \qquad \boxed{\tau = \frac{1}{2}[\eta - \sin\eta]} \tag{15.40}$$

This solution [see Figure 15.2], highlights the dynamical behavior of the scale-factor, unlike what happens in the Einstein static model [1] [see Exercise A.47]. The metric shows a singularity at $t = 0$, when $\xi = 0$. Afterwards, there is a decelerate expansion phase that ends when $\xi = 1$ or $a = a_{max}$. Since $a_{max} > 1$, this expansion phase will end in the future. At the present, $\eta_0 = 2\sin^{-1}\sqrt{\xi_0}$, $\cos(\eta_0/2) = \sqrt{1-\xi_0}$ and $\xi_0 = 1/a_{max}$. Then, Eq.(15.40) yields the predicted age of the universe

$$\boxed{t_0 = \frac{1}{H_0}\frac{\Omega_0}{(\Omega_0-1)^{3/2}}\left[\sin^{-1}\sqrt{\frac{\Omega_0-1}{\Omega_0}} - \frac{\sqrt{(\Omega_0-1)}}{\Omega_0}\right]} \tag{15.41}$$

as a function of the density parameter and of the Hubble constant. Note that $\lim_{\Omega_0\to+\infty} t_0 = 0$: the denser the universe, the younger it is. For $a > a_{max}$ the universe will collapse on itself, ending with a singularity at $\tau = \pi$, that is [see Exercise A.48]

$$t = \frac{\pi}{H_0}\frac{\Omega_0}{(\Omega_0-1)^{3/2}} \tag{15.42}$$

Again, $\lim_{\Omega_0\to+\infty} t_c = 0$: the denser the universe the faster it collapses.

15.10 THE OPEN FRIEDMANN UNIVERSE

In 1924, Friedmann discussed an open, low-density universe with a vanishing Cosmological Constant: $\Omega_0 < 1$ and $\Omega_k = 1 - \Omega_0$ [36]. For this model, the *rhs* of Eq.(15.36a) is positive defined. However, the relative importance of the density, $\Omega_0/a(t)^3$, and curvature, $\Omega_k/a(t)^2$, terms changes with time. There is an epoch when these two terms are equal. This happens for a specific value of the reduced scale-factor $a_c \equiv \Omega_0/(1-\Omega_0)$. Let's define two new variables: $\xi = a/a_c$ and $\tau = H_0 t(1-\Omega_0)^{3/2}/\Omega_0$. Then, Eq.(15.36a) becomes

$$\frac{d\xi}{d\tau} = \left(\frac{1}{\xi}+1\right)^{1/2} \tag{15.43}$$

After posing $\xi \equiv \sinh^2(\eta/2)$, we get a hyperbolic solution

$$\boxed{\xi = \frac{1}{2}(\cosh\eta - 1);} \qquad \boxed{\tau = \frac{1}{2}(\sinh\eta - \eta)} \tag{15.44}$$

As for the closed Friedmann model, the metric shows a singularity at $t = 0$, when $\xi = 0$. Then, there is a decelerate expansion phase that ends when $\xi \simeq 1$—that is, when $a(t) = a_c$. For $\eta \gg 1$, $\xi = \tau$—that is, $a(t) = H_0\Omega_k^{1/2}t$. The universe is so diluted that the curvature term dominates, providing a constant expansion velocity [see Figure 15.2 and Exercise A.49].

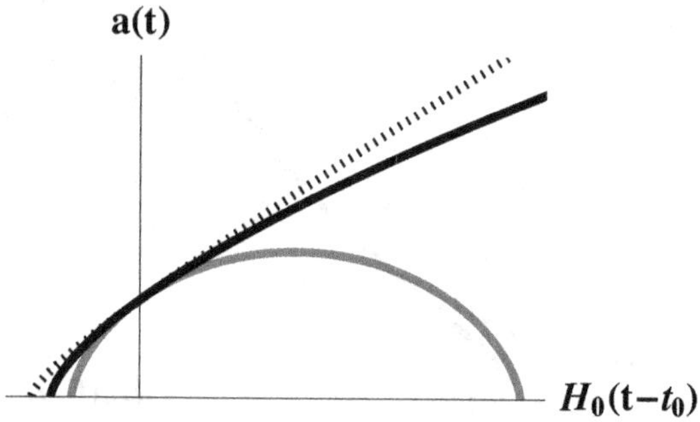

Figure 15.2 The reduced scale-factor $a(t)$ is plotted *vs.* time for the Friedmann closed model (gray line), the Einstein-de Sitter model (black line) and the open Friedmann model (dotted line). All the curves are normalized to have at the present the same expansion rate, $\dot{a}(t_0) = H_0$.

Since at the present $\eta_0 = 2\sinh^{-1}\xi_0^{1/2}$, $\cosh(\eta_0/2) = \sqrt{1+\xi_0}$ and $\xi_0 = 1/a_c$, Eq.(15.44) provides the predicted age of this model as a function of the density parameter and Hubble constant

$$\boxed{t_0 = \frac{1}{H_0}\frac{\Omega_0}{(1-\Omega_0)^{3/2}}\left[\frac{\sqrt{1-\Omega_0}}{\Omega_0} - \sinh^{-1}\sqrt{\frac{1-\Omega_0}{\Omega_0}}\right]} \tag{15.45}$$

15.11 THE EINSTEIN-DE SITTER UNIVERSE

Consider now the case of a flat universe, $\Omega_k = 0$, with a present matter density equal to the critical one, $\Omega_0 = 1$, and a vanishing Cosmological Constant, $\Omega_\Lambda = 0$. The dynamical behavior of this model was discussed in 1932 by Einstein and de Sitter [26]. For such a model, Eq.(15.36a) reduces to $(\dot{a}/a)^2 = H_0^2/a^3$, admitting a power law solution [see Figure 15.2]

$$\boxed{a(t) = \left(\frac{t}{t_0}\right)^{\frac{2}{3}}} \tag{15.46}$$

The integration constant

$$\boxed{t_0 = \frac{2}{3}\frac{1}{H_0}} \tag{15.47}$$

defines the age of the universe predicted by this model [see Exercise A.50].

15.12 A FLAT, Λ-DOMINATED UNIVERSE

There is an increasing number of observations that consistently call for a flat ($\Omega_k = 0$), low-density ($\Omega_0 < 1$) universe dominated today by a Cosmological Constant $\Omega_\Lambda = 1 - \Omega_0$. In this case, Eq.(15.36a) simplifies in

$$\left(\frac{\dot{a}}{a}\right)^2 = H_0^2\left[\frac{\Omega_0}{a^3} + \Omega_\Lambda\right] \tag{15.48}$$

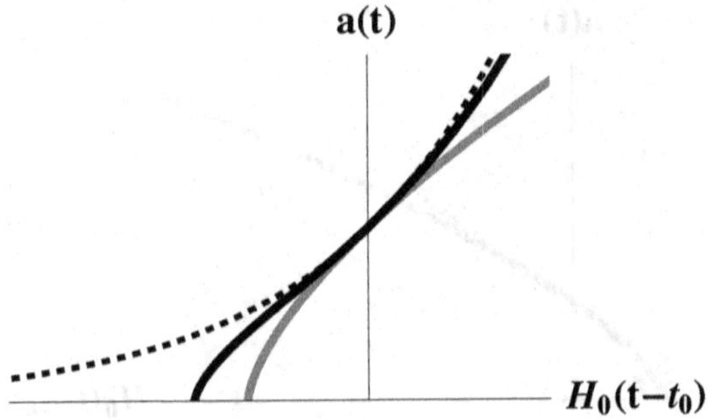

Figure 15.3 The reduced scale-factor $a(t)$ is plotted *vs.* time for three classes of models: the Einstein-de Sitter model (gray line), the concordance model (black line) and the de Sitter model (black, dashed line). All the curves are normalized to have at the present the same expansion rate, $\dot{a}(t_0) = H_0$.

The *rhs* of Eq.(15.48) is positive defined. However, the relative weight of the density, Ω_0/a^3, and Cosmological Constant, $\Omega_\Lambda = 1 - \Omega_0$, terms varies with time. They become equal when the scale-factor is equal to $a_\Lambda = (\Omega_0/\Omega_\Lambda)^{1/3}$. So, let's introduce a new variable $\xi \equiv a/a_\Lambda$ and rewrite Eq.(15.48) as follows:

$$\frac{\sqrt{\xi}\dot{\xi}}{\sqrt{1+\xi^3}} = H_0\sqrt{\Omega_\Lambda} \tag{15.49}$$

The integration of this equation provides $\sinh^{-1}\xi^{3/2} = 3\sqrt{\Omega_\Lambda}H_0 t/2$ or, equivalently,

$$\boxed{a(t) = \left(\frac{\Omega_0}{\Omega_\Lambda}\right)^{1/3} \sinh^{2/3}\left(\frac{3}{2}\sqrt{\Omega_\Lambda}H_0 t\right)} \tag{15.50}$$

[see Figure 15.3]. Note that $\lim_{\Omega_\Lambda\to 0} a(t) = (t/t_0)^{2/3}$, with $t_0 = 2/(3H_0\sqrt{\Omega_0})$, the result we would have obtained neglecting Ω_Λ in Eq.(15.48). The Ω_Λ term starts to dominate for $a \gtrsim a_\Lambda$. In this phase, Eq.(15.48) reduces to $\dot{a}/a = H_0\sqrt{\Omega_\Lambda}$, which admits an exponential solution *à la* de Sitter. The behavior of the solution given in Eq.(15.50) is plotted in Figure 15.3 together with the scale-factors of the Einstein-de Sitter and de Sitter models. Eq.(15.48) can be integrated to find the age of the universe

$$\boxed{t_0 = 2\frac{\sinh^{-1}\sqrt{\Omega_\Lambda/\Omega_0}}{3H_0\sqrt{\Omega_\Lambda}}} \tag{15.51}$$

Note that $\lim_{\Omega_0\to 1} t_0 = 2/(3H_0)$, as expected from Eq.(15.48) with $\Omega_0 = 1$.

15.13 H_0 AND THE AGE OF THE UNIVERSE

The history of the H_0 determination largely coincides with the history of cosmology. The precision on the H_0 measurements continuously improved over the years [see,

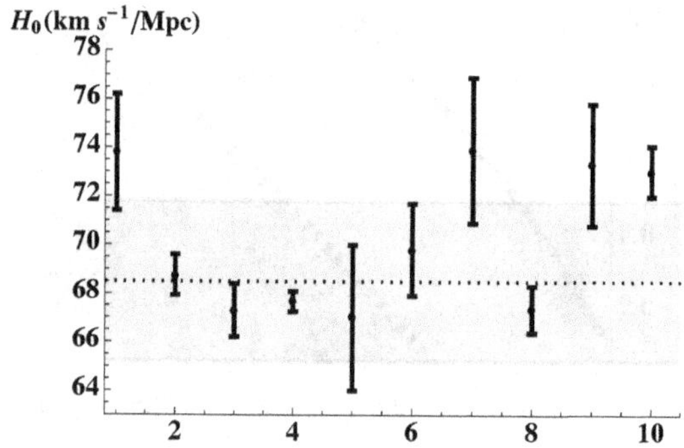

Figure 15.4 The more recent Hubble constant determinations: 1 [32]; 2 [4]; 3 [34]; 4 [10]; 5 [69]; 6 [33];7 [68]; 8 [84]; 9 [87]; 10 [77]. The gray shadowed area indicates a standard deviation of 3 kms^{-1}/Mpc around the weighted mean of 68.5 kms^{-1}/Mpc.

e.g., [31]]. This was/is due from one hand to an increasingly competitive technology and, on the other hand, to an increasingly in-depth knowledge of possible systematic errors.

We report recent H_0 determinations in Figure 15.4. We are still far from a "concordance" value, and there could be a tension between values derived by different groups with different experimental methods and techniques. However, for the goal of this book, we think it is sufficient to take "blindly" the best fit to the data points in Figure 15.4: $H_0 = (68.5 \pm 3.3)\, kms^{-1}/Mpc$. With this value of H_0, we can evaluate the age of different cosmological models.

The universe has to be older than their constituents, for example, old globular clusters. The age of these systems is well determined and constitutes a clear benchmark. Observations of Galactic globular cluster result in a lower limit on the age of the universe of 11.2 billion years (at the 95% c.l.) and in a best fit value of 13.5*Gyr* [54]. So, let's consider the predicted age of the universe in the $\Omega_0 - \Omega_\Lambda$ parameter space. This is done in Figure 15.5 where we fix the reduced Hubble constant to $h = 0.68$. Note that, for $\Omega_\Lambda = 0$, a lower limit of 11.2*Gyr* to the age of the universe implies an upper limit to the density parameter of an open universe: $\Omega_0 \lesssim 0.3$. Increasing Ω_Λ at fixed $\Omega_0 = 0.3$ increases the age of the universe quite considerably. Note that for a flat model with $\Omega_0 = 0.3$ and $\Omega_\Lambda = 0.7$ (and a reduced Hubble constant $h = 0.68$), the universe has an age $t_0 \simeq 13.5\,Gyr$ [*c.f.* Eq.(15.51)].

15.14 THE COSMOLOGICAL REDSHIFT

The field equations of General Relativity call for an expanding universe. It is then important to be able to evaluate/constrain abstract quantities as curvature and scale-factor with some observables. To do this, it is convenient to rewrite the FLRW metric

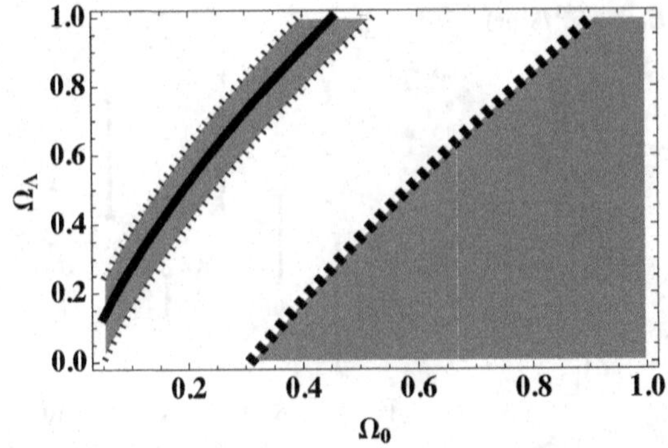

Figure 15.5 The dashed, black line refers to an age of $11.2Gyr$, taken to be a lower limit to the age of the universe: the gray-shaded region below it is then excluded at the 95% confidence level. The black continuous line corresponds to an age of $13.5Gyr$. The gray region around it identifies variations in the age of the universe by at most $\pm 5Gyr$. So, for a fixed value of Ω_0, the dotted line above (below) the black one corresponds to an age of $13Gyr$ ($14Gyr$). Here we assumed $h = 0.68$.

in the following form:

$$ds^2 = dx^{0 2} - a(t)^2 \left[dL_{\parallel}^2 + L_{\perp}^2 (d\theta^2 + \sin^2\theta d\phi^2\right] \tag{15.52}$$

where

$$dL_{\parallel} = \mathscr{R}_0 \frac{dr^2}{1 - kr^2}; \qquad L_{\perp} = \mathscr{R}_0 r; \tag{15.53}$$

The *proper* distances along the radial and transverse directions write

$$d\ell_{\parallel} = a(t) \times dL_{\parallel}; \qquad d\ell_{\perp} = a(t) \times L_{\perp} d\theta. \tag{15.54}$$

Note that, for the sake of simplicity, the transverse length is measured along a meridian—that is, at $\phi = const$. Note also that in the comoving frame, cosmic observers are at rest, whereas their proper distances change with time because of the cosmic expansion.

Consider a cosmic observer that sends us a light signal of frequency ν_{em} in the interval Δt_{em}. We observe the signal with a frequency ν_{obs} in the interval Δt_{obs}. The first wavefront, emitted at t_{em}, is observed at t_{obs}, whereas the last one (emitted at $t_{em} + \Delta t_{em}$) is observed at $t_{obs} + \Delta t_{obs}$. The condition $ds = 0$ [*c.f.* Eq.(15.30)] yields

$$\int_{t_{em}}^{t_{obs}} \frac{c\,dt}{a(t)} = \int_A^{us} dL_{\parallel} = \int_{t_{em}+\Delta t_{em}}^{t_{obs}+\Delta t_{obs}} \frac{c\,dt}{a(t)} \tag{15.55}$$

or, equivalently,

$$\int_{t_{em}}^{t_{em}+\Delta t_{em}} \frac{c\,dt}{a(t)} = \int_{t_{obs}}^{t_{obs}+\Delta t_{obs}} \frac{c\,dt}{a(t)}; \qquad \Rightarrow \qquad \boxed{\frac{\Delta t_{em}}{a(t_{em})} \simeq \Delta t_{obs}} \tag{15.56}$$

since $a \simeq const$ in both the time intervals Δt_{em} and Δt_{obs}, and $a(t_{obs}) = 1$. The last equality implies that the proper wavelength of the light is stretched because of the cosmic expansion, as all the other proper lengths [*c.f.* Eq.(15.54)]

$$\boxed{\lambda_{em} = a(t_{em})\lambda_{obs}} \tag{15.57}$$

This equation can be rewritten as follows:

$$\boxed{\frac{1}{a(t)} = 1 + z} \tag{15.58}$$

where the *cosmological redshift*, z, is defined as the fractional variation in wavelength experienced by the radiation during its travel from a source up to us: $z = (\lambda_{obs} - \lambda_{em})/\lambda_{em}$. This allows us to connect the abstract concept of a reduced scale-factor at a given time, $a(t)$, with the observed cosmological redshift of the light emitted at *that* time.

15.15 COMOVING DISTANCES

Observing far in space implies looking back in time: the light we receive today, *here and now*, was emitted *there and then*, at time t and redshift z. Light moves along null geodesics. So, Eq.(15.52) with the condition $ds = 0$ provides

$$\boxed{L_{\parallel}(z) = c\int_t^{t_0} \frac{dt'}{a(t')} = \frac{c}{H_0}\int_{a(z)}^{1} \frac{da}{\sqrt{\Omega_0 a + \Omega_k a^2 + \Omega_\Lambda a^4}}} \tag{15.59}$$

where the last equality comes first from changing integration variable ($dt = da/\dot{a}$) and then from using Eq.(15.36a). Because of Eq.(15.58), we can associate a specific line-of-sight comoving distance $L_{\parallel}(z)$ to a source observed at redshift z. The actual value of $L_{\parallel}(z)$ will obviously depend on the model, since the reduced scale-factor does. Let's see few examples.

The calculations for the closed and open Friedmann universes present several similarities. In fact, for these models, we can write

$$L_{\parallel}^{(closed)} = \frac{c/H_0}{\sqrt{\Omega_0 - 1}}\int_{\xi}^{\xi_0} \frac{d\xi'}{\xi'^{1/2}\sqrt{1-\xi'}} \tag{15.60a}$$

$$L_{\parallel}^{(open)} = \frac{c/H_0}{\sqrt{1 - \Omega_0}}\int_{\xi}^{\xi_0} \frac{d\xi'}{\xi'^{1/2}\sqrt{1+\xi'}} \tag{15.60b}$$

where we used $\xi = a(\Omega_0 - 1)/\Omega_0$ [$\xi = a(1-\Omega_0)/\Omega_0$] for $k = +1$ [$k = -1$]. The subscript "0" indicates the present—that is, $a(t_0) = 1$. The integral can be easily performed by defining $\xi = \sin^2(\eta/2)$ (for $\Omega_k < 0$) or $\xi = \sinh^2(\eta/2)$ (for $\Omega_k > 0$).

This leads to write

$$L_{\parallel}^{(closed)}(z) = \frac{2c/H_0}{\sqrt{\Omega_0 - 1}}\left(\sin^{-1}\sqrt{\frac{\Omega_0 - 1}{\Omega_0}} - \sin^{-1}\sqrt{\frac{\Omega_0 - 1}{\Omega_0(1+z)}}\right); \tag{15.61a}$$

$$L_{\parallel}^{(open)}(z) = \frac{2c/H_0}{\sqrt{1 - \Omega_0}}\left(\sinh^{-1}\sqrt{\frac{1 - \Omega_0}{\Omega_0}} - \sinh^{-1}\sqrt{\frac{1 - \Omega_0}{\Omega_0(1+z)}}\right); \tag{15.61b}$$

For the Einstein-de Sitter model, $a(t) = (t/t_0)^{2/3}$, $t_0 = 2/(3H_0)$ and $H(t) = H_0/a^{3/2}$ [see Section 15.11]. Thus, Eq.(15.59) provides

$$L_{\parallel}^{(flat)}(z) = 2\frac{c}{H_0}\left[1 - \frac{1}{\sqrt{1+z}}\right] \tag{15.62}$$

For the flat, Λ-dominated universe, the integral in Eq.(15.59) provides

$$L_{\parallel}^{(concordance)}(z) = \frac{c}{H_0}\int_{a_{em}}^{1} \frac{da}{\sqrt{\Omega_0 a + (1-\Omega_0)a^4}} \tag{15.63}$$

and it has to be performed numerically.

15.16 THE PROPER ANGULAR DIAMETER DISTANCE

Let's first define a new radial coordinate χ such that $r = \Sigma(\chi)$, where the definition of Σ depends on the curvature of the spatial hypersurface

$$\Sigma(\chi) \equiv \begin{cases} \sin\chi & k = +1 \\ \chi & k = 0 \\ \sinh\chi & k = -1 \end{cases} \tag{15.64}$$

In terms of this new variable, Eq.(15.53) can be written as follows:

$$dL_{\parallel}^{(c)} = \mathscr{R}_0 d\chi; \qquad L_{\perp} = \mathscr{R}_0\Sigma(\chi); \tag{15.65}$$

Let's now consider a ruler disposed perpendicular *w.r.t.* the line of sight at a given line-of-sight comoving distance, $L_{\parallel}(z)$. We can now extend the consideration given in Section 10.5 to the case of a FLRW metric. So, let's define the proper *angular diameter distance* as $\mathscr{D}_A(z) = \ell_{\perp}/\theta$, were θ is the angle subtended by the ruler and $\ell_{\perp} = L_{\perp}\theta/(1+z)$ is its proper length. Given Eq.(15.65), we can finally write

$$\boxed{\mathscr{D}_A(z) = \frac{\mathscr{R}_0\Sigma[\chi(z)]}{1+z}} \tag{15.66}$$

where $\chi(z) = L_{\parallel}(z)/\mathscr{R}_0$ [*c.f.* Eq.(15.65)]. We can now derive explicit analytical expressions for $\mathscr{D}_A(z)$ in a number of cosmological models.

Consider first a closed Friedmann universe with $\Omega_0 > 1$, $\Omega_k = \Omega_0 - 1$ and $\Omega_\Lambda = 0$. Let's use again the definitions given in Section 15.9: $\sin^2 \eta/2 = a/a_{max}$ and $\sin^2(\eta_0/2) = 1/a_{max} = (\Omega_0 - 1)/\Omega_0$. We can then write

$$1+z = \frac{\sin^2(\eta_0/2)}{\sin^2(\eta/2)} \tag{15.67}$$

Then, $(1-\cos\eta)(1+z) = 1-\cos\eta_0$—that is, $\cos\eta = z + \cos\eta_0/(1+z)$. It follows that [see Exercise A.51]

$$\sin\eta = \frac{2}{1+z}\sqrt{z+\cos^2(\eta_0/2)}\sin\frac{\eta_0}{2} \tag{15.68}$$

Along the light cone, $\chi = \eta_0 - \eta$ with $d\eta = c\,dt/\mathscr{R}(t)$ [see Exercises A.52 and A.53], implying $\sin\chi = \sin(\eta_0 - \eta)$. Thus, by using Eq.(15.66), we can finally obtain the wanted expression for the angular diameter distance [see Exercise A.54]. Interestingly enough we get the same expression also for an open Friedmann model [see Exercise A.55]

$$\boxed{\mathscr{D}_A(z) = \frac{2c/H_0}{\Omega_0^2(1+z)^2}\left\{\Omega_0 z + (2-\Omega_0)\left[1-\sqrt{1+\Omega_0 z}\right]\right\}} \tag{15.69}$$

For an Einstein-de Sitter universe, we have $\Omega_0 = 1$, $\Omega_k = \Omega_\Lambda = 0$, $\Sigma(\chi) = \chi$ and $L_\perp = \mathscr{R}_0\chi = L_\parallel$. It follows that angular diameter and comoving line-of-sight distances coincide. Then [*c.f.* Eq.(15.66)],

$$\boxed{\mathscr{D}_A(z) = 2\frac{c}{H_0}\frac{1}{1+z}\left[1-\frac{1}{\sqrt{1+z}}\right]} \tag{15.70}$$

Note that $\mathscr{D}_A(z)$ increases for small redshift as cz/H_0, to reach a maximum at $z = 1.25$ and to decrease for large redshifts as $2c/(H_0 z)$. Note also that Eq.(15.69) tends to Eq.(15.70) when $\Omega_0 \to 1$. Finally, for the flat, Λ-dominated universe, we still have $\Sigma(\chi) = \chi$ and

$$\boxed{\mathscr{D}_A(z) = L_\parallel^{(concordance)}/(1+z)} \tag{15.71}$$

However, we have to face again a numerical integration to find $L_\parallel$ [*c.f.* Eq.(15.63)].

The angular diameter distance is plotted in Figure 15.6. Note that, for redshift $z \lesssim 2$, the $\mathscr{D}_A$ values of the concordance model are larger than those predicted by an open model (with the same Ω_0) and by the Einstein-de Sitter model.

15.17 THE PROPER LUMINOSITY DISTANCE

Let's consider a source of intrinsic (bolometric) luminosity $\mathscr{L}$ at a given comoving distance, $L_\parallel(z) = \mathscr{R}_0\chi(z)$. For convenience, imagine the source at the center of a comoving shell with us on the border. The energy flux is given by the inverse square

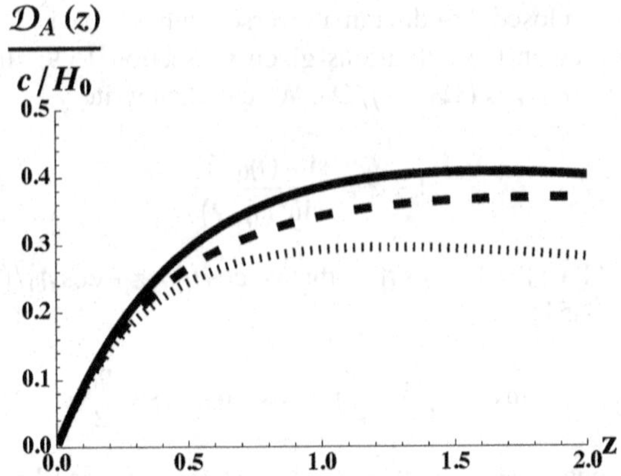

Figure 15.6 Angular diameter distance, in units of the Hubble radius, vs. redshift: for the concordance model ($\Omega_0 = 0.3$, $\Omega_\Lambda = 0.7$; continuous line); an open model ($\Omega_0 = 0.3$ and $\Omega_\Lambda = 0$; long dashed line); the Einstein-de Sitter model (dotted line).

law: $F = \mathscr{L}/S$, where $S = 4\pi\mathscr{R}_0^2\Sigma(\chi)^2$ is the surface of the comoving shell. Note, however, that this simple relation neglects two important cosmological effects. First, in an expanding universe, light is redshifted. Thus, the energy emitted by a source at redshift z is degraded by a factor $1+z$. Secondly, both luminosity and flux are given per unit time. The rates of emission and observation of photons result to be different. In fact, the number of photons, N_γ, received by the observer *per unit time* decreases, again by a factor $1+z$, *w.r.t.* the number of photons emitted per unit of time [*c.f.* Eq.(15.56) and Eq.(15.58)]

$$\frac{N_\gamma}{\Delta t_{obs}} = \frac{N_\gamma}{\Delta t_{em}} \times \frac{\Delta t_{em}}{\Delta t_{obs}} = \frac{N_\gamma}{\Delta t_{em}} \frac{1}{1+z} \tag{15.72}$$

Thus, the relation between observed energy flux and intrinsic luminosity must be correctly written as follows:

$$F = \frac{\mathscr{L}/(1+z)^2}{4\pi S} \tag{15.73}$$

In analogy with the classical case, the *proper luminosity distance* $\mathscr{D}_L$ is defined as the proper radial distance at which a source of given intrinsic luminosity and given redshift should be in order to explain the observed flux:

$$\boxed{F = \frac{\mathscr{L}}{4\pi\mathscr{D}_L(z)^2}} \tag{15.74}$$

Given Eq.(15.73), we then have

$$\boxed{\mathscr{D}_L(z) \equiv R_0\Sigma[\chi(z)](1+z)} \tag{15.75}$$

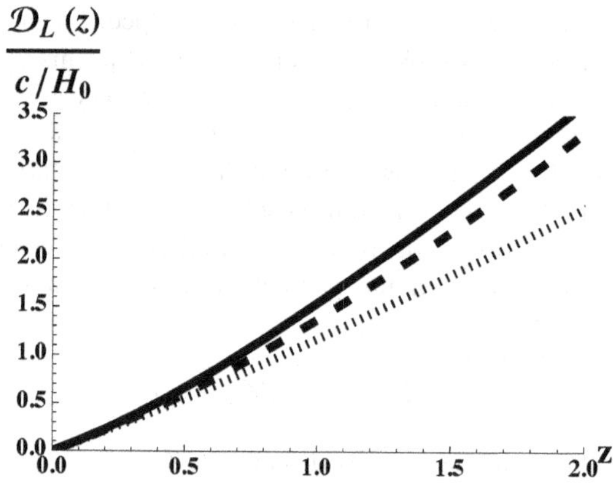

Figure 15.7 Luminosity distance, in units of the Hubble radius, for the concordance model ($\Omega_0 = 0.3$, $\Omega_\Lambda = 0.7$; continuous line), an open model ($\Omega_0 = 0.3$ and $\Omega_\Lambda = 0$; dashed line) and the Einstein-de Sitter model (dotted line)

Angular and luminosity distances are not independent [*c.f.* Eq.(15.66) and Eq.(15.75)]. Indeed,

$$\mathscr{D}_L = \mathscr{D}_A \times (1+z)^2 \tag{15.76}$$

So, we can immediately write the luminosity distance for cosmological models with $\Lambda = 0$ by multiplying Eq.(15.69) and Eq.(15.70) by $(1+z)^2$

$$\mathscr{D}_L = \frac{2c}{H_0} \begin{cases} \Omega_0^{-2} \left\{ \Omega_0 z + (2-\Omega_0)\left[1-\sqrt{1+\Omega_0 z}\right] \right\}; & \Omega_k \neq 0 \,\&\, \Omega_\Lambda = 0 \\ (1+z) - \sqrt{1+z}; & \Omega_k = 0 \,\&\, \Omega_\Lambda = 0 \end{cases} \tag{15.77}$$

[see Figure 15.7]. It is interesting to note that a Taylor expansion in z shows that, at the lowest order, the proper luminosity distance does not depend on the cosmological model

$$\boxed{\mathscr{D}_L \approx \frac{c}{H_0} z} \tag{15.78}$$

This is the well-known Hubble law, relating the cosmological redshift of a source with its (luminosity) distance. Note also that because of Eq.(15.77), at variance with $\mathscr{D}_A$, $\mathscr{D}_L$ doesn't have a maximum and it grows asymptotically as $2cz/(H_0\Omega_0)$. For flat cosmologies with $\Omega_\Lambda = 1 - \Omega_0$, the luminosity distance is given by $\mathscr{D}_L = L_\parallel(1+z)$, where $L_\parallel$ has to be numerically evaluated through Eq.(15.63). In Figure 15.7 we show the proper luminosity distance predicted by different cosmological models.

15.18 SNe La AND DARK ENERGY

Type Ia Supernovae (SNe Ia) have proved to be reliable "standardizable" candles, providing quite a power tool for constraining cosmological parameters. A breakthrough result was indeed obtained by two different collaborations: the more distant

SNe Ia are dimmer (and therefore further away) than expected in a matter-dominated universe with a vanishing Cosmological Constant [66, 76]. This was the first observational evidence in favor of a universe characterized by a late accelerated expansion driven by *dark energy*, able to make $\mathscr{D}_L$ larger than, *e.g.*, in an Einstein- de Sitter universe [*c.f.* Figure 15.7]. Although different theoretical scenarios have been proposed to explain this *dark* component, it must be stressed that the simplest case of a positive Cosmological Constant fits remarkably well the high-redshift SNe Ia data.

The standard relation between the apparent and absolute magnitudes for a single SN Ia in the B-band writes

$$m(z) = M_B + 5\log_{10}\mathscr{D}_L(Mpc) + 25 \tag{15.79}$$

Here $\mathscr{D}_L(Mpc)$ is the proper luminosity distance in Mpc. Since $\mathscr{D}_L(Mpc)$ is proportional to the Hubble radius $d_H = c/H_0$, it is convenient to define a *Hubble constant-free* luminosity distance $d_L \equiv \mathscr{D}_L/d_H$. Then, Eq.(15.79) becomes

$$m_B - \mathscr{M}_B = 5\log_{10} d_L(\Omega_0, \Omega_\Lambda) \tag{15.80}$$

where $\mathscr{M}_B = M_B + 5\log_{10} d_H + 25$. The dependence of d_L on Ω_k is not shown explicitly, as the latter is constrained by Eq. (15.34).

Consider the simple case of the two supernovae: one at low (z_1) and the other at high (z_2) redshift. Note that $\lim_{z\to 0} d_L \simeq z$ [*c.f.* Eq.(15.78)]. It follows that

$$d_L^{(2)} = z_1 10^{(m_2 - m_1)/5} \tag{15.81}$$

So, given the apparent magnitude of the two supernovae (m_2 and m_1) one can observationally evaluate $d_L(z_2)$ and compare it with the values predicted by different cosmological models. Note that this result is "Hubble constant-free" [see Exercise A.56].

In the real case, the quantity $\mathscr{M}_B$ in Eq.(15.80) can be fitted to the SNe Ia data together with the cosmological parameter Ω_0 and Ω_Λ. After marginalizing over $\mathscr{M}_B$, it is possible to define confidence level regions in the $\Omega_0 - \Omega_\Lambda$ plane. The best fit values indicate a low density ($\Omega_0 = 0.3$), flat ($\Omega_k = 0$) universe with a positive Cosmological Constant, $\Omega_\Lambda = 0.7$ [66, 76]. Being $\Omega_\Lambda = \Lambda c^2/(3H_0^2)$ [*c.f.* Eq.(15.33)], it is straightforward to verify that for $H_0 = 68\,kms^{-1}/Mpc$ the wanted value of the Cosmological Constant is $\Lambda \simeq 1.2 \times 10^{-56} cm^{-2}$. This is the so-called *Concordance Model* that also emerges from CMB anisotropy observations [see, *e.g.*, [70]].

16 The Hot Big-Bang

16.1 INTRODUCTION

The dust-filled cosmological models discussed in the previous chapter are mathematical solutions of the field equations of General Relativity that successfully confront with the observations of the *late* time universe. This can be definitely considered as a further test of General Relativity on cosmological scales. However, we have to understand how much we can trust and, possibly, test the theoretical predictions of these models at *early* times. The goal of this chapter is to discuss the experimental evidence that supports the General Relativity predictions of the early universe.

16.2 THE CMB

The discovery of the *Cosmic Microwave Background (CMB)* [65] constitutes a milestone in the study of the early universe. The CMB appeared soon to be a diffuse, highly isotropic radiation field. This corroborated the hypothesis of its cosmological origin [16]. However, more than isotropy, the key clue for the study of the early universe is the Planckian nature of the CMB. Its spectral properties were progressively tested over the years, and finally measured with an outstanding precision by the experiment FIRAS on board of the satellite COBE. The COBE/FIRAS experiment proved the CMB to be a blackbody of temperature $T_0 = (2.726 \pm 0.010)$ K, with deviations lower than 0.03% of the peak brightness over a wavelength range that goes from 0.5 to 5 mm [58, 59].

The energy density of an isotropic radiation field goes as $\varepsilon \propto a^{-4}$ if the cosmic expansion is adiabatic [*c.f.* Eq.(15.28) with $p = \varepsilon/3$]. This must be the case to preserve the universe homogeneity and isotropy [see Section 15.6]. It follows that the entropy of an isotropic, blackbody radiation field is conserved

$$S = \frac{4}{3}\mathscr{A}VT^3 \propto a^3(t)T^3(t) \tag{16.1}$$

where $\mathscr{A} = \pi^2 k^4/[15(\hbar c)^3]$ is the radiation density constant. All this leads to conclude that the CMB temperature varies with time

$$T(t) = \frac{T_0}{a(t)} = T_0(1+z) \tag{16.2}$$

Thus, the more we go back in time, the higher is the CMB temperature. It is worth noting that the predictions of Eq.(16.2) have been tested observationally, although only "locally"—that is, for redshifts $z \approx 1$. The CMB excites the rotational lines of carbon monoxide molecule. Thus, by studying the CO absorption lines in quasar spectra we can constrain $T(z)$ up to $z \sim 3$. Indeed, it is found that $T(z) = (2.725 \pm$

DOI: 10.1201/9781003141259-16

0.02) $\times (1+z)^{1-\beta}$ with $\beta = -0.007 \pm 0.027$, in perfect agreement with Eq.(16.2) [61].

To underline why the Planckian nature of the CMB is so important, let's look to its brightness

$$\boxed{I_\nu = \frac{1}{2\pi^2}\frac{(kT)^3}{(\hbar c)^2}\left[\frac{x^3}{e^x - 1}\right]} \tag{16.3}$$

Here the pre-factor, $\propto a^{-3}(t)$, indicates that the CMB brightness increases going back on time. In addition to this, note that the quantity $x = 2\pi\hbar\nu/kT$ is time-independent: because of the cosmic expansion, both $\nu(t)$ and $T(t)$ vary with time as $a^{-1}(t)$. It follows that the Planckian shape of the CMB spectrum—defined by the square parenthesis in Eq.(16.3)—is preserved during the cosmic expansion. In other words, the spectral properties of the CMB witness the period in the early universe when matter and radiation were in thermal contact, even if today they are clearly decoupled. This is a distinctive feature of the cosmological models expanding from a singularity, and this is why one can refer to the class of Friedmann models as to the *hot Big-Bang models*.

16.3 A RADIATION-DOMINATED UNIVERSE

A diffuse background of relativistic particles must have an impact on the cosmic expansion. We can still use Eq.(15.23), but we have now to interpret the mass density as the sum of the relativistic and non-relativistic components: $\rho(t) = \rho_{ER}(t) + \rho_{NR}(t)$, where $\rho_{ER}(t) = \rho_{ER}(t_0)/a^4$ and $\rho_{NR}(t) = \rho_{NR}(t_0)/a^3$. Following what was done in Section 15.7, we can then write Eq.(15.36a) as follows:

$$\frac{H^2(t)}{H_0^2} = \left[\frac{\Omega_{ER}}{a^4} + \frac{\Omega_0}{a^3} + \frac{\Omega_k}{a^2} + \Omega_\Lambda\right] \tag{16.4}$$

where $\Omega_{ER} \equiv \rho_{ER}(t_0)/\rho_{crit}$ and $\Omega_0 = \rho_{NR}(t_0)/\rho_{crit}$ are the density parameters of the relativistic and non-relativistic components, respectively. These two components are equally contributing to the cosmic expansion rate at the so-called *equivalence epoch* [see Exercise A.57]

$$a_{eq} \equiv \frac{\Omega_{ER}}{\Omega_0} \tag{16.5}$$

For $a \gtrsim a_{eq}$ the universe is *matter-dominated*, the relativistic component is not dynamically important, and in this regime all the considerations done in Chapter 15 remain valid. For $a \lesssim a_{eq}$ the reverse is true: the universe is *radiation-dominated*, the *rhs* of Eq.(16.4) is dominated by the first, relativistic term and the non-relativistic component can be ignored together with the curvature and the Cosmological Constant terms. In this regime, Eq.(16.4) simplifies in

$$\left(\frac{\dot{a}}{a}\right)^2 = H_0^2\frac{\Omega_{ER}}{a^4} \tag{16.6}$$

providing $a(t) = \sqrt{2\Omega_{ER}^{1/2} H_0} \times t^{1/2}$ [see Exercise A.58]. This allows to derive a very useful *time-energy* relation. In fact, remembering the definition of Ω_{ER}, we can write

$$\boxed{t = \sqrt{\frac{3c^2}{32\pi G \varepsilon_{ER}(t)}}} \tag{16.7}$$

where $\varepsilon_{ER}(t) \equiv \rho_{ER}(t)c^2$ is the total energy density due to all the relativistic components.

Let's restrict our discussion to the energy range between 1 and $100MeV$. In this phase, relativistic particles such as photons, electrons and positrons, neutrinos and antineutrinos all contribute to ε_{ER}. There are of course also non-relativistic particles such as neutron and protons, but they do not contribute to the expansion rate of a radiation-dominated universe. In this energy range, relativistic particles are kept in equilibrium by weak

$$n + e^+ \Longleftrightarrow p + \bar{\nu}_e; \qquad n + \nu_e \Longleftrightarrow p + e^-; \qquad n \Longleftrightarrow p + e^- + \bar{\nu}_e \tag{16.8}$$

and electromagnetic $e^+ + e^- \Longleftrightarrow \gamma\gamma$ interactions. So, for $1 \lesssim E(MeV) \lesssim 100$, we can write $\varepsilon_{ER} = g_*(\pi^2/30)k^4T^4/(\hbar^3c^3)$. The effective statistical weight is defined as follows:

$$g_* = \left[\sum_{i,bosons} g_i + \frac{7}{8}\sum_{i,fermions} g_i\right] = g_\gamma + \frac{7}{8}\left(g_{e^-} + g_{e^+} + g_\nu \times N_{eff}\right) = 10.75 \tag{16.9}$$

where the g_i is the statistical weight of the $i-th$ component: $g_\gamma = 2$, $g_{e^-} = 2$, $g_{e^+} = 2$, $g_\nu = 2$ for photons, electrons, positrons, neutrino, and antineutrino, respectively. In Eq.(16.9) N_{eff} is the effective number of neutrino families, taken to be equal to three [91]. The contributions of fermions and bosons to the total energy density are different because of the different statistics they obey to. In the range $1 \lesssim E(MeV) \lesssim 100$, Eq.(16.7) provides

$$\boxed{t = \frac{0.74s}{T^2(MeV)}} \tag{16.10}$$

Clearly, this relation is valid for a radiation-dominated universe *and* for a very specific energy interval, $1 \lesssim E(MeV) \lesssim 100$.

16.4 THE NEUTRON-TO-BARYON RATIO

In the energy range between 1 and $100MeV$, neutron and protons are not relativistic, their mass being $\approx 1GeV$. However, they continuously transform the ones in the others due to the weak reactions of Eq.(16.8). In equilibrium conditions, the neutron-to-baryon ratio would be given by the Boltzmann factor $n_n^{(0)}/n_p^{(0)} = \mathrm{e}^{-\Delta m/T}$ where $\Delta m = m_n - m_p = 1.293MeV$ is the difference between the neutron and the proton masses. It would then seem that the neutron-to-proton ratio decreases exponentially.

If this were the case, today we shouldn't expect any survived neutron to be around. The point is that, in general, the reactions given in Eq.(16.8) can occur also in non-equilibrium conditions. In this case, there are two competing rates to be taken into account. From one hand, we have the total neutron-to-proton conversion rate that takes into account all the three reactions of Eq.(16.8). For the temperature of interest here, there is a good fit to this rate [17,47,96]: $\lambda_{n\to p}(x) \simeq 255\left(12+6x+x^2\right)/(\tau_n x^5)$, where $x \equiv \Delta m/T$ and $\tau_n = (885.7 \pm 0.8)s$ is free-neutron decay time scale [100]. The other competing effect is provided by the cosmic expansion rate, $H(x) = 1.13s^{-1}/x^2$ [see Exercise A.59].

These two rates, $\lambda_{n\to p}$ and H, are equal for $x = 1.9$, which corresponds to $T \simeq 0.7\,MeV$. It follows that for $T \gtrsim 0.7\,MeV$, $\lambda_{n\to p}(x) > H(x)$, the neutron-to-proton conversion is quite efficient, and it occurs on time scales much shorter than those of the cosmic expansion. For $T \lesssim 0.7\,MeV$, $\lambda_{n\to p}(x) < H(x)$, the conversion becomes less and less effective and the neutron-to-baryon ratio progressively freezes out to the value

$$X_n = 0.15 \tag{16.11}$$

where $X_n = n_n/(n_n+n_p)$ is the *neutron-to-baryon* ratio.

16.5 A ν'S COSMIC BACKGROUND

After the freezing out of the weak interactions, neutrinos decoupled from the rest of the other cosmological constituents. We are then left with a homogeneous Cosmic Neutrinos Background-CνB, homogeneously distributed over the space, with an energy density $\varepsilon_\nu = g_\nu/(2\pi^2)\,(kT_\nu)^4\,(\hbar c)^{-3}\,7\pi^4/120$. The C$\nu$B temperature decreases as a^{-1}, because of the adiabatic nature of the cosmic expansion. Before decoupling ($T \gtrsim 0.7\,MeV$), neutrinos share the same temperature with all the other relativistic particles. However, at $T \lesssim 0.5\,MeV$ this is not true anymore. In fact, at that temperature, electrons and positrons annihilate. Being already decoupled, neutrinos are basically unaffected by this process that instead produces a heating of the photon gas. Let's evaluate the relevance of such a process. Note that before the freezing of the weak reactions photons, electrons, positrons, and N_{eff} families of neutrinos contribute to the entropy of the relativistic gas

$$S_- \propto \left[g_\gamma + \frac{7}{8}\left(g_{e^-} + g_{e^+} + g_\nu N_{eff}\right)\right] T_-^3 a^3(t_) \tag{16.12}$$

On the contrary, after the e^+e^- annihilation, only the CMB and the CνB contribute to the total entropy

$$S_+ \propto \left[g_\gamma T_{\gamma+}^3 + \frac{7}{8} g_\nu N_{eff} T_{\nu+}^3\right] a^3(t_+) \tag{16.13}$$

Electrons and positrons clearly contribute to S_-, but not to S_+. Before the e^+e^- annihilation, there is a single equilibrium temperature, T_-, for all the relativistic cosmic components. After annihilation, the photon ($T_{\gamma+}$) and the neutrino ($T_{\nu+}$)

temperatures are different. Since the cosmic expansion is adiabatic, entropy is conserved and, then, for $N_{eff} = 3$, we have

$$[T_- \times a(t_-)]^3 \frac{43}{4} = \left\{ 2\left(\frac{T_{\gamma+}}{T_{\nu+}}\right)^3 + \frac{42}{8} \right\} [T_{\nu+} \times a(t_+)^3] \tag{16.14}$$

where we assumed an instantaneous annihilation of the e^+e^- pairs. Exploiting again the adiabatic nature of the cosmic expansion, we can also conclude that $T_{\nu+} \times a(t_+) = T_{\nu-} \times a(t_-) = T_- \times a(t_-)$: before annihilation, $T_{\nu-}$ is the temperature of all the other relativistic components, photons included. Then, Eq.(16.14) provides

$$\boxed{T_{\nu+} = \left(\frac{4}{11}\right)^{1/3} T_{\gamma+}} \tag{16.15}$$

The CνB is cooler than the CMB, with a present temperature of $T_\nu(t_0) = 1.97K$. As mentioned above, after the $e + e^-$ annihilation, the only relativistic particles contributing to ε_{ER} are neutrinos and photons. In this regime the effective statistical weight results to be [*c.f.* Eq.(16.9)]

$$g_\star \equiv \left[g_\gamma + \frac{7}{8}\left(\frac{4}{11}\right)^{4/3} g_\nu N_{eff} \right] = 3.36 \tag{16.16}$$

which provides the right time-temperature relation for a radiation-dominated universe with temeperature $T \lesssim 0.5MeV$

$$\boxed{t = 5.3s \left[\frac{0.5MeV}{T(MeV)}\right]^2} \tag{16.17}$$

16.6 PRIMORDIAL NUCLEOSYNTHESIS

For an appropriate energy range, we do expect that the synthesis of light nuclei could occur in the early universe The neutron-to-proton ratio is the key ingredient to provide reliable predictions. Let's stress that the value X_n given at the end of Section 16.4 [*c.f.* Eq.(16.11)] doesn't take into account two important and competing physical processes. From one hand, free-neutrons decay on a time scale τ_n in protons, electrons, and electron anti-neutrinos [see Eq.(16.8)]. On the other hand, the synthesis of light elements stabilizes neutrons in atomic nuclei. To evaluate the relative importance of these two processes, we must first know when primordial nucleosynthesis actually starts. Deuterium (D) is clearly the first isotope to be synthesized via the reaction

$$p + n \Longleftrightarrow D + \gamma \tag{16.18}$$

The binding energy of D nucleus is $B_D \equiv m_n + m_p - m_D = 2.22MeV$. For temperatures higher than this, Deuterons can form at high rate, but they are immediately photo-dissociated by the high energy CMB photons. This is an important point. How many CMB photons do we have in the universe? To answer to this question, let's evaluate their number density $n_\gamma(t) = 2\zeta(3)\,[kT(t)]^3\,[\hbar c]^{-3}/\pi^2$, where ζ is the Riemann function. At the present $T(t_0) = 2.726K$ and, then, $n_\gamma(t_0) \simeq 410\,cm^{-3}$. Note that today the baryonic number density is much lower: $n_b(t_0) \simeq 1.1\cdot 10^{-5}\Omega_b h^2\,cm^{-3}$, $\Omega_b = \rho_{b0}/\rho_{crit}$ being the baryonic density parameter [*c.f.* Eq.(15.35)]. As a result, the baryon-to-photon ratio, which is constant in time, results to be

$$\boxed{\eta_b \equiv \frac{n_b}{n_\gamma} = 6\times 10^{-10}\left(\frac{\Omega_b h^2}{0.022}\right)} \tag{16.19}$$

Because of their large abundance, the number of CMB photons with an energy $\gtrsim B_D$ equals the number of baryons only when the universe cooled down to a temperature of $82keV$, almost a factor 3 smaller than B_D [94]. It follows that the reaction $n+p \Rightarrow D+\gamma$ becomes effective in forming Deuterons only when $T \lesssim 82keV$ [see Exercise A.60]. This occurs when the universe had an age of [*c.f.* Eq.(16.17)]

$$t_{nuc} = 5.3s\left[\frac{0.5MeV}{0.082MeV}\right]^2 \simeq 196s \tag{16.20}$$

So, at the onset of the primordial nucleosynthesis—roughly $3'$ after the Big-Bang–free-neutron decay lowered the abundance of free-neutrons by a factor $e^{-t_{nuc}/\tau_n} \simeq 0.80$ Thus, a more reliableat estimate of the neutron-to-baryon ratio is given by

$$\boxed{X_n(T_{nuc}) \approx 0.123} \tag{16.21}$$

16.7 PRIMORDIAL ABUNDANCE OF LIGHT NUCLEI

Once Deuterons formed at $T \sim 82keV$, other light nuclei—such as ^3H, ^3He, ^4He, ^7Li, and ^7Be—can be synthesized on a very short time scale, accordingly to the scheme shown in Figure 16.6.

It is worth remembering two facts: i) the ^4He nuclei have the largest binding energy among the light elements; ii) there are not tightly bound isotopes with mass number between 5 and 8. Hence, starting with neutrons and protons, it is impossible to build elements heavier than ^4He. This occurs in stars, via the triple-α reaction $^4He+^4He+^4He \to ^{12}C$. However, this reaction requires high temperatures *and* high densities. It is interesting to note that the last condition is not satisfied at the onset of primordial nucleosynthesis. Indeed, at $T \simeq 82keV \simeq 10^9K$ the baryonic density is $\rho_b(t_{nuc}) \simeq 10^{-3}\Omega_b h^2 g\,cm^{-3}$, lower by at least three order of magnitudes *w.r.t.* the water density.

On the basis of this consideration, it is not unreasonable to assume that, to zeroth order, all neutrons are captured in ^4He nuclei. Under this hypothesis, the neutron abundance is simply twice the abundance of ^4He nuclei: $n_n = 2n_{^4He}$. By convention,

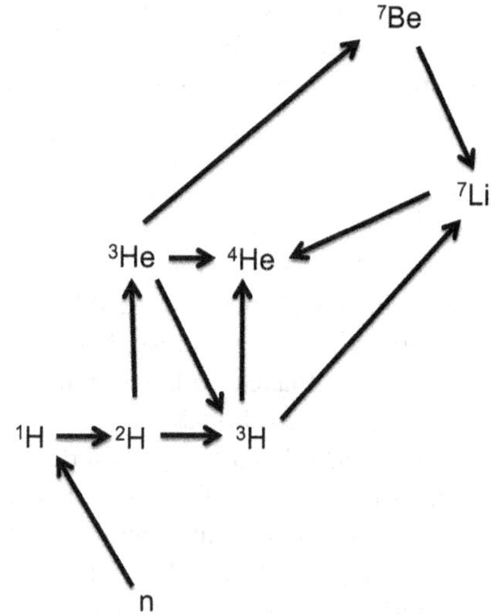

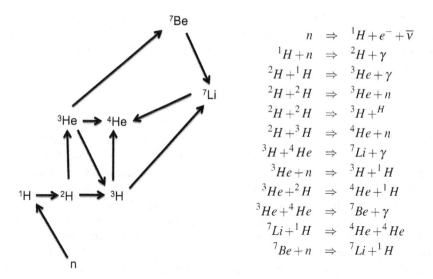

$$\begin{array}{rcl} n & \Rightarrow & {}^1H+e^-+\overline{\nu} \\ {}^1H+n & \Rightarrow & {}^2H+\gamma \\ {}^2H+{}^1H & \Rightarrow & {}^3He+\gamma \\ {}^2H+{}^2H & \Rightarrow & {}^3He+n \\ {}^2H+{}^2H & \Rightarrow & {}^3H+{}^H \\ {}^2H+{}^3H & \Rightarrow & {}^4He+n \\ {}^3H+{}^4He & \Rightarrow & {}^7Li+\gamma \\ {}^3He+n & \Rightarrow & {}^3H+{}^1H \\ {}^3He+{}^2H & \Rightarrow & {}^4He+{}^1H \\ {}^3He+{}^4He & \Rightarrow & {}^7Be+\gamma \\ {}^7Li+{}^1H & \Rightarrow & {}^4He+{}^4He \\ {}^7Be+n & \Rightarrow & {}^7Li+{}^1H \end{array}$$

Figure 16.1 The main reactions involved in the primordial nucleosynthesis process

the Helium mass fraction, Y_p, is defined as the mass in ${}^4He$ over the total mass:

$$Y_p \equiv \frac{N_{{}^4He} m_{{}^4He}}{N_n m_n + N_p m_p} = \frac{4 n_{{}^4He}}{n_n + n_p} = 2X_n \qquad (16.22)$$

For $X_n = 0.123$ [*c.f.* Eq.(16.21)], we can predict that $Y_p \simeq 0.246$. This is consistent with the conservative upper limit $Y_p \leqslant 0.251 \pm 0.002$ [38].

If all available neutrons end up in ${}^4He$ nuclei, we don't expect any surviving Deuterons from the early universe. This is clearly an oversimplification. In fact, proper and detailed calculations of primordial nucleosynthesis show that Deuteron abundance first decreases and then freezes out to a non-vanishing value [38]. Interestingly enough, the surviving Deuterium abundance—measured in terms of Deuteron-to-proton ratio, D/H—is very sensitive to the baryon abundance or, equivalently, to the baryon-to-photon ratio, η_b. This is not unexpected: the higher the value of η_b, the more efficient the burning of D in heavier nuclei and, then, the lower its residual abundance. An accurate numerical fit for $5.7 \times 10^{-10} \lesssim \eta_b \lesssim 6.7 \times 10^{-10}$ provides [38]

$$\left.\frac{D}{H}\right|_{th} = 2.68 \times 10^{-5}(1 \pm 0.03)\left(\frac{6 \times 10^{-10}}{\eta_b}\right)^{1.6} \qquad (16.23)$$

with a quite steep dependence on η_b. This allows the D/H abundance to be a very effective "baryometer": indeed, any measure of the primordial D/H ratio constrains η_b and, then, $\Omega_b h^2$ [*c.f.* Eq.(16.19)]. Clearly, the residual fraction of primordial Deuterium will be destroyed by the nuclear reactions occurring in stars. Therefore, any measure of the D/H ratio represents only a lower limit to the primordial Deuterium abundance. However, observations of either low-metallicity or high redshift systems could provide reasonable estimates of the primordial D/H ratio. A conservative measure of the D/H ratio is provided by [38]:

$$\left.\frac{D}{H}\right|_{obs} = \left(2.68^{+0.27}_{-0.25}\right) \times 10^{-5} \tag{16.24}$$

This value is consistent with $\eta_b \simeq 6 \times 10^{-10}$ and, then, with $\Omega_b h^2 \simeq 0.022$ [*c.f.* Eq(16.19), Eq.(16.23), and Eq.(16.24)]. It is very important to stress that this result is fully consistent with the results obtained by the Planck collaboration: $\eta_b = (6.01 \pm 0.06) \times 10^{-10}$, corresponding to $\Omega_b h^2 = 0.02205 \pm 0.00028$ [69].

Let's conclude this Section by reminding that the SNe Ia data constrain the non-relativistic content of the universe to have a density parameter $\Omega_0 \simeq 0.3$. For $h = 0.68$, the primordial D/H ratio constrains the baryonic component to be $\Omega_b \simeq 0.04$. This is one of the strongest arguments for requiring that, beside baryons, a non-relativistic *dark matter* component actually contribute to the non-relativistic mass density of the universe. What is dark matter is an open question, a forefront research in astrophysics and particle physics.

16.8 THE RECOMBINATION OF THE PRIMORDIAL PLASMA

After primordial nucleosynthesis ended, baryons and photons remain thermally coupled together via Compton and Coulomb scattering processes. Both these physical mechanisms act over a quite large period of time, from about $200s$ up to roughly $400,000y$ after the Big-Bang. During this period, there is a sufficiently large number of photons with energies larger than the binding energy of hydrogen atoms, $B_H = 13.6eV$. As for the Deuterium bottle-neck, we have to wait for the universe to cool until the number of photons with energy $2\pi\hbar\nu > B_H$ is equal to the number of protons. When this happens, there is the transition from the primordial plasma phase to a neutral one, which leads to the formation of hydrogen

$$p + e^- \Rightarrow H + \gamma \tag{16.25}$$

This process is referred to as the *recombination* of the primordial plasma. Helium atoms formed earlier and, for sake of simplicity, we neglect their contribution to the overall recombination process.

Let's define the free-electron fraction as the fractional abundance of the free electrons *w.r.t.* electrons and hydrogen atoms:

$$X_e \equiv \frac{n_e}{n_e + n_H} = \frac{n_p}{n_p + n_H} \tag{16.26}$$

the last equality following from the requirement that the universe has not a net electric charge. To zeroth order, we could study how $X_e(t)$ changes with time by using the Saha approximation. However, there are two reasons for the Saha solution not to work properly. First, the Saha equation assumes direct ground state recombination. Secondly, the $13.6\,eV$ photon released during an hydrogen atom formation immediately reionize another hydrogen atom already formed. Therefore, recombination to the ground state is ineffective. Thirdly, and more important, when X_e decreases, the rate of hydrogen formation drops and the reaction of Eq.(16.25) progressively gets out of equilibrium. Then, we have to be back to proper, non-equilibrium calculations.

The most important outcome of these calculations is to identify which redshift layer(s) is (are) directly observable *via* the CMB. In order to answer this question, let's first evaluate the optical depth for Thomson scattering

$$\tau = \int_t^{t_0} n_e \sigma_T c dt \sigma_T c \int_a^1 n_b X_e \frac{1}{H(a)} d\ln a \tag{16.27}$$

where $n_b(t) = n_b(t_0)/a^3(t)$ [*c.f.* Eq.(16.6)] and $H(a) = H_0\Omega_0^{1/2} a^{-3/2}$, a good approximation for the period of interest. In terms of redshift,

$$\boxed{\tau(z) = 8.75 \times 10^{-4} \frac{\Omega_b h^2}{0.022} \sqrt{\frac{0.15}{\Omega_0 h^2}} \int_0^{1+z} (1+x)^{1/2} X_e(x) dx} \tag{16.28}$$

Given the optical depth, we can evaluate the probability density for a CMB photon to be last-scattered by a free electron. This is given by the *visibility function*

$$\boxed{g(z) = \mathrm{e}^{-\tau(z)} \frac{d\tau}{dz}} \tag{16.29}$$

Proper calculations show that the visibility function has a peak at $z = 1064$, weakly dependent on the considered cosmological model [13, 50]. It is then common to talk about the *last scattering surface*, as the surface where most of the CMB photons experience their last scatter [see Figure 16.2]. It must be noted, however, that the visibility function has a non-negligible width around its maximum and it is quit skew. The half width half maximum (HWHM) of the visibility function for the concordance model is about 110, corresponding to a Gaussian dispersion $\sigma = HWHM/(\sqrt{2\ln 2}) \approx 80$. This means that the last scattering surface samples a redshift shell centered on $1 + z_{rec} \simeq 1064$ (corresponding to $a_{rec} \approx 10^{-3}$) with a width $\Delta z \approx 80$ [see Exercise A.61].

To conclude, the importance of the recombination process stands on the fact that all free electrons disappear to form neutral hydrogen atoms. It follows that photons can't Compton scatter against free-electrons, as it happened before recombination in the so-called plasma phase. This exactly why photons can free-stream from the last scattering surface up to now, bringing with them a wealth of information about the universe at roughly $400,000\,yr$ after the Big-Bang.

16.9 THE PUZZLES OF THE STANDARD MODEL

The Friedmann models are nowadays supported by quite a number of independent observations. The universe is, on the average, very homogeneous and isotropic, and

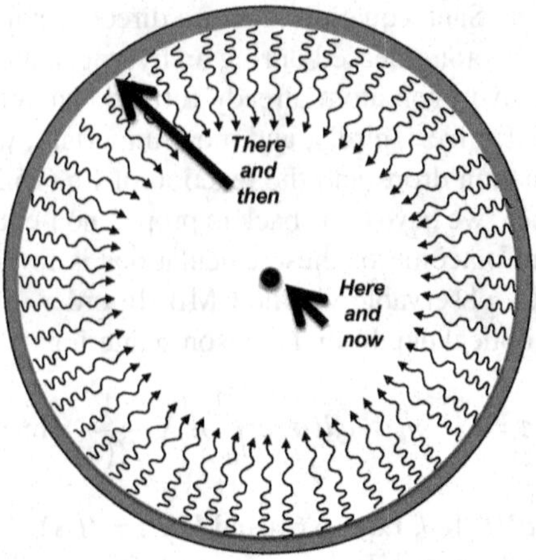

Figure 16.2 Most of the CMB photons were last scattered *there and then*, at a redshift $1+z \simeq 1064 \pm 80$. The last scattering surface (gray thick line) is a shell that surrounds any cosmic observer and from which the CMB photons can free-stream freely toward the observer, *here and now*, see text.

it expands at a rate which has only recently become constant. However, beyond all its successes, the hot Big Bang model leaves a number of open questions, such as the nature of dark matter and dark energy, and, at an even more basic level, a couple of very known "puzzles".

Let's start from the so-called *horizon problem*. On the last scattering surface, an angle θ subtends a comoving length $\ell_\perp^{(c)} = \ell_\perp(1+z)$. The high redshift limit of Eq.(15.69), written for a vanishing Cosmological Constant, provides $\mathscr{D}_A = (2c/H_0)\Omega^{-1}(1+z)^{-1}$. Then, $\lim_{z\to\infty}\theta = \Omega_0 \ell_\perp^{(c)}/(2c/H_0)$. After substituting the numerical values, we can write

$$\theta(deg) \simeq \Omega_0 \left[\frac{\ell_\perp^{(c)}(h^{-1}Mpc)}{100h^{-1}Mpc}\right] \tag{16.30}$$

It follows that in an Einstein- de Sitter universe $\approx 1^\circ$ subtends on the last scattering surface a comoving scale of $\approx 100h^{-1}Mpc$. On the other hand, the comoving size of the region casually connected at the epoch of last scattering is instead given by

$$L_H^{(c)}(a_{rec}) = \int_0^{t_{rec}} \frac{cdt}{a(t)} = \int_0^{a_{rec}} \frac{cda}{a^2 H(a)} \tag{16.31}$$

where $H(a) \simeq H_0\sqrt{\Omega_{ER}a^{-4} + \Omega_0 a^{-3}}$, as at early times we can neglect the contribution to the expansion rate of the curvature and of the Λ terms [*c.f.* Eq.(16.4)]. The integration of Eq.(16.31) provides

$$L_H^{(c)}(a_{rec}) = 2\frac{c}{H_0}\frac{1}{\sqrt{\Omega_0}}\left[\sqrt{a_{eq}+a_{rec}} - \sqrt{a_{eq}}\right] \tag{16.32}$$

where $a_{eq} = \Omega_{ER}/\Omega_0 = 4.14 \times 10^{-5}/(\Omega_0 h^2)$ and $a_{rec} \simeq 10^{-3}$ [*c.f.* Eq.(16.5) and Section 16.8]. It turns out that, for $h = 0.68$, Eq.(16.32) is well fitted by $L_H^{(c)}(a_{rec}) \simeq 210\Omega_0^{-0.28} Mpc$. Then, the angle subtended by the causally connected region on the last scattering surface is [*c.f.* Eq.(16.30)]

$$\theta_H \approx 1.5\Omega_0^{0.72} deg \tag{16.33}$$

So, if we observe the CMB sky in two different directions, more than $\approx 1.5\Omega_0^{0.72} deg$ apart, we probe regions that *were not* in causal contact at the last scattering epoch. On the other hand, the CMB is observed to have the same intensity all over the sky, by one part over 10^5. So, how did two unconnected regions manage to reach the same physical properties? This seems to be an obvious paradox, the so-called *horizon problem.*

The other puzzle is known as the *curvature problem.* We can obviously use the definition given in Eq.(15.33) at an arbitrary time. Then, in terms of the today curvature parameter we can write

$$\Omega_k(t) \equiv -\frac{kc^2}{H^2(t)\mathscr{R}^2(t)} \simeq \Omega_k(t_0)\frac{a^2}{\Omega_{ER}} \tag{16.34}$$

where in the Friedmann equation we consider only the contribution of the relativistic backgrounds [*c.f.* Eq.(16.4)]. Note that the sign of the curvature is conserved, as we should have expected from the derivation of the Friedmann metric [*c.f.* Eq.(15.14)]: if the universe is flat, it is flat forever. If $k \neq 0$, the magnitude of the curvature parameter tends to zero as $a(t) \to 0$, irrespectively of the value of $\Omega_k(t_0)$ at the present time. So, having in the very early universe $\Omega_k(t) = \varepsilon$ (with ε arbitrarily small) produces either a universe which expands forever ($\varepsilon > 0$) or a universe which recollapses on itself ($\varepsilon < 0$). In other words, we need to *fine tune* the initial conditions to an incredible precision to obtain today something that resembles the universe we observe. This is the so-called *curvature problem.*

16.10 AN EARLY ACCELERATED PHASE?

Both the horizon and the curvature problems arise because of an implicit assumption. By neglecting the contribution of the Cosmological Constant term at early times, we forced the universe to expand in a *decelerated* way. Indeed, if $\Lambda = 0$, Eq.(15.26) provides

$$\boxed{\frac{1}{H_0^2}\frac{\ddot{a}}{a} \simeq -\frac{\Omega_{ER}}{a^4} - \frac{1}{2}\frac{\Omega_0}{a^3}} \tag{16.35}$$

Let's instead assume that in the very early phases the cosmic expansion was *accelerated*, rather than decelerated. Let's also assume that this accelerated phase started when $a(t_i) = a_i$ and ended when $a(t_f) = a_f$. Consider, for sake of simplicity, a simple exponential accelerated expansion *à la* de Sitter [*c.f.* Section 15.8]. In this case

$H = const$ and, at the end of the exponential expansion, the size of the comoving horizon writes

$$\boxed{L_H^{(c)}(a_f) = \frac{c}{H}\left[-\frac{1}{a_f} + \frac{1}{a_i}\right]} \tag{16.36}$$

Now, let's assume that this primordial accelerated expansion started very early, and it ends when $a_f \gg a_i$. Then, for $a_i \to 0$, *all* the presently observable Universe was once in causal contact, and there is no need of invoking *ad hoc* initial conditions.

Under the same assumption of an early accelerated expansion phase, also the curvature problem disappears. The de Sitter model provides a constant expansion rate, $H = H(t_{in}) = H(t_f)$, and a scale factor $\mathscr{R}(t) = \mathscr{R}(t_{in})\exp[H(t - t_{in})]$. Under this conditions, we can write Eq.(16.34) as follows:

$$\boxed{\Omega_k(t) \equiv \Omega_k(t_{in})\exp[-2H(t - t_{in})]} \tag{16.37}$$

Thus, we get exactly the opposite behavior of Eq.(16.34): $\Omega_k(t)$ is now decreasing with increasing time. If this exponential expansion is lasting long enough, we get $\lim_{t \to t_f \gg t_{in}} \Omega_k(t) = 0$, from "above" ("below") for $\Omega_k(t_{in}) > 0$ ($\Omega_k(t_{in}) < 0$), independently on the actual value of $\Omega_k(t_{in})$. So, it is possible to have a smooth transition between the early, accelerated phase, and the standard Friedman decelerated expansion without any fine-tuning of the *initial conditions*.

16.11 COSMIC INFLATION

Let's assume that the early universe evolution has been driven, for a period of time, by a scalar field ϕ. Field equations and equations of motion can be written considering the Einstein-Hilbert action plus the scalar field contribution: $S = S_{EH} + S_\phi$ [*c.f.* Sections 8.6 and 8.7].

For sake of simplicity, consider the case of a constant scalar field, ϕ_0 say, which corresponds to a minimum of the potential: $V'(\phi)|_{\phi_0} = 0$. In this case $V(\phi_0) = \Lambda c^4/(8\pi G)$ [*c.f.* Eq.(8.48)] So, we recover an exponential accelerated expansion *à la* de Sitter. However, we also need an exponential expansion lasting long enough to ensure that $a(t_f) \gg a(t_{in})$. Let's then measure the duration and the effectiveness of such primordial exponential expansion in terms of the number of e-foldings

$$e^N \equiv \frac{a(t_f)}{a(t_{in})} \tag{16.38}$$

Note that this definition is quite general. For the de Sitter solution, $N = H(t_f - t_i)$. For other models, as the Friedmann models, $N = \int_{t_i}^{t_f} H(t)dt$.

To estimate how long the inflationary phase should last to be effective, let's consider again the curvature problem. To resolve it, we want the present value of the curvature parameter much less than the corresponding value at the onset of inflation [see Exercise A.62]:

$$\frac{\Omega_k(t_0)}{\Omega_k(t_{in})} \equiv \frac{H^2(t_{in})R^2(t_{in})}{H^2(t_0)R^2(t_0)} = e^{-2N}\Omega_0\left(\frac{T_f}{T_{eq}}\right)^2\frac{T_{eq}}{T_0} \ll 1 \tag{16.39}$$

Neglecting $\Omega_0 \approx 1$, this provides

$$N \gg \ln \frac{T_f}{T_{eq}} + \frac{1}{2} \ln \frac{T_{eq}}{T_0} \tag{16.40}$$

Now, considering $T_f \simeq 10^{16} GeV$, $T_{eq} \simeq 1eV$ and $T_0 \simeq 10^{-4} eV$ one gets

$$N \gg 60 \tag{16.41}$$

So, 60 is, so to speak, the minimum number of e-foldings necessary for resolving the standard problems of the Big-Bang. It is interesting to note that this implies an incredibly huge amplification of the scale factor, in a very tiny time interval

$$\Delta t \gtrsim \frac{60}{H(t_f)} \simeq 60 \sqrt{\frac{3}{8\pi G \rho_{ER}}} \left(\frac{T_0}{T_f}\right)^2 \approx 10^{-37} s \tag{16.42}$$

Many models have been considered for explaining such an early accelerated expansion phase [9]. Inflation is nowadays a paradigm of modern cosmology. However, there is a way to test observationally this paradigm. Indeed, the discovery of B-modes in the CMB polarization pattern would reveal the existence of a primordial, stochastic background of gravitational waves. These are expected to be generataed by tensorial fluctuations of the FLRW space-time. LiteBIRD is a satellite designed to hunt for this elusive signal, and it is expected to fly at the end of this decade [56].

A Exercises

Chapter 1

Exercise A.1. *Expand the following expressions assuming a 3D-Euclidean space with Cartesian coordinates* $x^j \equiv \{x, y, z\}$

$$\frac{\partial f}{\partial x^i}\frac{\partial x^i}{\partial x^k}; \qquad \frac{\partial^2 f}{\partial x^i \partial x^j}\frac{\partial x^i}{\partial x^m}\frac{\partial x^j}{\partial x^n}$$

Exercise A.2. *Derive the explicit expression of the metric of a 3D Euclidean space using cylindrical coordinates.*

Exercise A.3. *Verify that the sum of two vectors is a vector*

Exercise A.4. *Show that Eq.(1.17) effectively provides the covariant components of a vector.*

Exercise A.5. *Prove that the inner product of two vectors is invariant under coordinate transformation.*

Exercise A.6. *Use the definition given in Eq.(B1.2.f) to derive the non vanishing Christoffel symbols given in Eq.(B1.2.e).*

Chapter 2

Exercise A.7. *Find the explicit expression of the* $L^{\alpha}{}_{\mu}$ *and* $M^{\mu}{}_{\alpha}$ *matrixes for a rotation of the reference frame around the* ξ^1*- and* ξ^2*- axis. Show that the condition given in Eq.(2.11) is fulfilled by these rotations.*

Exercise A.8. *Find the explicit expression of the* $L^{\alpha}{}_{\mu}$ *and* $M^{\mu}{}_{\alpha}$ *matrixes for a hyperbolic rotation of the reference frame in the* ξ^i*-*ξ^0 *plane for* $i = 2$ *and* $i = 3$*. Show that the condition given in Eq.(2.11) is fulfilled by these hyperbolic rotations.*

Exercise A.9. *Show that the inner product of two vectors is an invariant*

Exercise A.10. *Show that two hyperbolic rotations do not commute unless they occur around the same axis.*

Exercise A.11. *Consider the lab. frame* $\mathscr{K}_0$ *and a series of other frame,* $\mathscr{K}_i$*, each moving w.r.t. to the previous one,* $\mathscr{K}_{i-1}$*, with the same drag velocity V along the same directions, the x-axis of* $\mathscr{K}_0$*, say. Find the velocity of the frame* $\mathscr{K}_N$ *w.r.t. the lab.frame.*

DOI: 10.1201/9781003141259-A

Exercise A.12. *Show that in the non-relativistic limit, two consecutive boosts along different direction do indeed commute.*

Chapter 3

Exercise A.13. *Given Eq.(3.2) and Eq.(3.4), find the metric coefficients in the rotating reference frame.*

Exercise A.14. *Given Eq.(3.8) and Eq.(3.11), verify that when passing from the rotating to the lab. frame we recover the Minkowski metric.*

Exercise A.15. *Consider a 2D Euclidean space and choose a polar coordinate system. Derive the Christoffel symbols given in Eq.(B1.2.e)*

Exercise A.16. *Derive Eq.(3.32)*

Exercise A.17. *Derive Eq.(3.37)*

Exercise A.18. *Two reference frames $\mathscr{X}$ and $\mathscr{X}'$ rotate around the z-axis of the lab. with equal, but opposite angular velocities. Observer A is at rest in $\mathscr{X}$, while observer B is at rest in $\mathscr{X}'$. They are at the same distance from the rotation axis. When they first meet, their watches indicate the same time. Discuss if and why this is still the case when they meet again after one round.*

Chapter 4

Exercise A.19. *Derive Eq.(4.2)*

Exercise A.20. *Verify that the sum of two vectors is a vector.*

Exercise A.21. *Prove that the inner product of two vectors is invariant under coordinate transformation.*

Exercise A.22. *Under which conditions a given second order tensor, $T^{\alpha\beta}$, can be written as the outer product of two vectors?*

Exercise A.23. *Prove that the properties of from $\phi = 0$ to $\phi = 2\pi$, symmetry or skew-symmetry are conserved under coordinate transformation.*

Exercise A.24. *Does the relation*

$$A_{\alpha\beta}B^{\beta\gamma} = A_{\alpha\beta}C^{\beta\gamma}$$

imply that $B^{\beta\gamma} = C^{\beta\gamma}$?

Exercise A.25. *A two index object is defined as follows: $T^{\alpha\beta} = U^{\alpha} + V^{\beta}$, where U^{α} and V^{β} are vectors. Discuss if $T^{\alpha\beta}$ can be a tensor.*

Chapter 5

Exercise A.26. *A unit vector A on a 2D sphere is bound to the point $\theta = \theta_0$ and $\phi = 0$. It is initially oriented along the meridian: $\vec{A} = \vec{e}_\theta$. Evaluate the vector $\vec{A}$ after parallel transporting it along the parallel $\theta = \theta_0$ from $\phi = 0$ to $\phi = 2\pi$.*

Exercise A.27. *Derive Eq.(5.21)*

Chapter 6

Exercise A.28. *Derive the expressions given in Eq.(6.33) and Eq.(6.34)*

Chapter 10

Exercise A.29. *Derive Eq.(10.11) and Eq.(10.12)*

Chapter 11

Exercise A.30. *Derive Eq.(11.3a)*

Exercise A.31. *Derive Eq.(11.3b)*

Exercise A.32. *Derive Eq.(11.4)*

Exercise A.33. *Derive Eq.(11.5)*

Exercise A.34. *Derive Eq.(11.7)*

Exercise A.35. *Derive Eq.(11.17)*

Exercise A.36. *Derive Eq.(11.17)*

Exercise A.37. *Derive the components of the field equations $G^\alpha{}_\beta = 0$*

Chapter 13

Exercise A.38. *Derive Eq.(13.11)*

Exercise A.39. *Evaluate the Ricci tensor for the stationary metric given in Eq.(13.41)*

Chapter 14

Exercise A.40. *Evaluate the quantity* $\Gamma^{\sigma}_{1\mu} T_{\sigma}{}^{\mu}$.

Exercise A.41. *Derive Eq.(14.4).*

Exercise A.42. *Derive Eq.(14.68)*

Chapter 15

Exercise A.43. *Given the metric of Eq.(15.14), find the corresponding Christoffel symbols*

Exercise A.44. *Find the* R^3_3 *components of the Ricci tensor.*

Exercise A.45. *Evaluate the maximum distance from an arbitrary chosen point on a hypersphere of radius* $\mathscr{R}$. *Evaluate also the hypersphere proper volume.*

Exercise A.46. *Find a static form for the metric of a de Sitter space-time.*

Exercise A.47. *Consider a static closed universe with a non-vanishing cosmological constant. Give an estimate of the radius of the hypersphere, the magnitude of the Cosmological Constant, and the mass of universe assuming a density of* $\approx 10^{-30} g\, cm^{-3}$

Exercise A.48. *Discuss the consistency of Eq.(14.69) with Eq.(15.42) in light of the homogeneity of the closed Friedmann universe.*

Exercise A.49. *Consider the Milne universe:* $\Omega_0 = 0$, $\Omega_k = 1$ *and* $\Omega_\Lambda = 0$. *Derive and discuss the solution of the first Friedmann equation. Explain the difference between this solution and the asymptotic behavior of an open Friedmann model.*

Exercise A.50. *Hubble original estimate of* H_0 *was* $= 500\, km\, s^{-1}/Mpc$. *Evaluate the corresponding age of the universe. Is it consistent with the age of the Earth* (4.5×10^9)*? Evaluate the upper bound on* H_0 *that renders the age of the universe consistent with the age of the Earth.*

Exercise A.51. *Derive Eq.(15.68)*

Exercise A.52. *Show the relation between the conformal time and the parameter* η *used in Eq.(15.40) and Eq.(15.44).*

Exercise A.53. *Consider a closed Friedmann universe. How long does it take for a light signal to travel along a geodesic to return to the emission point? Ignore any scattering process and assume that the photons free-stream all the time.*

Exercise A.54. *Derive Eq.(15.69) for a closed Friedmann universe.*

Exercise A.55. *Derive Eq.(15.69) for an open Friedmann universe.*

Exercise A.56. *Consider two SNe Ia: 2001cp, at redshift* $z = 0.0224$ *and distance modulus* $\mu_{01cp} = 34.9 \pm 0.3$*; 2004Eag, at redshift* $z = 1.02$ *and distance modulus* $\mu_{04Eag} = 44.2 \pm 0.3$*. Estimate first the Hubble constant and then the luminosity distance of 2004Eag.*

Chapter 16

Exercise A.57. *Evaluate the equivalence epoch under the hypothesis that the CMB is the only relativistic component in the universe*

Exercise A.58. *Consider a flat, radiation-dominated universe:* $\Omega_{ER} = 1$*. Estimate the age of this model and compare it with what was expected in an Einstein-de Sitter universe. Discuss the reason of the difference between the two ages.*

Exercise A.59. *Derive the expansion rate* $H(x)$

Exercise A.60. *The cross section for Deuterium formation is* $\sigma \approx 10^{-29} cm^2$*. Knowing that the baryonic density is today* $\approx 10^{-7} cm^{-3}$*, give an order-of-magnitude estimate for the CMB temperature you expect today.*

Exercise A.61. *Assume that the universe never recombined. Discuss if and when the CMB photons can decouple form the matter.*

Exercise A.62. *Derive Eq.(16.39)*

References

1. A. Einstein. Kosmologische Betrachtungen zur allgemeinen Relativitätstheorie. *Preussische Akademie der Wissenschaften, Sitzungsberichte*, Part 1:142–152, 1917.
2. J. Aasi et al. Advanced LIGO. *Classical and Quantum Gravity*, 32(7):074001, 2015.
3. B. P. Abbott et al. (LIGO Scientific Collaboration and Virgo Collaboration). Observation of Gravitational waves from a binary black hole merger. *Physical Review Letters*, 116(6):061102, February 2016.
4. S. Casertano et al. A 3% solution: Determination of the Hubble Constant with the Hubble Space Telescope and Wide Field Camera 3. *The Astrophysical Journal*, 730:119, April 2011.
5. W. S. Adams. The relativity displacement of the spectral lines in the companion of Sirius. *PNAS*, 117:382–387, 1925.
6. ALMA Partnership et al. The 2014 ALMA long baseline campaign: Observations of the strongly lensed submillimeter galaxy HATLAS J090311.6+003906 at z = 3.042. *The Astrophysical Journal Letters*, 808(1):L4, July 2015.
7. G. B. Arfken and H. J. Weber. *Mathematical Methods for Physicists*. Academic Press, 2005.
8. M. A. Barstow et al. Hubble Space Telescope Spectroscopy of the Balmer lines in Sirius B. *Monthly Notices of the Royal Astronomical Society*, 362(4):1134–1142, August 2005.
9. D. Baumann. TASI Lectures on Inflation. *arXiv e-prints*, page arXiv:0907.5424, July 2009.
10. C. L. Bennett et al. Nine-year Wilkinson Microwave Anisotropy Probe (WMAP) Observations: Final maps and results. *The Astrophysical Journal Supplement*, 208:20, October 2013.
11. B. Bertotti, L. Iess, and P. Tortora. A test of general relativity using radio links with the Cassini spacecraft. *Nature*, 425:374–376, 2003.
12. H. Bondi. Spherically symmetrical models in general relativity. *M.N.R.A.S.*, 107:410, 1947.
13. S. Bonometto et al. Ionization curves and last scattering surfaces in neutrino-dominated universes. *Astronomy & Astrophysics*, 123:118–120, June 1983.
14. G. M. Clemence. The relativity effect in planetary motions. *Reviews of Modern Physics*, 19(4):361–364, Oct 1947.
15. T. E. Cranshaw, J. P. Schiffer, and A. B. Whitehead. Measurement of the gravitational red shift using the Mössbauer effect in Fe^{57}. *Physical Review Letters*, 4:163–164, February 1960.
16. R. H. Dicke et al. Cosmic black-body radiation. *The Astrophysical Journal*, 142:414–419, July 1965.
17. S. Dodelson. *Modern Cosmology*. Academic Press, 2003. ISBN: 9780122191411.
18. A. Einstein. Zur elektrodynamik bewegter körper. *Annalen der Physik*, 17:891–921, 1905.
19. A. Einstein. Uber das relativitiitsprinzip und die aus demselben gezogenen folgerungen. *Jahrbuch fiir Radioaktivitiit und Elektronik*, 4:411, 1907.
20. A. Einstein. Die feldgleichungen der gravitation. *Sitzungsberichte der Preussischen Akademie der Wissenschaften zu Berlin:*, pages 844–847, 1915.

21. A. Einstein. Erklärung der perihelbewegung des merkur aus der allgemeinen relativitätstheorie. *Preussische Akademie der Wissenschaften, Sitzungsberichte*, pages 831–839, 1915.
22. A. Einstein. Grundgedanken der allgemeinen relativitätstheorie und anwendung dieser theorie in der astronomie. *Preussische Akademie der Wissenschaften, Sitzungsberichte*, 315, 1915.
23. A. Einstein. Zur allgemeinen relativitätstheorie. *Preussische Akademie der Wissenschaften, Sitzungsberichte*, pages 778–786, 799–801, 1915.
24. A. Einstein. Näherungsweise integration der feldgleichungen der gravitation. *Preussische Akademie der Wissenschaften, Sitzungsberichte*, Part 1:688–696, 1916.
25. A. Einstein. Gravitationswellen. *Preussische Akademie der Wissenschaften, Sitzungsberichte*, Part 1:154–167, 1918.
26. A. Einstein and W. de Sitter. On the relation between the expansion and the mean density of the universe. *Proceedings of the National Academy of Sciences*, 18:213–214, 1932.
27. R. Eötvös. On the gravitation produced by the earth on different substances. *http://zelmanov.ptep-online.com/papers/zj-2008-02.pdf.*
28. R. Eötvös. Uber die anziehung der erde auf verschiedene substanzen. volume 8, page 65. Mathematische und Naturwissenschaftliche Berichte aus Ungarn, 1890.
29. R. Eötvös. In G. Reiner, editor, *Verhandlungen der 16 Allgemeinen Konferenz der Internationalen Erdmessung*, volume 319, 191.
30. D. Finkelstein. Past-future asymmetry of the gravitational field of a point particle. *Physical Review*, 110(4):965, 1958.
31. W. L. Freedman and B. F. Madore. The Hubble Constant. *Annual Review of Astronomy and Astrophysics*, 48:673–710, September 2010.
32. W. L. Freedman et al. Final results from the Hubble Space Telescope key project to measure the Hubble Constant. *The Astrophysical Journal*, 553(1):47–72, May 2001.
33. W. L. Freedman et al. The Carnegie-Chicago Hubble Program. VIII. An Independent Determination of the Hubble Constant Based on the Tip of the Red Giant Branch. *The Astrophysical Journal*, 882(1):34, September 2019.
34. W. L. Freedman et al. Carnegie Hubble program: A mid-infrared calibration of the Hubble Constant. *The Astrophysical Journal*, 758:24, 2012.
35. A. Friedmann. Über die Krümmung des Raumes. *Zeitschrift für Physik*, 10:377–386, 1922.
36. A. Friedmann. Über die Möglichkeit einer Welt mit konstanter negativer Krümmung des Raumes. *Zeitschrift für Physik*, 21:326–332, 1924.
37. L. R. Miller, G. E. Moss, and R. L. Forward. Photon-Noise-Limited Laser Transducer for Gravitational Antenna. *Applied Optics*, 10:2495, 1971.
38. G. Steigman. Primordial nucleosynthesis in the precision cosmology era. *Annual Review of Nuclear and Particle Science*, 57:463–91, 2007.
39. A. Genova et al. Solar system expansion and strong equivalence principle as seen by the nasa messenger mission. *Nature Communications*, 9(1):289, 2018.
40. M. E. Gertsenshtein and V. I. Pustovoit. On the detection of low-frequency gravitational waves. *Soviet Physics JETP*, 16:433, 1962.
41. J. B. Oke, H. L. Shipman, and J. L. Greenstein. Effective temperature, radius, and gravitational redshift of Sirius B. *The Astrophysical Journal*, 169:563, 1971.
42. K. Hentschel. Measurements of gravitational redshift between 1959 and 1971. *Annals of Science*, 53(3):269–295, 1996.

43. N. S. Hetherington. Sirius B and the gravitational redshift - An historical review. *Royal Astronomical Society, Quarterly Journal*, 21:246–252, 1980.
44. J. N. Hewitt et al. Unusual radio source mg1131+0456: A possible Einstein ring. *Nature*, 333(6173):537–540, 1988.
45. D. Hilbert. Die grundlagen der physik. *Nachrichten von der Gesellschaft der Wissenschaften zu Göttingen – Mathematisch-Physikalische Klasse*, pages 1395–407, 1915.
46. R. A. Hulse and J. H. Taylor. Discovery of a pulsar in a binary system. *The Astrophysical Journal Letters*, 195:L51–L53, January 1975.
47. J. Bernstein. *Kinetic Theory in the Expanding Universe*. Cambridge monograph on mathematical physics. Cambdrige University Press, 1988.
48. J. T. Jebsen. On the general spherically symmetric solutions of Einstein's gravitational equations in vacuo. *General Relativity and Gravitation*, 37(12):2253–2259, 2005.
49. J. T. Jebsen. Über die allgemeinen kugelsymmetrischen lösungen der einsteinschen gravitationsgleichungen im vakuum. *Arkiv for matematik, astronomi och fysik*, 15(18), 1921.
50. B. J. T. Jones and R. F. G. Wyse. The ionisation of the primeval plasma at the time of recombination. *Astronomy & Astrophysics*, 149:144–150, August 1985.
51. J. Kardontchik. Riemann tensors in 3D isotropic spaces. *https:// www. researchgate. net/ publication/ 341522167*, 2020.
52. R. P. Kerr. Gravitational field of a spinning mass as an example of algebraically special metrics. *Physical Review Letters*, 11:237–238, Sep 1963.
53. M. D. Kruskal. Maximal extension of Schwarzschild metric. *Physical Review*, 119:1743–1745, Sep 1960.
54. L. M. Krauss and B. Chaboyer. Age estimates of globular clusters in the Milky Way: Constraints on cosmology. *Science*, 299:65–69, January 2003.
55. U. J. Le Verrier. Theorie du mouvement de Mercure. *Annales de l'Observatoire de Paris*, 5:1, January 1859.
56. E. Allys et al. (LiteBIRD Collaboration). Probing cosmic inflation with the LiteBIRD cosmic microwave background polarization survey. *arXiv e-prints*, page arXiv:2202.02773, February 2022.
57. J. J. Condon et al. The megamaser cosmology project. IV. A direct measurement of the Hubble Constant from UGC 3789. The Astrophysical Journal, 767:154, April 2013.
58. J. C. Mather et al. Measurement of the Cosmic Microwave Background spectrum by the COBE FIRAS instrument. *The Astrophysical Journal*, 420:439–444, January 1994.
59. J. C. Mather et al. A preliminary measurement of the Cosmic Microwave Background spectrum by the Cosmic Background Explorer (COBE) satellite. *The Astrophysical Journal Letters*, 354:L37–L40, May 1990.
60. R. L. Mossbauer. Kernresonanzfluoreszenz von Gammastrahlung in Ir^{191}. *Zeitschrift fur Physik*, 151(2):124–143, April 1958.
61. P. Noterdaeme et al. The evolution of the Cosmic Microwave Background temperature. Measurements of T_{CMB} at high redshift from carbon monoxide excitation. *A&A*, 526:L7, February 2011.
62. R. B. Orellana and H. Vucetich. The principle of equivalence and the Trojan asteroids. *Astronomy and Astrophysics*, 200(1–2):248–254, July 1988.
63. R. B. Orellana and H. Vucetich. The Nordtvedt effect in the Trojan asteroids. *Astronomy and Astrophysics*, 273:313–317, June 1993.
64. R. Penrose and R. M. Floyd. Extraction of rotational energy from a black hole. *Nature Physical Science*, 229(6):177–179, 1971.

65. A. A. Penzias and R. W. Wilson. A measurement of excess antenna temperature at 4080 Mc/s. *The Astrophysical Journal*, 142:419–421, July 1965.
66. S. Perlmutter et al. Measurements of Ω and Λ from 42 high-redshift supernovae. *The Astrophysical Journal*, 517:565–586, June 1999.
67. S. Pireaux. Solar quadrupole moment and purely relativistic gravitation contributions to Mercury's perihelion advance. *Astrophysics and Space Science*, 284(4):1159–1194, 2003.
68. D. W. Pesce et al. The Megamaser Cosmology Project. XIII. Combined Hubble Constant Constraints. *Ap.J.Lett*, 891(1):L1, March 2020.
69. P.A.R. Ade et al. (Planck Collaboration). Planck 2015 results. XIII. Cosmological parameters. *Astronomy and Astrophysics*, 594, A13 (2016).
70. N. Aghanim et al. (Planck collaboration). Planck 2018 results. VI. Cosmological parameters. *Astronomy and Astrophysiscs*, 641:A6, September 2020.
71. R. V. Pound and G. A. Rebka. Apparent weight of photons. *Physical Review Letters*, 4:337–341, Apr 1960.
72. R. V. Pound and J. L. Snider. Effect of gravity on gamma radiation. *Physical Review*, 140:B788–B803, Nov 1965.
73. R.V. Pound and A. Rebka, Glen. Resonant absorption of the 14.4keV γ-ray from 0.10-μsec Fe^{57}. *Physical Review Letters*, 3:554, 1959.
74. M. M. G. Ricci and T. Levi-Civita. Méthodes de calcul différentiel absolu et leurs applications. *Mathematische Annalen*, 54(1):125–201, 1900.
75. G. Ricci Curbastro. Resume de quelques travaux sur les systemes variables de fonctions associes a une forme differentielle quadratique. *Bulletin des Sciences Mathematiques*, 2:167.
76. A. G. Riess et al. Observational Evidence from Supernovae for an Accelerating Universe and a Cosmological Constant. *The Astrophysical Journal*, 116:1009–1038, September 1998.
77. A. G. Riess et al. A Comprehensive Measurement of the Local Value of the Hubble Constant with 1 km s^{-1} Mpc^{-1} uncertainty from the Hubble Space Telescope and the SH0ES Team. *Ap.J.Lett.*, 934(1):L7, July 2022.
78. D. S. Robertson and W. E. Carter. Relativistic deflection of radio signals in the solar gravitational field measured with VLBI. *Nature*, 310(5978):572–574, August 1984.
79. P.G. Roll, R. Krotkov and R.H. Dicke. The equivalence of inertial and passive gravitational mass. *Annals of Physics*, 26:442, 1964.
80. K. Schwarzschild. Über das gravitationsfeld einer kugel aus inkompressibler flüssigkeit nach der einsteinschen theorie. *Sitzungsberichte der Königlich Preussischen Akademie der Wissenschaften*, pages 424–434, 1916.
81. K. Schwarzschild. Uber das gravitationsfeld eines massenpunktes nach der einsteinschen theorie. *Sitzungsberichte der Königlich Preussischen Akademie der Wissenschaften*, pages 189–196, 1916.
82. K. Schwarzschild. On the gravitational field of a mass point according to Einstein's theory. *arXiv e-prints*, page physics/9905030, May 1999.
83. K. Schwarzschild. On the gravitational field of a sphere of incompressible fluid according to einstein's theory. *arXiv e-prints*, December 1999.
84. T. M. Sedgwick et al. The effects of peculiar velocities in SN Ia environments on the local H0 measurement. *Monthly Notices of the Royal Astronomical Society*, 500(3):3728–3742, 11 2020.

85. I. I. Shapiro. Fourth test of general relativity. *Physical Review Letters*, 13:789–791, Dec 1964.
86. I. I. Shapiro et al. Fourth test of general relativity: Preliminary results. *Physical Review Letters*, 21:266–266, Jul 1968.
87. J. Soltis, S. Casertano, and A. G. Riess. The Parallax of ω Centauri Measured from Gaia EDR3 and a Direct, Geometric Calibration of the Tip of the Red Giant Branch and the Hubble Constant. *Ap.J.Lett.*, 908(1):L5, February 2021.
88. G. Szekeres. On the singularities of a Riemannian manifold. *Publicationes Mathematicae Debrecen 7*, 7:285, January 1960.
89. J. H. Taylor, L. A. Fowler, and P. M. McCulloch. Measurements of general relativistic effects in the binary pulsar psr1913 + 16. *Nature*, 277(5696):437–440, 1979.
90. J. H. Taylor and J. M. Weisberg. A new test of general relativity-gravitational radiation and the binary pulsar psr 1913+ 16. *The Astrophysical Journal*, 253:908–920, 1982.
91. The ALEPH, DELPHI, L3, OPAL, SLD Collaborations, the LEP Electroweak Working Group, the SLD Electroweak and Heavy Flavour Groups. Precision Electroweak Measurements on the Z Resonance. *Physics Report*, 427:257, September 2006.
92. P. Touboul et al. Microscope mission: First results of a space test of the equivalence principle. *Physical Review Letters*, 119:231101, 2017.
93. R. F. C. Vessot et al. Test of relativistic gravitation with a space-borne hydrogen maser. *Physical Review Letters*, 45:2081–2084, Dec 1980.
94. N. Vittorio. *Cosmology*. Series in Astronomy and Astrophysics. CRC Press, 2017.
95. D. F. Watson, A. S. Eddington, and C. Davidson. IX. A determination of the deflection of light by the sun's gravitational field, from observations made at the total eclipse of May 29, 1919. *Philosophical Transactions of the Royal Society of London. Series A, Containing Papers of a Mathematical or Physical Character*, 220:291–333, January 1920.
96. S. Weinberg. *Gravitation and Cosmology: Principles and Applications of the General Theory of Relativity*. John Wiley and & Sons, Inc., July 1972.
97. S. Weinberg. The cosmological constant problem. *Review of Modern Physics*, 61:1–23, 1989.
98. J. M. Weisberg and Y. Huang. Relativistic measurements from timing the binary pulsar PSR B1913+16. *The Astrophysical Journal*, 829(1):55, September 2016.
99. J. G. Williams et al. New test of the equivalence principle from lunar laser ranging. *Physical Review Letters*, 36(11):551–554, March 1976.
100. W-M Yao et al. Review of particle physics. *Journal of Physics G: Nuclear Physics*, 33:1–1232, July 2006.

Index

Printed in the United States
by Baker & Taylor Publisher Services